Quantitative Drug Safety and Benefit-Risk Evaluation

Practical and Cross-Disciplinary Approaches

Chapman & Hall/CRC Biostatistics Series

Series Editors

Shein-Chung Chow, Duke University School of Medicine, USA
Byron Jones, Novartis Pharma AG, Switzerland
Jen-pei Liu, National Taiwan University, Taiwan
Karl E. Peace, Georgia Southern University, USA
Bruce W. Turnbull, Cornell University, USA

Recently Published Titles

Signal Detection for Medical Scientists
Likelihood Ratio Test-based Methodology
Ram Tiwari, Jyoti Zalkikar, and Lan Huang

Single-Arm Phase II Survival Trial Design
Jianrong Wu

Methodologies in Biosimilar Product Development
Sang Joon Lee, Shein-Chung Chow (eds.)

Statistical Design and Analysis of Clinical Trials
Principles and Methods, Second Edition
Weichung Joe Shih, Joseph Aisner

Confidence Intervals for Discrete Data in Clinical Research
Vivek Pradhan, Ashis Gangopadhyay, Sandeep Menon, Cynthia Basu, and Tathagata Banerjee

Statistical Thinking in Clinical Trials
Michael A. Proschan

Simultaneous Global New Drug Development
Multi-Regional Clinical Trials after ICH E17
Edited by Gang Li, Bruce Binkowitz, William Wang, Hui Quan, and Josh Chen

Quantitative Drug Safety and Benefit-Risk Evaluation: Practical and Cross-Disciplinary Approaches
Edited by William Wang, Melvin Munsaka, James Buchanan and Judy X. Li

For more information about this series, please visit: https://www.routledge.com/Chapman--Hall-CRC-Biostatistics-Series/book-series/CHBIOSTATIS

Quantitative Drug Safety and Benefit-Risk Evaluation

Practical and Cross-Disciplinary Approaches

Edited by
William Wang, Melvin Munsaka,
James Buchanan and Judy X. Li

CRC Press
Taylor & Francis Group
Boca Raton London New York

CRC Press is an imprint of the
Taylor & Francis Group, an **informa** business

A CHAPMAN & HALL BOOK

First edition published 2022
by CRC Press
6000 Broken Sound Parkway NW, Suite 300, Boca Raton, FL 33487-2742

and by CRC Press
2 Park Square, Milton Park, Abingdon, Oxon, OX14 4RN

CRC Press is an imprint of Taylor & Francis Group, LLC

ISBN: 978-1-138-59406-7 (hbk)
ISBN: 978-1-032-19111-9 (pbk)
ISBN: 978-0-429-48880-1 (ebk)

DOI: 10.1201/9780429488801

Typeset in Palatino
by SPi Technologies India Pvt Ltd (Straive)

Contents

Editors.. vii
List of Contributors.. viii
Introduction Chapter.. xiii

Part A Regulatory Landscape and Interdisciplinary Collaboration in Safety Evaluation

1. The Emergence of Aggregate Safety Assessment across the
    Regulatory Landscape..3
    James Buchanan and William Wang

2. Monitoring of Aggregate Safety Data in Clinical Development:
    A Proactive Approach 15
    Lothar Tremmel and Barbara Hendrickson

3. Safety Signaling and Causal Evaluation...................................45
    James Buchanan and Mengchun Li

4. Safety Monitoring through External Committees......................77
    Amit Bhattacharyya, William Wang and James Buchanan

Part B Statistical Methodologies for Safety Monitoring

5. An Overview of Statistical Methodologies for Safety
    Monitoring and Benefit-Risk Assessment...............................101
    William Wang, Ed Whalen, Melvin Munsaka and Judy X. Li

6. Quantitative Methods for Blinded Safety Monitoring 119
    Melvin Munsaka and Kefei Zhou

7. Bayesian Safety Methodology...133
    *Michael Fries, Matilde Kam, Judy X. Li, Satrajit Roychoudhury,
    Mat Soukup and Brian Waterhouse*

8. Likelihood-Based Methods for Safety Monitoring............................167
    *Ed Whalen, Krishan Singh, Li-An Lin, Melvin Munsaka and
    William Wang*

9. Meta-Analysis of Drug Safety Assessment 187
    Rositsa B. Dimova, Sourav Santra and Melvin Munsaka

**Part C   Design and Analysis Considerations
           in Randomized Controlled Trials and Real-World
           Evidence for Safety Decision-Making**

10. Pragmatic Trials with Design Considerations for
    Cardiovascular Outcome Trials 215
    Li-An Lin, Ed Whalen, Melvin Munsaka and William Wang

11. Post-Market Safety Assessment Using Observational Studies
    and the Food and Drug Administration Sentinel System 233
    Yong Ma, Rima Izem, Ellen Snyder and Wei Liu

12. Analysis Considerations for Real-World Evidence and
    Clinical Trials Related to Safety 253
    *Richard C. Zink, Melvin Munsaka, Birol Emir, Yong Ma,
    Judy X. Li, William Wang and Dimitri Bennett*

**Part D   Safety/Benefit-Risk Evaluation and
           Visualization**

13. Trends and Recent Progress in Benefit-Risk Assessment
    Planning for Medical Products and Devices 285
    Lisa R. Rodriguez, Cheryl Renz, Brian Edwards and William Wang

14. Estimands in Safety and Benefit-Risk Evaluation 317
    Shuai S. Yuan, Yodit Seifu, William Wang and Michael Colopy

15. Visual Analytics of Safety and Benefit-Risk from
    Clinical Trial Data 351
    *Melvin Munsaka, Kefei Zhou, Krishan Singh, Michael Colopy,
    Chen Chen and Meng Liu*

Index 375

Editors

William Wang, PhD, is currently Executive Director in Clinical Safety Statistics and Late Develop Statistics (Asia Pacific region), Biostatistics and Research Decision Sciences (BARDS), Merck Research Laboratories. He has over 25 years of global clinical development experiences. He has co-chaired the ASA Safety Working Group and served as the deputy topics leader for the ICH E17 Multi-Regional Clinical Trial Working Group. He is an elected ASA Fellow.

Melvin Munsaka, PhD, is currently Senior Director and Head of Safety Statistics at AbbVie. He has more 25 years of industry experience working in various capacities including strategic leadership roles. He has participated in multiple global submissions covering a wide range of therapeutic areas and is a member of the ASA Biopharmaceutical Safety Scientific Working Group (WG), DIA Bayesian Scientific WG, the PHUSE Safety Analytics and the Data Visualization and Open-Source Technology WGs.

James Buchanan, PharmD, is President of Covilance LLC, a drug safety consulting firm. He teaches the Advanced Signal Detection course for DIA, is a co-lead of the American Statistical Association Biopharmaceutical Safety Working Group and also co-lead on the Interactive Safety Graphics task force that is developing novel, open-source interactive graphical tools to identify and evaluate safety issues during drug development.

Judy X. Li, PhD, is currently Senior Director and Biostatistics Lead for San Diego site at Bristol Myers Squibb. Before joining the industry, she worked at the US Food and Drug Administration (FDA) as a master statistical reviewer and supervisory mathematical statistician in multiple centers. She is an active member of the ASA Biopharmaceutical Section Executive Committee, as the co-chair of 2019 ASA Biopharmaceutical Section Regulatory Industry Statistics Workshop, and the founding co-chair of the ASA Safety Working Group.

Contributors

Dimitri Bennett, MD, MPH, is the Director Pharmacoepidemiology at Takeda Pharmaceuticals, where his key responsibility includes leadership for the development of global safety pharmacoepidemiology programs for Gastroenterology and Cardiovascular products. He has 20 years of experience in epidemiology and is a Fellow of the International Society of Pharmacoepidemiology and the American College of Epidemiology.

Amit Bhattacharyya, PhD, is currently the Senior Director of Biostatistics at Alexion, where he leads statistical support for the Amyloidosis program. He has more than 25 years of experience in various statistical leadership roles globally in drug discovery and development in R&D. He is an elected Fellow of the American Statistical Association.

Chen Chen, PhD, is a Senior Biostatistician at UCB. She previously worked for AbbVie in clinical development and in regulatory submissions. She is involved in research on visual analytics of safety data. Since joining UCB, she has been engaged in the development of Bayesian methods for blinded safety signal detection and visualization tools for benefit-risk assessment.

Michael Colopy, PhD, is a Safety Statistics and Decision Scientist at UCB with over 25 years of industry experience. His areas of expertise include statistical methods in clinical trials and epidemiology, signal detection, benefit-risk analysis and Monte Carlo simulation, Patient Preference and Structured Benefit-Risk Network, predictive analytics, visualization and Bayesian decision science. He is an active collaborator on the ASA Safety Monitoring Work Stream and EFSPI Benefit-Risk Special Interest Group.

Rositsa B. Dimova, PhD, is affiliated with the Center for Biologics Evaluation and Research at the U.S. Food and Drug Administration. She has more than 13 years of experience in the biopharmaceutical research area. Dr. Dimova received her PhD degree in Biostatistics from Columbia University and is a member of the ASA Safety Working Group.

Brian Edwards, MD, spent 14 years in hospital medicine and clinical research before joining the UK Medicines Control Agency for 5 years. Over the last 20 years, he has held senior roles in pharmacovigilance in Parexel, Johnson & Johnson and latterly NDA group. Currently he consults for his own company, Husoteria Ltd.

Birol Emir, PhD, is Senior Director and RWE Statistics Group Lead in the Statistical Research and Data Science Center at Pfizer. He is also Adjunct Professor of Statistics at Columbia University in New York and External PhD Committee Member in the Graduate School of Arts and Sciences at Rutgers, The State University of New Jersey. He is co-author of the book *Interface Between Regulation and Statistics in Drug Development*.

Michael Fries, PhD, is Head, Biostatistics at CSL Behring which provides statistical support for clinical trials, statistical innovation, and product development. He has over 20 years of experience in Pharmaceuticals and academics. Mike has research interests in safety biostatistics, quantitative decision making including go/no go criteria and assurance, and Biomarkers in clinical trials.

Barbara Hendrickson, MD, is Immunology Therapeutic Area Head in Pharmacovigilance and Patient Safety, AbbVie. She is a subspecialist in Pediatric Infectious Diseases with over 15 years in academic medicine and 17 years of pharmaceutical industry experience. She has been involved in multiple safety statistics initiatives and co-leads the DIA-ASA Aggregate Safety Assessment Planning Task Force.

Rima Izem, PhD, is an Associate Director at Novartis where she supports development and implementation of novel statistical methods using RWD or hybrid designs in clinical development across therapeutic areas. Her experience includes pioneering work in causal inference, signal detection, and survey research at FDA, comparative effectiveness in rare diseases at Children's National Research Institute, and dimension reduction methods at Harvard University.

Maria Matilde Kam, PhD, is Associate Director for Analytics and Informatics for the Office of Biostatistics (OB) at FDA/CDER. She has strategic and oversight responsibility for data standards, data integrity, data science, scientific computing and statistical programming activities in OB. She has over 25 years of industry and regulatory experience.

Mengchun Li, MD, MPA, is the Director of Pharmacovigilance at TB Alliance, where she chairs the multidisciplinary safety management team and takes the leading role on clinical safety. She has more than 13 years of drug development experience in the pharmaceutical industry. She co-leads the DIA-ASA Interdisciplinary Safety Evaluation workstream under the ASA Biopharmaceutical Safety Working Group.

Li-An Lin, PhD, is the Associate Principal Biostatistician at Merck, where he leads the statistical support for early oncology safety. He has more than 6 years of experience in safety statistics with research interest in the areas of adaptive trial design, blinded safety monitoring, meta-analysis, Bayesian methods, and causal inference methods.

Meng Liu, PhD, is a Statistics Manager at AbbVie. She joined AbbVie where she has been since 2017. She has had an increasing role in provide statistical support in product safety, clinical trial development and regulatory submissions. She is also an active participant in the activities of the ASA Biopharm Safety Working Group focusing on visual analytics.

Wei Liu, PhD, is a Senior Epidemiologist and team lead at FDA CDER's Office of Surveillance and Epidemiology where he oversees epidemiological reviews of study protocols, finalized study reports, and provides epidemiological recommendations for regulatory decisions. He is also involved in the design, execution and interpretation of real-world epidemiological studies conducted under FDA's Sentinel System.

Yong Ma, PhD, is a lead mathematical statistician at CDER, FDA, where she covers the review and surveillance of post-market drug safety. She has more than 20 years of working experience in government and academia and over 30 publications in peer-reviewed journals. She co-leads the Workstream 3 of the ASA Biopharm Safety Working Group.

Cheryl Renz, MD, is former Head of the Benefit Risk Management group at AbbVie, responsible for benefit-risk assessment efforts and risk management strategies. She also leads pioneering digital health innovation initiatives. She is involved in several CIOMS working groups and the ASA Biopharmaceutical Safety Working Group.

Lisa R. Rodriguez, PhD, is a Deputy Division Director in the Office of Biostatistics at the Center for Drug Evaluation and Research, U.S. Food and Drug Administration. She has more than 15 years of experience in the biopharmaceutical environment and is currently serving as co-lead of the Benefit-Risk Assessment Planning Taskforce within the ASA Safety Working Group.

Satrajit Roychoudhury, PhD, is a Senior Director and a member of the Statistical Research and Innovation group in Pfizer Inc. He has 15 years of extensive experience in different phases of clinical trials. His research areas include survival analysis, model-informed drug development, and Bayesian methodologies.

Sourav Santra, PhD, is currently the Senior Director, Biostatistics, at Sarepta Therapeutics. He leads biostatistics for medical affairs and real-world evidence studies on a rare disease, Duchenne Muscular Dystrophy at Sarepta. He has more than 14 years of experience in the pharmaceutical industry at various pharmaceutical companies with experience in several therapeutic areas.

Yodit Seifu, PhD, is a Director at Amgen Biosimilar unit where she leads the statistical support for two biosimilar products. She has more than 20 years working experience in the pharmaceutical industry. Primary therapeutic experience includes gastroenterology, fertility, oncology, endocrinology, urology, transplantation and cardiovascular diseases. She is interested in win-statistics and Bayesian inference.

Krishan Singh, PhD, Senior Director (retired) at GlaxoSmithKline. He has 32 years of experience in leading statistical support for the development of drugs across therapeutics areas, and 12 years of experience in academia and government. He co-leads statistical methodology and interactive safety graphics work streams for the ASA Biopharmaceutical Safety Working Group.

Ellen Snyder, PhD, is a Senior Director at Merck Research Laboratories and has over 28 years of industry experience. She currently works in the Clinical Safety Statistics group on innovative approaches for monitoring and characterizing drug safety. She has published over a hundred journal articles and abstracts in the medical and statistical literature.

Mat Soukup, PhD, is Deputy Director, Division of Biometrics 7 in the Office of Biostatistics at FDA/CDER which provides statistical expertise to the quantitative assessment of drug safety across FDA/CDER. He has over 15 years of regulatory experience with research interests in signal detection, statistical graphics, and safety outcome trial design.

Lothar Tremmel, PhD, leads the statistical, epidemiological, PRO, and medical writing functions in clinical research at CSL Behring. He has over 20 years of leadership experience in clinical development, with a long-standing interest in safety statistics. He currently leads Workstream 1 of the ASA Safety Working Group and co-leads the Aggregate Safety Assessment Planning Task Force.

Brian Waterhouse, MS, is a senior principal scientist in clinical safety statistics at Merck Research Laboratories. He has more than 20 years of pharmaceutical industry experience analyzing clinical trial data across different therapeutic areas. His research interests include safety biostatistics and interactive graphics as investigative and communication tools.

Ed Whalen, PhD, is a Senior Director, Clinical Statistics, at Pfizer Inc. He leads the statistical support for pain and CNS clinical studies. He has more than 30 years of experience in clinical trials statistics with research interest in safety evaluation and communication, causal inference, and quantitative benefit-risk analysis.

Shuai S. Yuan, PhD, is a director in oncology statistics of GSK. He has more than 12 years of industry experience in clinical trials across different therapeutic areas and phases. His research interests include adaptive designs, Bayesian statistics and BRA. He co-led the ASA-Biopharm Safety and BR Estimand taskforce and contributed to the BRA planning task force.

Kefei Zhou, PhD, is currently Senior Director at Bristol Myers Squibb and has some work experience at JAZZ and Amgen. She has over 15 years of industry experience with extensive work in the Oncology area. She is an active member of the ASA Biopharm Safety Work Group.

Richard C. Zink, PhD, is Vice President of Data Management, Biostatistics, and Statistical Programming at Lexitas Pharma Services. Richard is a contributor to 9 books on statistical topics in clinical trials. He holds a Ph.D. in Biostatistics from UNC-Chapel Hill, where he serves as Adjunct Professor of Biostatistics. He is a Fellow of the American Statistical Association.

Introduction Chapter

Safety assessment, monitoring and safety surveillance, also referred to as pharmacovigilance, are key components of a pharmaceutical product's development life cycle. Different types of data, such as clinical trial data and real-world data (RWD), can all be used for pharmacovigilance. In the past, most pharmacovigilance departments at biopharmaceutical companies focused on the handling of individual adverse event reports (called individual case safety reports). In recent years, there has been a shift in focus from individual cases toward aggregate analysis to identify potential adverse drug reactions. While individual adverse event case handling remains relevant for understanding the specifics of each case, the safety assessment based on aggregate data and quantitative evaluation has been increasingly used in drug development, life cycle management and post-marketing surveillance to better understand the scope of product risks.

Statistical methodology for safety assessment for monitoring and reporting will need to be further developed to match that for efficacy evaluation. In addition to evolving improved methodologies to support safety evaluation, our new complex world will require a multidisciplinary approach to effectively address safety assessment efforts. No longer will the 'safety physician in a project team' be alone in determining and evaluating safety issues with the simple case-based analysis used in the past years. That said, statisticians also need to enhance their knowledge of the clinical perspective of safety evaluation so that they can assist in the interpretation of the safety concerns in drug development. What is needed is a true medical-statistical joint venture in order to develop the models in the 21st century for safety evaluation.

In 2014, the American Statistical Association (ASA) Biopharmaceutical (BIOP) Section started a Safety Working Group (SWG) including active members and advisors from regulatory agencies and pharmaceutical companies. Since then, the working group has expanded its workstreams and task forces from memberships purely in the US region to a multiregional collaboration, from designing cardiovascular safety outcome trials (CVOTs) to safety monitoring methodologies, from pre-marketing safety to randomized clinical trial/real-world evidence (RCT/RWE) integration, from sole safety assessment to benefit-risk evaluation. Throughout its discussions with advisors and thought-leaders, the working group has identified the following four key components to enable the application of statistics in safety evaluation: cross-disciplinary scientific engagement, efficient operational processes, intelligent data architecture, and visual/analytical tools and methodologies.

Four Key Components of Quantitative Safety Evaluation

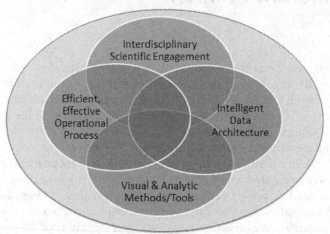

Consequently, the working group has set its major mission to cultivate interdisciplinary collaboration. To achieve that, the working group has established close collaboration with the drug information association safety communities in the form of a dedicated workstream (i.e., workstream 1 on interdisciplinary safety evaluation) and its multiple task forces. Since 2017, the workstream has performed a systematic review of the global regulatory landscape and conducted an industry survey and thought-leader interviews on the current practices and future direction of safety and benefit-risk evaluation. Based on the findings of these efforts, workstream 1 has been developing interactive safety graphics and general frameworks/processes on aggregate safety assessment planning (ASAP) and benefit-risk assessment planning (BRAP).

In general, the overall safety working group is organized in the following three workstreams:

- **Workstream 1 on Interdisciplinary Safety Evaluation**: the workstream is focused on developing interdisciplinary frameworks for aggregate safety assessment and benefit-risk planning and open-source interactive visual tools to enable/enhance cross-disciplinary collaboration for safety evaluations.

- **Workstream 2 on Safety Statistical Methodology**: the workstream is focused on methods for quantitative analysis of safety data from various data sources, including clinical trials, post-marketing and other healthcare data sources.

- **Workstream 3: Integration and Bridging the RWE and RCT for Safety Decision-Making**: the workstream is focused on investigating structured and principled approaches for the integration and bridging of data from RCTs and RWE for risk and benefit decision-making.

This book will be largely based on the work of the ASA safety working group. The book will consist of the following four parts (Part A to Part D), which align with the working streams and associated task forces:

Part A covers the overall regulatory landscape, interdisciplinary safety processes, safety signaling and the data monitoring committee. It includes the following four inter-related chapters:

- **Chapter 1** provides an overview from both global and regional perspectives on aggregate safety monitoring and evaluation. The chapter offers a historical perspective with a focus on recent developments and identifies important long-term trends which include (1) a shift from a single-case medical review of serious adverse events to an interdisciplinary evaluation of aggregate data, (2) closer interdisciplinary collaboration for bridging the gap between pre- and post-marketing surveillance and (3) a more proactive determination of patient populations with a positive benefit-risk profile for product use. In addition to these trends, the chapter also highlights specific regulations at the US FDA, EU EMA, Japan PMDA, and China NMPA. These regulatory landscapes motivate our workstreams and task forces whose works are summarized in the following chapters.

- **Chapter 2** discusses Aggregate Safety Assessment Planning (ASAP), a framework of drug safety monitoring and evaluation proposed by the working group. In the last decades, there has been a shift in focus from post- to pre-marketing surveillance, as well as a shift from case-oriented analysis toward aggregate assessments of increased event frequencies in order to identify safety signals. Sponsors are faced with new expectations to use their own analyses and interpretation before reporting pre-marketing safety events to health authorities. This can only be performed by looking at all accumulating clinical data within a project in a planned, systematic, interdisciplinary way. The ASAP establishes a framework for interdisciplinary, systematic safety planning, data review, and characterization of the emerging product safety profile. It is aimed at aligning the safety topics of interest, standardizing product-level safety assessments, identifying

safety knowledge gaps, planning ongoing aggregate safety evaluation, and preparing for safety communications. The ASAP document
is intended to be modular and flexible in nature, depending on the
program complexity, phase of development and coexisting sponsor
processes. Patient safety in ongoing clinical trials needs the utmost
attention from all stakeholders. Safety monitoring through a DMC,
independent of the sponsor, provides the ultimate source of unbiased and critical oversight of ethical conduct of clinical trials and
ensuring the safety of the participants in the trials.

- **Chapter 3** focuses on safety signaling and causal inference evaluation. The concept of a safety signal – an event with an unknown
  causal relationship to treatment that is recognized as worthy of further exploration – has engendered a degree of confusion due to the
  misuse of the term to mean actual product risk as well as the multiple types of signals based on the varied data sources from which
  it may arise. Then having identified the safety event of interest, the
  reviewer is challenged with deciding what information constitutes
  sufficient evidence to conclude a causal association with the drug.
  Chapter 3 clarifies how a safety signal can be identified, how it may
  be the result of confounding or bias within the source data, and the
  means by which it can be evaluated to determine if it is actually the
  result of exposure to the drug of interest. In particular, this chapter
  offers unique perspectives on the causal evaluation of a safety signal using the Hill's criteria on causality. Since tools are as important as appropriate methodology for these evaluations, the chapter
  describes innovative work in the realm of interactive safety graphics.
  The Hepatic Explorer is one such example that offers a powerful way
  to examine liver toxicity.

- **Chapter 4** provides an overview of the critical roles of data monitoring committees, their composition and operations during a
  trial and the importance of their recommendations with respect
  to the ongoing trial throughout its critical phases. The recent FDA
  IND safety reporting draft guidance recommendation to establish
  a safety assessment committee (SAC) is also discussed in relation
  to the DMC. The authors hope that the reader will appreciate the
  impact of the DMC and SAC as significant mechanisms within the
  overall safety monitoring processes that sponsors are expected to
  implement in order to meet expectations from regulators and fulfill
  responsibilities for patient safety.

**Part B is constructed from the perspective of statistical methodologists
to enable holistic safety monitoring and evaluation. Chapter 5 provides
an overview of safety monitoring methodologies but also connects safety**

assessment with the benefit-risk evaluation – this chapter also serves as a precursor for Part D on benefit-risk evaluation. Chapters 6 to 9 each provide summaries on a specific approach of analyzing drug safety.

- **Chapter 5** presents a holistic overview of safety and benefit-risk methodologies. Using the "elephant" metaphor, existing individual methods usually focus on a singular, specific perspective of the "elephant". This chapter tries to visualize and draw the whole elephant by reviewing the following key perspectives of defining and contrast: Bayesian versus frequentist methods, blinded versus unblinded safety monitoring, individual versus aggregate data meta-analyses, pre- and post-marketing methods, static versus dynamic safety reviews, and safety assessment vs benefit-risk evaluation. These perspectives may serve as a background for future statistical work, both in methodology development and its application. Since safety methodologies should be connected with benefit-risk evaluation, we provide toward the end of this chapter a high-level summary of the benefit-risk methodologies. This not only serves to bridge but also to prepare our expanded discussion on benefit-risk evaluation in Part D of this book

- **Chapter 6** is built on the premise that many ongoing RCTs, even when blinded, may hold important clues to the safety profile. The key is to utilize existing knowledge and observed data to assess the likelihood of potential safety risk elevation. This chapter provides a summary of various approaches in evaluating safety while the trial is still blinded. These include both Bayesian and frequentist approaches. Such proactive safety signal monitoring will inevitably raise logistical questions regarding monitoring patient safety while at the same time maintaining the study blind - this chapter will offer practical points for consideration during the blinded review.

- **Chapter 7** leverages the properties of Bayesian statistics to address some of the challenges found in analyzing safety data. After a short orientation to some basic concepts of Bayesian statistics, we look at three key applications. The first application is in safety monitoring of accumulating blinded data, leveraging the methods ability to incorporate historical information and handle accumulating data. The second is in assessing adverse events with unblinded data. Here, we take advantage of the method's ability to model complex structures and borrow information within groupings and apply that to the hierarchy of adverse events defined by MedDRA. The third section shows how we can incorporate external information into safety signal detection in a way that accounts for trial to trial variability - even dynamically determining the amount of information to borrow in case the observed data do not align with the historic information.

- **Chapter 8** discusses the application of the likelihood ratio test (LRT) methodology for safety monitoring. It first discusses the general likelihood principles for dynamic or continuous safety monitoring. The implementation of these principles includes Bayesian and frequentist approaches. Most of the chapter will focus on the frequentist likelihood-based methods. In addition to the classical use of LRT in safety monitoring for clinical trials and spontaneous safety reporting, the chapter also covers longitudinal LRT, sequential LRT, and its special application in the Tree-scan method as used in the Sentinel system. A case study on the safety of the Rotavirus vaccine is used to illustrate the LRT application.

- **Chapter 9** provides an overview and discussion on the use of meta-analysis approaches for safety assessments. Meta-analyses offer the advantage of looking at specific safety events of interest by aggregating data from multiple clinical trials. This is particularly useful when the safety event of interest is rare but critical. The major challenges to the conduct of meta-analysis using safety data include the lack of consistency of the definitions of the adverse events across the different studies, the variability in the exposure and observation times, as well as the poor reporting, or lack of it, in the published literature. This chapter discusses the main components of the process of conducting a meta-analysis for safety assessments, including a discussion on the US FDA (2018) "Draft Guidance: Meta-Analyses of Randomized Controlled Clinical Trials to Evaluate the Safety of Human Drugs or Biological Products".

Part C discusses randomized and real-world evidence for safety decision-making.

Safety evaluation is a continual and iterative process throughout the drug development life cycle. While clinical trials provide high-quality data for an initial assessment of the safety profile of a new drug, they alone cannot fully characterize the safety profile. Thus, post-marketing and RWD sources play a critical role in further understanding the safety profile of a drug. Regulatory agencies have become increasingly open to incorporating these sources of safety evidence for their fit-for-purpose safety decision making.

- **Chapter 10** provides a summary of the design of pragmatic trials for safety evaluation. The CVOT for type II diabetes drug development is used as a good illustration example. This chapter will discuss key design considerations in randomization, safety endpoint determination, choice of comparators and safety monitoring in the pragmatic real-world settings. The use of PRECIS-2 is discussed as a key tool for the design considerations. The FDA CVOT draft guidance (2020) is used to illustrate how to potentially fulfill regulatory requirements with a pragmatic trial design.

- **Chapter 11** discusses the design considerations for post-marketing observational safety data surveillance. It particularly focuses on the observational RWD through the FDA's Sentinel initiative. Sentinel is the FDA's national electronic system which has transformed the way researchers monitor the safety of FDA-regulated medical products, including drugs, vaccines, biologics and medical devices. This chapter introduces the history and structure of Sentinel as well as its common data model and distributed database. It further discusses the design considerations and statistical methods used in Sentinel. The unique approaches in safety signal detection, signal refinement and signal evaluation in the Sentinel system will be elaborated with examples.

- **Chapter 12** summarizes analysis considerations for real-world data alone and in conjunction with data from randomized controlled trials with applications to safety. The 21st Century Cures Act passed by the US Congress in 2016 empowers the US FDA to accelerate the approval process by allowing evidence from real-world data sources to support new indications for approved drugs or to support or satisfy post-approval regulatory requirements. In this chapter, we provide an overview of real-world data (RWD), discuss methods to account for confounding of observational sources, describe ways to utilize RWD alone or in conjunction with data from randomized controlled trials to answer important safety questions, and discuss the challenges and benefits of their inclusion into an analysis to support the regulatory decision-making process.

Part D connects and expands safety discussions into the benefit-risk evaluation.

- **Chapter 13** provides a summary of the work of our task force for BRAP. A structured approach to benefit-risk evaluation is essential for pharmaceutical product development throughout its product life cycle. Integrating / connecting proactive safety evaluation with structured benefit-risk evaluation is one of the key efforts for the ASA safety working group. In this chapter, we will provide a general overview of the current global and regional regulatory guidance on benefit-risk evaluation and summarize the scientific literature on planning for benefit-risk evaluation in connection with the safety evaluation. We will also summarize the BRAP industry survey that has been performed by the task force. A discussion will be provided on the future trends and potential points for consideration in benefit-risk assessment planning

- **Chapter 14** is built on the fundamental principle that getting the questions right is a critical part of any scientific pursuit. The ICH-E9 R1 provides a good framework to clarify the question but has focused mainly on efficacy evaluation. This chapter explores the estimand framework for safety and benefit-risk evaluations. Safety estimands are impacted by intercurrent events differently from efficacy estimands. The estimands for benefit-risk evaluation include population-level and subject-level considerations. Two case studies will be used to illustrate these concepts

- **Chapter 15** discusses the role of graphical methods in safety monitoring and benefit-risk and assessment. Safety data from all sources present many challenges with regard to analysis and interpretation. It is well recognized that visual analytics present a useful alternative to tabular outputs for exploring safety data and present a great opportunity to enhance evaluation of drug safety. Similarly, the use of visualization in benefit-risk assessment can help enhance transparency and clarity of tasks and decisions made. The visualizations can be static and/or dynamic and interactive to allow for efficient exploration for communicating benefits and risks. The chapter discusses how visual analytics, coupled with a question-based approach, can help get maximum gain from using graphics in safety monitoring and benefit-risk assessment. We will show through some examples. We will also discuss enhancements that can be applied to visual analytic tools to help the assessment of safety monitoring and risk-benefit.

There is a wealth of opportunity to move the safety discipline forward with new methods, new tools and new interdisciplinary practices. This book provides a systematic review of these practices, methods and tools. Through our various multidisciplinary efforts, the ASA BIOP-SWG is changing how drug safety and benefit-risk shall be incorporated into clinical development programs and into post-marketing surveillance. Altogether this will bring into fruition a coherent drug safety and benefit-risk assessment based on sound clinical judgment and statistical rigor. Supplemental information (eg programming code, working examples, working templates, white papers etc) to book chapters can be found at https://community.amstat.org/biop/workinggroups/safety/safety-home

Part A

Regulatory Landscape and Interdisciplinary Collaboration in Safety Evaluation

1

The Emergence of Aggregate Safety Assessment across the Regulatory Landscape

James Buchanan
Covilance, LLC, Belmont, CA, USA

William Wang
Merck & Co, Inc, Kenilworth, NJ, USA

CONTENTS

1.1 Introduction 3
1.2 Food and Drug Administration (FDA) Regulations 5
1.3 EMA Regulations 7
1.4 Japan Pharmaceuticals and Medical Devices Agency (PMDA) Regulations 8
1.5 China National Medical Products Administration (NMPA) Regulations 9
1.6 Post-Marketing Aggregate Safety Analysis 11
1.7 Real-World Health Data 12
1.8 Discussion 13
References...13

1.1 Introduction

The evolution of safety evaluation practices has been guided in large part by advances in methods promulgated by expert working groups, such as the Council for International Organizations of Medical Sciences (CIOMS), the International Council for Harmonization (ICH) and guidances and regulations from health authorities. Initially, these processes were focused on managing adverse events that met criteria for expedited reporting to health authorities. From a process standpoint, this led safety professionals within the industry to focus their attention on individual serious adverse event (SAE) reports. Safety databases within the safety department were

DOI: 10.1201/9780429488801-2

only populated with SAE cases since the principle purpose of such databases was to meet regulatory reporting obligations, with safety evaluation a secondary consideration. As SAE data accumulated in the safety database, there was periodic reconciliation with the clinical adverse event database, but for the most part, the collective AE database was not subject to routine evaluation until the safety section of the clinical study reported was written at the conclusion of the clinical trial. This focus on individual case safety reports also carried over into post-marketing safety practices where, again, attention was focused on those serious cases subject to expedited reporting. In the US, the non-serious cases were generally not evaluated until the crafting of the quarterly Periodic Adverse Drug Experience Report and in Europe with the writing of the Periodic Safety Update Report.

It can be appreciated that focusing attention on a subset of the complete safety experience precludes an early appreciation of the emerging risk profile of a drug and the ability to implement effective risk mitigation strategies. A clear understanding of which adverse events a drug can cause is required to create a list of expected events from which decisions to expedite a given case are made. Thus, regulatory authorities had interest in companies performing more comprehensive safety evaluations on a regular basis so that the submission of uninformative expedited reports would be minimized. This has resulted in a variety of communications to encourage this practice.

In order to perform a comprehensive analysis of aggregate safety data, professionals from a variety of disciplines are required, including not only safety clinicians but also data managers, biostatisticians and programmers. Such an interdisciplinary approach is a step beyond the more traditional approach of the safety clinician reviewing individual SAE reports in isolation of the broader data. The interdisciplinary concept can be expanded even within the safety department where it was not unusual in large organizations to functionally separate clinical development safety functions from post-marketing safety evaluations. The need for communication between the pre-marketing and post-marketing safety clinicians is as important as the inclusion of other disciplines. Thus, the communications from regulatory authorities emphasized the need for such a multidisciplinary approach to aggregate data review.

In 2005, the CIOMS Working Group VI released the report "Management of Safety Information from Clinical Trials". As described by Ball et al. (2020), the core idea of CIOMS VI was the implementation of product-specific, multidisciplinary safety management teams for the regular review of program-level safety information from ongoing blinded trials, completed unblinded trials and other data sources to make decisions on safety and risk management in a more timely and informed manner. An unfortunate consequence of organizational structures within pharmaceutical companies has been the separation of the safety functions for pre- and post-marketing data. An important concept articulated by CIOMS VI was to help bridge the gap between pre- and post-approval activities to better understand and manage risk.

The principles of this systematic approach were the basis for the Program Safety Analysis Plan (PSAP) proposed by the Pharmaceutical Research and Manufacturers of America (PhRMA) Safety Planning, Evaluation and Reporting Team (Crowe et al. 2009). The PSAP authors supported the CIOMS VI recommendation to conduct periodic aggregate safety data reviews, including review of blinded data for ongoing trials and un-blinded data for completed trials. The aggregate safety reviews were recommended to be conducted by a multidisciplinary safety management team in line with the CIOMS VI proposal. The PSAP has since been updated to reflect current regulatory guidance in the form of the Aggregate Safety Assessment Plan as described in Chapter 3.

The importance of a planning framework for conducting safety evaluations was described in the ICH E2E (2005) guidance on pharmacovigilance planning activities. The guidance described the creation of a Safety Specification, a summary of the important identified risks of a drug, important potential risks, and important missing information. The safety specification is created from a comprehensive review of non-clinical and clinical experience as well as a consideration of the epidemiology of medical outcomes in the population under study and the safety profile in vulnerable populations such as the young, the elderly and those with various organ insufficiencies. Based on the findings in the safety specification, a pharmacovigilance plan is prepared. In addition to describing how regulatory reporting obligations will be met, the Plan defines the process for continuous monitoring of the safety profile of the drug including signal detection and evaluation. Although not specifically stated, it is clear that the optimal approach to such aggregate safety evaluation processes is via a multidisciplinary team.

1.2 Food and Drug Administration (FDA) Regulations

The use of aggregate assessment of safety data was further conveyed in the FDA guidance on pre-market risk assessment (FDA 2005). While covering a variety of safety issues to consider during clinical development, the guidance stated

> Used appropriately, pooled analyses can enhance the power to detect an association between product use and an event and provide more reliable estimates of the magnitude of risk over time. Pooled analyses can also provide insight into a positive signal observed in a single study by allowing a broader comparison.

The application of statistical approaches mentioned in the guidance requires a team with the requisite statistical skills.

The recommendation that a multidisciplinary approach to aggregate safety assessment should be utilized was not limited to the pharmaceutical industry. Indeed, FDA's internal guidance for medical reviews recommended a similar approach (FDA 2005). The guidance cautioned the medical reviewer that

> Consideration of the safety findings in individual studies, without a thoughtful integration of the overall safety experience, is not adequate for a safety review. Although much of the guidance in this document is directed primarily toward the clinical reviewer and toward the analysis of particular events, the evaluation of safety data also involves analyses of event rates, estimation of risk over time, exploration of possible subgroup differences and identification of risk factors associated with serious events, all analyses that involve substantial knowledge of methods of validly quantifying risk and providing measures of uncertainty. Clinical reviewers should therefore collaborate with their biostatistical colleagues when necessary in the preparation of reviews and consider when it may be appropriate to conduct a joint statistical and clinical review for particularly important safety issues.

The reference to aggregate analyses then appeared in documents required to update regulatory authorities on programs in clinical development. The Development Safety Update Report (DSUR) (2011), which harmonized the previous US IND Annual Report and the European Union (EU) Annual Safety Report, included a section to summarize relevant safety information from any other clinical trial sources that became available to the sponsor during the reporting period, such as results from pooled analyses.

The FDA Final Rule on Safety Reporting Requirements for INDs and BA/BE Studies (2011) described the circumstances in which sponsors are required to submit expedited safety reports to the Agency. One of such circumstance was when

> an aggregate analysis of specific events observed in a clinical trial (such as known consequences of the underlying disease or condition under investigation or other events that commonly occur in the study population independent of drug therapy) that indicates those events occurring more frequently in the drug treatment group than in a concurrent or historical control group.

An aggregate approach to data analysis is also needed in order to identify increases in the frequency of expected SAEs, which would also trigger an expedited report. These aggregate assessments should be part of "ongoing safety evaluations, including periodic review and analyses of their entire safety database, not only for IND safety reporting purposes." To meet the requirements for IND safety reporting based on aggregate data assessment, the guidance stresses the importance of having in place a systematic approach for evaluating the accumulating safety data.

Following the release of the FDA Final Rule, a subsequent guidance was released in 2015 to further clarify the Agency's position on expedited reporting and to emphasize the importance of aggregate safety assessments entitled "Industry Safety Assessment for IND Safety Reporting" (FDA 2015). Aggregate analyses are recommended for the comparison of adverse event rates across treatment groups from accumulating safety data collected across multiple studies (completed and ongoing). Whereas prior guidances did not specifically describe a recommended process for systematically reviewing aggregate safety data, this guidance drew attention specifically to the concept of a Safety Assessment Committee (SAC), a group of individuals chosen by the sponsor to review safety information in the development program. The FDA clearly recognized the importance of interdisciplinary involvement in the SAC, stating

> A safety assessment committee should be multidisciplinary. It should include at least one physician who is familiar with the therapeutic area for which the investigational drug is being developed as well as clinicians who have general or specific (e.g., cardiology, hepatology and neurology) safety experience. Other disciplines should be considered on a regular or an ad hoc basis (e.g., epidemiology, clinical pharmacology, toxicology, chemistry and biostatistics).

The SAC described in this guidance was considered distinct from a Data Management Committee (DMC). A more thorough discussion of the roles and responsibilities of the DMC and SAC is provided in Chapter 4.

1.3 EMA Regulations

Consistent with the FDA's approach to aggregate safety assessment, the EU has also devised requirements that are based on aggregate assessments. For instance, as in the US, the finding of an increase in the frequency of an expected SAE, which would be identified from an aggregate data analysis, would trigger an expedited report to the competent authorities (Detailed Guidance 2006, Regulation (EU) No 536/2014). The EU has also adopted the ICH E2F DSUR to serve as the annual report during clinical development (ICH 2010), which includes the section on relevant safety information from pooled analyses. The Reference Safety Information section of the Investigator's Brochure requires that the frequency of expected serious adverse reactions be calculated on an aggregated basis (CTFG 2017).

The importance of aggregate safety assessment has bridged from clinical development to the post-marketing environment. The CIOMS Working Group V report (2005) included multiple references to aggregate data analyses, in the context of observational studies, registries and safety databases. When the Good Vigilance Practice (GVP) modules were released, they represented a major update to the approach to post-marketing safety processing in the EU. Module IX on safety signal management refers to the use of aggregated data in the signal detection process (EMA 2017). The availability of post-marketing safety reports from the FDA Adverse Event Reporting System (FAERS) database, the Uppsala Monitoring Centre's Vigibase and the EU's Eudravigilance database affords the ability to perform disproportionality analyses across these aggregated datasets.

1.4 Japan Pharmaceuticals and Medical Devices Agency (PMDA) Regulations

The PMDA is Japan's independent administrative organization for overseeing all consultation and review work from the preclinical stage through marketing approval and into post-marketing surveillance. The PMDA and the Ministry of Health, Labor and Welfare (a cabinet level ministry of the Japanese government) were founding regulatory members of the ICH.

The expectations for the aggregate safety data review in Japan are not as explicitly stated as in other regions. In fact, while CIOMS VI promoted the use of aggregate safety analysis, risk evaluation during clinical development in Japan was based solely on individual case safety reports. This resulted in an initiative by the Japan Pharmaceutical Manufacturers Association to recommend that safety evaluation be based on an aggregate data analysis and that benefit-risk balance also be based on an aggregate analysis, in order to align Japanese regulations with those in effect in the US and the EU (Ohshima 2009).

Since then, Japan has adopted many of the ICH guidances such as the Development Safety Update Report (ICH E2F) during clinical development and Pharmacovigilance Planning (ICH E2E) and the Periodic Benefit-Risk Evaluation Report (ICH E2C) in the post-marketing period (Pharmaceutical and Food Safety Bureau 2012, Japan Pharmaceutical Manufacturers Association 2020). Each of these contains safety assessments that depend on the analysis of aggregate safety data. Marketing applications are submitted in the format of the Common Technical Document, including the Summary of Clinical Safety (Section 2.7.4) and the Clinical Overview (Section 2.5). Each of these requires an aggregate analysis of clinical trial safety data.

Japan follows ICH technical requirements to evaluate the long-term safety of pharmaceutical products and to accommodate multiregional clinical trial data. However, data obtained from other regions must meet Japan's regulatory requirements and demonstrate that the data can be extrapolated to Japan's population.

GVP guidelines establish standards for post-marketing safety management related to the collection, evaluation and assessment of safety information and are one of the licensing requirements for the marketing authorization holder. Of the 17 articles that comprise the GVP, available post-marketing data sources are identified in Article 7, and the examination of the aggregated data is described in Article 8. The findings from that evaluation may result in one or more safety assurance measures, including the development of a Risk Management Plan (Article 9).

The PMDA requires confirmation that the recommended dosing regimen does not introduce any additional safety concerns for the Japanese population. In accordance with ICH E2A, sponsors must report ADRs within a specified time frame. Also, as part of early post-marketing phase vigilance, sponsors are required to collect adverse events from health clinics and hospitals for the first six months after marketing approval and to report relevant safety information to health authorities.

Marketed drugs are subject to a process of reexamination (Article 14-4 of the Pharmaceutical Affairs Law) to reconfirm the continued efficacy and safety of the product at four to ten years after initial marketing authorization depending on the type of approval. One component of the safety reevaluation is to identify any marked increase in the incidence of ADRs and changes in the incidence of ADRs due to factors such as dosage, duration of administration, complications or concomitant medication, which requires an analysis of aggregated data in order to do so.

1.5 China National Medical Products Administration (NMPA) Regulations

Over the past 30 years, the China NMPA, formerly known as the China Food and Drug Administration, has been establishing a comprehensive PV system. The Center of Drug Evaluation (CDE) and the Center for Drug Re-evaluation, under the China NMPA, handle pre-marketing and post-marketing drug evaluation, respectively. In the last five years, the NMPA has undertaken major changes in drug regulation overall and drug safety evaluation. Most notable has been the accelerated adoption of the ICH regulatory guidances. China has joined the ICH as its eighth regulatory member in 2017 pledging to gradually implement the international coalition's technical standards and guidelines. On November 12, 2019, NMPA issued the Announcement on the Application of 15 ICH Guidelines in order to align the technical standards for drug registration with international standards. Relevant to the aggregate review of safety data, this includes the adoption of ICH E2F (Development Safety Update Report) and ICH E2E (Pharmacovigilance Planning). Of note, the 15 ICH guidances adopted did not include ICH E2C on the Periodic Benefit-Risk Evaluation Report. However, the principles of the post-marketing safety review are described in the Drug Administration Law (China National People's Congress Network 2019). Specifically, Article 77 states "holder of a drug marketing authorization shall formulate a post-marketing risk management plan, actively carry out post-marketing research, further confirm the safety, effectiveness and quality controllability of the drug and strengthen the review of the drug on the market," Article 80 states "Holders of drug marketing permits shall carry out post-market adverse reaction monitoring, actively collect, track and analyze information on suspected adverse drug reactions and take timely risk control measures for drugs with identified risks" and Article 83 states

> holder of a drug marketing authorization shall conduct regular post-marketing evaluations on the safety, effectiveness and quality controllability of marketed drugs. When necessary, the drug regulatory authority under the State Council may order the drug marketing license holder to carry out post-marketing evaluation or directly organize and carry out post-marketing evaluation.

Although not specifically articulated, in order to meet the evaluations broadly described in these Articles, an aggregate approach to safety data analysis would logically be required. Considering the rapid pace of adoption of international standards, it can be anticipated that additional guidances may be released to align post-marketing safety activities with evolving international standards.

The NMPA issued the official Chinese version of "M4: Common Technical Document for human drug registration in 2019 in order to bring the applications for drug marketing authorization in line with international standards." At the time of this writing, it was anticipated that all applications for chemical and biological products will be required to follow this guideline beginning October 2020. This includes the Summary of Clinical Safety (Section 2.7.4) and the Clinical Overview (Section 2.5), both requiring an aggregate analysis of clinical trial safety data.

With the adoption of these guidelines, only SUSARs as defined in ICH E2A (instead of all SAEs) will need to be reported to the CDE in an expedited manner; individual case safety reports will be reported using E2B (R3) format (instead of line listings) and MedDRA will be used as the standard dictionary. These changes, substantially advancing and broadening the global harmonization of safety monitoring and reporting, will have a major impact on the efficiency of PV practices for clinical development. However, as with Japan, China has not provided any specific guidance regarding ongoing aggregate safety assessment and reporting.

The CDE of the NMPA released the Good Safety Information Evaluation and Management Practice during Drug Clinical Trials guidance on July 1, 2020, including the Good Management Practice for the Development Safety Update Report. One of the requirements is that sponsors shall conduct risk monitoring, identification, and evaluation which will require an aggregate assessment of safety data. In addition, as noted previously, the DSUR contains safety assessments that depend on the analysis of aggregate safety data.

1.6 Post-Marketing Aggregate Safety Analysis

The preceding discussion of regulatory frameworks has dealt primarily with aggregate safety assessments in the clinical development phases. In contrast, aggregate assessment of post-marketing data has complemented the individual case safety report review since the availability of large public databases, for example, the FAERS and the WHO Drug Vigibase. Methodologies based on identifying disproportionate reporting rates, such as frequentist and Bayesian approaches, have been applied to these collections of spontaneous adverse event reports to seek signals of emerging product risks (Ahmed et al. 2015, Suling & Pigeot 2012). However, until 2017, there was no regulatory requirement that large post-marketing databases be analyzed. An FDA Guidance for Industry, Good Pharmacovigilance Practices and Pharmacoepidemiologic Assessment, released in 2005, included a section on data-mining applied to large adverse event databases, such as FAERS or the FDA Vaccine Adverse Event Reporting System, to identify unusual or unexpected product-event combinations warranting further investigation. However, the guidance specifically stated that the "use of data mining techniques is not a required part of signal identification or evaluation." Nevertheless, it was common industry practice to at that time include disproportionality analyses of databases like FAERS to augment individual case reviews.

The regulatory landscape changed in 2017 with the revision to the EU Good Pharmacovigilance Practices Module IX Signal Management (EMA 2017) which instituted a legal requirement for marketing authorization holders (MAHs) to monitor safety data pertaining to their products in EudraVigilance. Up until this time, MAHs were required to submit post-marketing reports of adverse events to EudraVigilance, but there was no external access to the system to perform disproportionality analyses. The revision to Module IX led not only to access to EudraVigilance but also provided the MAH with the results of a disproportionality analysis via the EudraVigilance Data Analysis System.

1.7 Real-World Health Data

Analysis of post-marketing databases can identify signals of potential safety risks but are limited to the cases that are voluntarily reported into these systems. Electronic healthcare records have proved to be a rich source of "real-world" data that can be used to both identify safety signals and also to evaluate them. Several examples are worth highlighting. The FDA established the Sentinel system in 2008 to provide a means to interrogate patterns in medical billing information and electronic health records in order to investigate safety questions. Within Sentinel, the Active Postmarket Risk Identification and Analysis (ARIA) System was created in 2016. The ARIA system consists of two components: a common data model of the healthcare data formatted in the Sentinel and the analytical tools. A variety of statistical methods are available to identify signals, including propensity score-based TreeScan, self-controlled TreeScan, temporal TreeScan and sequency symmetry analysis. Having identified a safety signal, ARIA can be used to estimate the size of the risk and study the potential relationship to the drug in question.

The Medical Information for Risk Assessment Initiative (MIHARI) was established by the PMDA in 2009 to use electronic health records and insurance claims data to monitor product safety. After developing a framework for data access and testing various pharmacoepidemiolgic methods, the MIHARI became operational in 2014. Similar to Sentinel, the MIHARI serves to identify safety signals and seed evidence of a causal relationship to the drug of interest. It has also been used to evaluate the impact of regulatory actions. Analytical methods employed include segment regression analysis and propensity scoring.

Such "real-world" data will be increasingly important to serve as a source of safety signals and a means by which signals can be evaluated. The reader is referred to Chapter 12 for a discussion of integrating information from randomized clinical trials and real-world evidence to improve the evaluation of safety data.

1.8 Discussion

Aggregate safety data analysis has been the subject of a number of recommendations from international organizations. In response to these, regulatory agencies in the US, the EU, Japan and China and other regulatory authorities have also promulgated guidances or regulations addressing aggregate safety data analysis in both the pre-marketing and post-marketing environments. Aggregate data analysis has been more extensively employed by the pharmaceutical industry in the post-marketing setting, but the trend is increasingly to apply the principles of aggregate data analysis during the clinical development of drugs. The infrastructure for large-scale data analysis in electronic medical records systems has been established by several regulatory agencies and offers a window into the next phase of aggregate safety data analysis. Beyond the aggregate safety evaluation, there is a growing trend of connecting aggregate safety evaluation with benefit-risk evaluation that is most noticeable in the ICH M4E (R2), ICH E2C (R2) and the FDA benefit-risk framework. More discussion will be provided in BRAP Chapter 13.

References

Ahmed I, Begaud B, Tubert-Bitter P. Evaluation of post-marketing safety using spontaneous reporting databases. In *Statistical Methods for Evaluating Safety in Medical Product Development*, First Edition. John Wiley & Sons, Ltd., 2015. pp. 332–344.

China National People's Congress Network. Drug Administration Law of the People's Republic of China. August 26, 2019. http://www.npc.gov.cn/npc/c30834/2019 08/26a6b28dd83546d79d17f90c62e59461.shtml

Clinical Trial Facilitation Group (CTFG). Q&A document—reference safety information. November 2017.

Council for International Organizations of Medical Sciences. *Management of Safety Information From Clinical Trials—Report of CIOMS Working Group VI*. Geneva, Switzerland: CIOMS, 2005.

Crowe BJ, Xia -HA, Berlin JA, et al. Recommendations for safety planning, data collection, evaluation and reporting during drug, biologic and vaccine development: a report of the safety planning, evaluation, and reporting team. *Clin Trials*. 2009;6(5):430–440.

Detailed guidance on the collection, verification and presentation of adverse reaction reports arising from clinical trials on medicinal products for human use. ENTR/ CT 3, April 2006.

European Medicines Agency. Guideline on good pharmacovigilance practices (GVP) Module IX – Signal management (Rev 1). EMA/827661/2011 Rev 1, November 2017.

Food and Drug Administration. Reviewer guidance—conducting a clinical safety review of a new product application and preparing a report on the review. February 2005a.

Food and Drug Administration. Premarketing risk assessment, 2005b.

Food and Drug Administration. Good Pharmacovigilance Practices and Pharmaco-epidemiologic Assessment, 2005c.

Food and Drug Administration. Guidance for Industry E2F Development Safety Update Report, 2011.

Food and Drug Administration. Guidance for Industry Safety Assessment for IND Safety Reporting, 2015.

ICH E2E Note for Guidance on Planning Pharmacovigilance Activities (CPMP/ICH/5716/03), 2005.

ICH Harmonised Tripartite Guideline, Development Safety Update Report, E2F, August 2010.

Japan Pharmaceutical Manufacturers Association. Information on Japanese Regulatory Affairs, Regulatory Information Task Force, 2020. http://www.jpma.or.jp/english/parj/pdf/2020.pdf.

Ohshima H. Japanese New Regulation for Periodic 6-monthly Safety Report During Clinical Development. Presented at the Drug Information Association, 45th Annual Meeting, 2009.

Pharmaceutical and Food Safety Bureau, Ministry of Health, Labour and Welfare. Risk Management Plan Guidance. PFSB/SD Notification No. 0411-1, PFSB/ELD Notification No. 0411-2, April 11, 2012. https://www.pmda.go.jp/files/000153333.pdf.

Regulation (EU) No 536/2014 of the European Parliament and of the Council of 16 April 2014 on Clinical Trials on Medicinal Products for Human Use, and Repealing Directive 2001/20/EC.

Suling M, Pigeot I. Signal detection and monitoring based on longitudinal healthcare data. *Pharmaceutics* 2012;4:607–640.

2

Monitoring of Aggregate Safety Data in Clinical Development: A Proactive Approach

Lothar Tremmel

CSL Behring, King of Prussia, PA, USA

Barbara Hendrickson

AbbVie, Inc., North Chicago, IL, USA

CONTENTS

2.1 Background: The Need for Pre-Marketing Safety Surveillance 16
2.2 The Aggregate Safety Assessment Plan (ASAP) 18
    2.2.1 Section 1: Value Proposition and Governance 19
    2.2.2 Section 2: Safety Topics of Interest and Pooling Strategy 20
    2.2.3 Section 3: Data Analysis Approaches 22
    2.2.4 Section 4: Analysis of Key Gaps 24
    2.2.5 Section 5: Ongoing Aggregate Safety Evaluation (OASE) 25
        2.2.5.1 Multiplicity in Safety Data 25
        2.2.5.2 Some Methods for Monitoring and Signal
              Detection 26
    2.2.6 Section 6: Communication of Safety Information 29
2.3 The ASAP as a Modular Template-Driven Approach 30
2.4 The ASAP as a Safety Planning Tool to Respond to a Pandemic
    Such as COVID-19 30
    2.4.1 General Effect on Data Collection 30
    2.4.2 Unanticipated Infections of Subjects in the Clinical
         Trials Due to the Pandemic 31
    2.4.3 Using the ASAP To Respond to a Pandemic: 32
2.5 Anticipated Hurdles toward Implementation of the ASAP 33
2.6 Foundation: Intelligent Data Architecture 34
    2.6.1 Consistent Implementation of Derivation Rules 34
    2.6.2 Consistent MedDRA Coding 35
    2.6.3 Data Traceability and Single Source of Truth 36

DOI: 10.1201/9780429488801-3

2.7   Foundation: Background Rates..37
2.8   Trends Likely to Influence the Further Direction of the ASAP 40
Note..41
References...41

Sponsors are responsible for continuously assessing the safety of treatments in development. Regulations require pre-marketing analysis of serious adverse events (SAEs) in order to determine causality and the need for expedited SAE reporting. During the last few years, the interdisciplinary sub-team of the Drug Information Association-American Statistical Association Safety Working Group embarked on a systematic effort to survey current working practices and emerging expectations of health authorities in the most important markets worldwide. The results from these efforts were summarized in two publications (Colopy et al. 2019; Ball et al. 2020). Based on those results, the interdisciplinary sub-team developed a planning tool for pre-marketing safety surveillance called the Aggregate Safety Assessment Plan (ASAP). The ASAP will be the focus of this chapter, which is structured as follows: First, we provide some background about regulatory developments that led to the need for more quantitative monitoring of aggregate safety during the development of new treatments. Then we introduce the ASAP as a planning tool. A discussion follows of potential hurdles of ASAP implementation. We then describe the foundations that a company needs to have in place before they can implement effective pre-marketing safety monitoring. We conclude with an analysis of current trends that are likely to influence how pre-marketing safety surveillance will be performed in the future.

2.1 Background: The Need for Pre-Marketing Safety Surveillance

One important goal of the safety evaluation of new medical treatments is to identify important adverse events caused by the treatment. Traditionally, such determinations were conducted based on scrutinizing single or few "sentinel" cases, i.e., cases for which other reasons for the event could reasonably be excluded. An example where such a single-case review played an important role in regulatory decision-making was the highly publicized removal of the herbal supplement Ephedra from the market (Morton 2005). The decision to ban Ephedra was made based not only on a meta-analysis of clinical trial data but also on a single-case review of sentinel cases of deaths and other serious events which occurred in persons within 24 hours of consuming Ephedra with no other cause identified. After the decision, the Food and Drug Administration (FDA) stated that the conclusion was not based solely on those single-case reviews but that those reviews provided supportive evidence.

During the last decades, there has been a trend away from the single-case analysis approach and toward a more statistical viewpoint based on increased frequency analysis. The simple and scientifically compelling argument behind this change is that if more events appear under drug exposure, all other things being equal, the drug must be the culprit. The origins of this trend are often attributed to the CIOMS think tank (CIOMS 2005). However, the same year the FDA issued a "Reviewer's guidance" that expressed skepticism about the importance of investigator's subjective assessments of drug relatedness (FDA 2005). For further details regarding this and other trends in regulations regarding drug safety, we refer to the study by Ball et al. (2020).

One essential tool for assessing events from multiple clinical trials is *Aggregate Assessment*, which we define loosely as evaluating the totality of the product safety data before a final judgment is made about the role of product administration given an observation of increased frequency of events under exposure. In the publication by Crowe et al. (2009), a framework was established for planning incremental meta-analyses of the growing body of evidence from an ongoing clinical development program (Crowe et al. 2009). In 2010, after much public discussion, the FDA issued their long-expected "Final Rule" regarding IND safety reporting (FDA 2010). Herein, FDA re-emphasized the importance of aggregate analysis to identify increased frequencies of events under drug treatment, compared to the reference treatment or background reference rate. The FDA also clarified that the occurrence of certain single events and one or more events atypical for the study population could establish causality between a drug and an adverse event. Specifically, they distinguish between different types of events: (A) *"single occurrences of an uncommon event that is often associated with drugs (for example, hepatic injury),"* (B) *"A few examples of an event that is uncommon in the population exposed (for example, tendon rupture),"* and (C) everything else.

One main point of the Final Rule is that in addition to completed studies, accumulated data from ongoing studies need to be included in the aggregate analysis of Type C events. As discussed in more detail below, through subsequent guidance (FDA 2012, 2021) and public discussion, the FDA advocated an unblinded review of the accumulating safety data as one approach. The guidance acknowledged the potential challenges in "firewalling" study teams from this review and the governance complexity involved (FDA 2021). With another approach, referred to as the "trigger approach", an unblinded analysis is performed only after the results of blinded analyses demonstrate that the rate of events in the pooled treatment groups substantially exceeds a pre-specified predicted rate.

Type (A) events are sometimes also called *"Designated Medical Events."* There are compilations of such events; see elsewhere in this book (Section A,

Safety Signaling Process). As pointed out by Wittes et al. (2015), the boundary between Type (B) and (C) events is somewhat fluid. For suspected Type (B) events, presumably, the background rate in the study population is very low. For Type (C) events, that is not the case, and therefore, causality assessments need to be grounded in comparison of event frequencies between exposed and unexposed subjects.

In addition to Type (A), (B), and (C) events, it might be helpful to consider a fourth Type, D: *Safety topics of interest*. These events represent concerns that arise from several different sources and have the potential to impact the product's benefit-risk profile. Safety topics of interest may emerge from the product's mechanism of action, reported risks of products of the same class, or traditional regulatory concerns (e.g., drug-induced liver injury). Type (D) events include events with some reason to believe there is a causal relationship to treatment – *a priori*, e.g., based on mechanistic considerations or known product class risk, or *a posteriori* because a couple of exposed patients have shown the events. Typically, a sponsor would have a lower threshold to conclude that such events when they occur have a reasonable possibility of being related to product administration. Alternatively, safety topics of interest could include certain Type (C) events, which are known to occur at an increased incidence in the study population (e.g. cardiovascular events in an elderly population) and for which there is interest to show that these events are not increased under treatment.

Type (D) events will need to be carefully characterized. The need to debunk the belief or silent assumption of likely causality could impact the clinical trial sample size chosen and the approach to data collection and assessment. A case study to underline this point is REBINYN, a pegylated (PEG) Factor IX product for the treatment of Hemophilia B. The FDA Advisory Committee raised concerns about a preclinical signal of PEG accumulation in the CNS. Although not a single specific CNS event was reported in humans, the FDA reviewer stated that "conclusions in this regard cannot be made as the studies were not designed to specifically evaluate neurological function."[1]

Methods of identifying and thoroughly validating safety signals are described elsewhere in the book (Chapter 3 on Safety Signaling). The purpose of this chapter is to explain how to establish the technological, logistical, interdisciplinary, and quantitative framework to characterize known risks and detect unknown ones.

2.2 The Aggregate Safety Assessment Plan (ASAP)

Pre-marketing assessment of a new drug's emerging safety profile is a complex endeavor, involving blinded and unblinded analyses by various bodies and integrating evidence from prior and ongoing studies. The proposed

ASAP acknowledges this complexity and delineates a proactive approach to deal with it upfront.

Similar efforts were undertaken before, with the Program Safety Analysis Plan (PSAP) proposal (Crowe et al. 2009), published shortly before the Final Rule was issued. This proposal emphasized cumulative meta-analysis of the studies of a program and advocated the creation of sponsor interdisciplinary Safety Management Teams (SMTs), as proposed by CIOMS VI (CIOMS 2005). The PSAP was implemented in some major pharmaceutical companies, and experience gathered was published (Chuang-Stein and Xia, 2013).

In many ways, the ASAP builds on the PSAP and so could be viewed as "PSAP 2.0". Both documents propose an ongoing, integrated look at accumulating safety data by an interdisciplinary team, early clarity about the safety topics of interest, and standardization of important aspects of study design, safety data collection, and assessment. However, the ASAP addresses in more detail safety governance, identification of safety topics of interest that need further scrutiny, methodologies for signal detection and analysis of key events, and harmonized safety messaging. Also the ASAP is meant to be modular, which means that depending on the program's phase and size and the processes that a company already has in place, only the relevant components of the proposed content may be implemented.

A detailed proposal for the ASAP is described elsewhere (Hendrickson et al. 2021). What follows is a summary of the proposed structure and content.

2.2.1 Section 1: Value Proposition and Governance

This section of the ASAP describes the governance model used for assessments of the accumulating pre-marketing safety data. A scheme such as that in Figure 2.1 may be included to depict the relationship between different entities involved in signal detection. In this example, based on a real-life study, the Internal Safety Committee consists of persons with relevant expertise selected by the sponsor to review unblinded clinical trial safety data. Per FDA's 2015 Draft Guidance, the purpose of the SAC is to make recommendations about whether particular safety events based on the aggregate review meet a threshold for reporting to the FDA in an IND safety report. This objective differs from an independent Data Monitoring Committee (IDMC), which recommends to a sponsor a modification or halting of a study based on an unfavorable benefit-risk profile of the investigational product as currently administered.

This section should also contain pointers to related documents informing the ASAP, including the target product profile, development risk management plan, product-specific regulatory feedback, and related pre-existing corporate guidances and SOPs. Finally, the section defines the frequency and triggers for the aggregate assessments and for updating the plan.

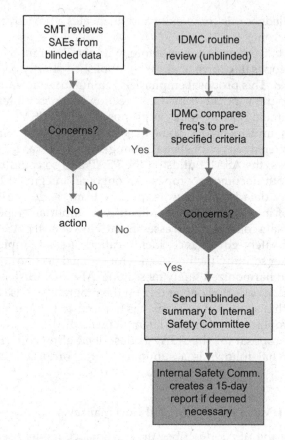

FIGURE 2.1

Example of a governance structure for evaluating safety signals

Note: SMT=Safety Management Team; IDMC = Independent Data Monitoring Committee

2.2.2 Section 2: Safety Topics of Interest and Pooling Strategy

The purpose of this section is two-fold: (a) Identify safety topics of interest for the drug, based on current knowledge and (b) proactively develop a data pooling plan. In general, safety topics of interest are events that require specialized data collection, monitoring, or analyses (i.e., actions beyond "routine"). Those would typically contain identified or potential risks. Other safety topics of interest are anticipated events commonly observed in the target population, as well as events of routine interest for the drug class or the reviewing division, for which concerns may have to be debunked. This information should be summarized in a table such as the one below (Table 2.1).

Subsection (b) contains a proactive plan that specifies how the data will be pooled across different sets of studies. Reference will be made to the

TABLE 2.1

Example Table of Safety Topics of Interest

Safety Topic of Interest	Basis for Inclusion	Identification of Events for Review	Use of Event Adjudication	Special Collection Forms or Safety Studies	Relevant Restrictions[b]
Identified Risks					
Risk #1 Infusion reactions[a]	Temporally associated events with IV administration reported in clinical trials	PTs of infusion-related reaction and infusion-related hypersensitivity reaction	No	Infusion reaction adverse event supplemental case report form	Subjects excluded with Prior history of severe hypersensitivity reaction to drug in the same class
Potential risks					
Risk #1 Hepatotoxicity[a]	Reported in products of the same class; Increases from baseline in ALT/AST noted on drug in Phase 1 studies	Drug-related hepatic disorders-comprehensive search (Narrow) Laboratory values of ALT/AST$\geq$5xULN and/or Total Bilirubin$\geq$2xULN	External expert adjudication (procedures per pre-specified charter)	Hepatic Adverse Event supplemental case report form	Study exclusion: ALT or AST>3xULN
Other safety topics of interest					
Risk #1 Congestive heart failure	Increased risk in the study population	Cardiac failure Standardized MedDRA Query (SMQ) Narrow Cardiac echo results	Blinded case review by internal cardiologist quarterly	None	Subjects excluded with Class III and IV Heart Failure

[a] Considered Important Risks

[b] e.g., protocol exclusion criteria limiting data on certain patient populations

company's data architecture, including steps to be taken to make the data "poolable" and specification of target MedDRA version and frequency of MedDRA upgrades.

2.2.3 Section 3: Data Analysis Approaches

This section also has two subsections: The "Foundational Analysis Rules" spell out basic analysis, coding, and data processing rules to harmonize the approach to safety data analyses across the program. The section contains rules for calculating exposure, determining "treatment-emergent" adverse events, and exposure adjusted analyses. It may also include criteria for abnormal values considered potentially clinically significant for vital signs, ECG, and laboratory data. If a standardized grading system is used, such as CTCAE (Common Terminology Criteria for Adverse Events: U.S. Department of Health and Human Services 2017), that system should be specified here, with specifics such as version number and modifications that may be needed to enable numerical analysis. This section should also spell out rules for safety-modifying data such as baseline conditions and prior and concomitant medications. The dictionary used should be specified (e.g., WHO-Drug), with version identifier, and rules should be stated for distinguishing prior from concomitant medications, what types of medications are of interest, and how they could be grouped.

These rules would typically be listed in study-specific statistical analysis plans (SAPs). Having all of them together here, in one project-wide central place, serves the dual purpose of streamlining the single SAPs (which could point to that central place) and of driving harmonization of all safety-related statistical outputs for a particular product.

The second subsection is about the Analysis of *Key Safety Topics of Interest*, which are those safety topics of interest that are important for determining benefit and risk. The key consideration is that the "crude risk," defined as

the number of subjects affected / number of subjects exposed

is not always the proper way to summarize safety data. For instance, for the topic "Serious infections," this simplistic approach would treat a patient with one episode the same way as a patient who experienced many episodes. Since every episode comes with significant risk to the patient, it is important to process all episodes to estimate the hazard of suffering such an episode under exposure and control.

We propose to extend the "estimand approach" to the key safety topics of interest. The estimand approach was developed with efficacy parameters in mind (ICH 2020). Based on this guidance, an estimand has the following attributes:

A. *The treatment condition of interest (could be a complex regimen)*

B. *The population, that is, the patients targeted by the scientific question*

C. *The variable (or endpoint) to be obtained for each patient*

D. *The treatment of intercurrent events, as far as A–C have not already addressed this topic*

E. *The population-level summary for the variable which provides a basis for the estimation of the estimand*

The estimand approach is described in more detail in Chapter 14 of this book. It imposes a rigor of thinking that helps with getting the safety question right: What is the proper quantity (or "*estimand*") to be estimated? For adverse events that typically happen early on, the crude risk may well be a suitable approach. For episodic events such as Serious Infections, the proper estimand is more likely the hazard of experiencing an event of interest, be it the first one or a subsequent one. For events that happen with certainty or at least a high probability, time-to-event may be the variable we are interested in (e.g., death; kidney failure). The table below gives some guidance on what estimand may be proper for what type of adverse event.

Clarity about the estimand drives the statistical analysis that will provide the estimate for the estimand. For instance, in the example of serious infections, the proper estimand is the hazard to experience an episode. The estimate could be the Poisson parameter of a statistical model that will process

TABLE 2.2

Estimands to Consider Based on Event Characteristics

Features	Estimand	Estimate
Short exposure time Fixed follow-up time Risk is relatively constant Repeat events not important	Crude risk	# patients affected / # patients exposed
Individual follow-up time varies Risk is relatively constant Time is continuous Repeat events not important	Exposure-adjusted incidence rate	Counts per total person-time at risk
Risk is relatively constant Repeat events are important	Poisson parameter	# events / exposure
Individual follow-up time Risk can vary over time	Cumulative Incidence Rate	Kaplan-Meier method
Individual follow-up time Risk can vary over time, but hazard ratios are constant Covariates are important	Hazard ratio	Ratio of two hazard rates, e.g. obtained from a Cox proportional hazard model
Risk varies over time Hazard ratio not constant over time	Difference in restricted mean survival time	Difference in areas under the survival curve

all data, not just first events. In the time-to-event example, the estimand is the probability of experiencing the event as a function of time, which would be estimated by a cumulative incidence function based on standard non-parametric survival analysis.

The question of how to treat intercurrent events has received a lot of regulatory attention recently: How should data be treated that are "contaminated" by preceding events such as rescue medications or treatment interruptions or discontinuations? The ICH guidance discusses how clarity about the estimand determines how the analyst should treat these events. For instance, if the estimand is the risk of incurring an event under an assigned *treatment regimen*, data after treatment interruption or discontinuation should be included, as the word "regimen" implies the consideration of all real-world scenarios that would happen under assigned treatment. This approach contrasts with estimating the event's risk under treatment ("as-treated"), where we would only consider data from times when the patient was exposed.

Of note, there is an asymmetry between efficacy and safety: For efficacy, the "treatment regimen" estimand is conservative and tends to be unfavorable for the drug because its effect will be attenuated by including data when the drug was not received. For safety, the opposite may be true: Including such data segments will increase the risk denominator, although patients are not exposed to treatment-related risk during these segments.

2.2.4 Section 4: Analysis of Key Gaps

This section's focus is to acknowledge the "known unknowns," i.e., aspects of the new drug's safety profile that we may not know but should know. This includes the following questions:

- Do we understand the safety in special populations, according to standard health authority expectations? (e.g., safety in renally impaired patients)

- Do the safety data match the target product profile? (e.g., that a key risk does not exceed the stated frequency occurrence or severity of this risk for a competitor)

- Structured benefit-risk considerations: How well have we characterized the risks of key importance for risk-benefit evaluation?

- Are there any theoretical risks, based on preclinical findings or pharmacological considerations, that require special attention or data collection to be addressed and hopefully debunked?

- For global submissions, are there any regional safety concerns that need to be addressed?

2.2.5 Section 5: Ongoing Aggregate Safety Evaluation (OASE)

The main purpose of this section is to define procedures for the detection of new, emerging safety concerns, i.e., detection of new potential risks or potentially the elevation of potential risks to identified risks (i.e., adequate evidence of an association with the medicinal product of interest has been established). Referring back to the FDA Final Rule as discussed at the beginning of this chapter, medically designated events and serious events unanticipated for the study population and without a clear alternative etiology should be assumed drug-related and reported in an expedited fashion. For all other events, an association with exposure must be based on some kind of increased frequency analysis. Both Bayesian and frequentist approaches toward increased frequency analysis are available, as well as methods for blinded and unblinded analysis, as further outlined below. Many of these approaches are based on a firm understanding of background rates of anticipated events in the target population. The ASAP or its appendix should include, where available, estimates for these rates. Also, the ASAP may specify an appropriate database from which background rates could be obtained within a fast turn-around time if an entirely unexpected safety signal emerges ("standing cohort" approach).

2.2.5.1 Multiplicity in Safety Data

The fundamental problem of signal detection in safety is the relatively massive multiplicity problem from considering all possible events over multiple timepoints. Left uncontrolled, this multiplicity will inevitably lead to false signals, which may lead to unnecessary IND serious adverse event (SAE) reporting and inclusion of adverse events as adverse reactions in product labeling that are unlikely to be removed in the future. Methods for multiplicity control are largely developed for efficacy. These controls assume much more modest multiplicity and are designed to be 'conservative,' meaning they err on the side of overlooking valid efficacy signals. Of course, in the realm of safety analyses, such a potential overcorrection of valid signals is considered 'anticonservative' and deemed unacceptable.

Those methods also may be overly ambitious because they strive to control the so-called family-wise error rate, which is the probability that any of the many tested hypotheses results in a false signal. In settings with massive multiplicity, methods may be preferable that strive for the less ambitious goal of controlling the "False Discovery Rate" or FDR (Benjamini and Hochberg 1995), which is the percentage of falsely rejected hypotheses. Since in that approach, a certain rate of false discoveries is accepted, it is clear that signals thus detected would have to go through a subsequent thorough signal validation, such as described in Chapter 3 about Safety Signal Evaluation.

An important note of caution is that a signal for an unanticipated safety issue can come from any source, whether pre-specified to be measured or not. Mathematically, this means that the dimensionality of the multiplicity problem is not even known. Therefore, many of the methods discussed below do not even attempt to adjust for multiplicity directly; rather, they try to define reasonably calibrated thresholds for signal detection. Unadjusted p values may serve as a simple "flagging device."

The following is a classification of adverse events that is often made (e.g., Crowe et al. 2009; Mehrotra and Heyse 2004); it is somewhat helpful in the context of multiplicity control as follows:

- Tier 1: Associated with specific hypotheses. Typically, these would be the safety topics of interest; the dimensionality is similar to efficacy, so methods for multiplicity control developed for efficacy may be considered. Techniques to adjust for multiple safety outcomes are rarely used in practice because of the already mentioned "anti-conservativeness" of such controls. Sequential monitoring methods to enable proper tracking of known or suspected safety issues over time are more commonly used; they are described below under "dynamic" methods.

- Tier 2: Common events. We can use methods based on FDR or Bayesian hierarchical methods to adjust for multiplicity of multiple safety outcomes, as further described below as "holistic, static" methods.

- Tier 3: Rare events. Unless they are "Medically Designated Events," which would be considered *a priori* drug-related, apparent signals would need to be thoroughly evaluated medically and epidemiologically by the principles described in Chapter 3 about Safety Signal Evaluation.

2.2.5.2 Some Methods for Monitoring and Signal Detection

The table below gives an overview of some common methods that may be used for Tier 1 and Tier 2 events, separately for scenarios of blinded and unblinded monitoring. The rows for "specific event" focus on one event at a time, while "holistic" means that the method looks at all events reported simultaneously and allows the whole picture to influence conclusions for single events. The methods for specific events considered here will not adjust for multiplicity, while the "holistic" approaches may address this important issue one way or another. The table also distinguishes between *"dynamic methods,"* which are proper for a sequence of looks over time, and *"static methods,"* which focus on a snapshot of safety data over time.

TABLE 2.3

Some Quantitative Frameworks for Screening for New Safety Signals

	Static	Dynamic
Blinded or single arm-specific event	Bayesian methods with binomial or Poisson likelihood	SPRT (Sequential Probability Ratio Test) versions SGLT (Sequential Generalized Likelihood Ratio Test) Bayesian methods, e.g., Thall-Simon beta-binomial model.
Unblinded – specific events	"Wittes rule" Beta-binomial model	GSDs (Group seq designs) SPRT versions; SGLT
Unblinded – holistic	Bayesian Hierarchical Models FDR-based methods	Chen's generalization of the Bayesian Hierarchical model

Blinded, static methods: Various Bayesian methods are available based on the same basic idea: The probability that the combined (or single-arm) event frequency exceeds what would be expected based on prior background rate data was evaluated. Beta-binomial models can be used for the simple case that repeated events can be ignored, and exposure adjustments are unnecessary. The Thall–Simon model, as described in the study by Yao et al. (2013), falls in that category, as well as the model by Ball (2011). For exposure-adjusted events or repeated events, Poisson / Gamma models can be used, as proposed by Schnell and Ball (2016), Kashiwabara et al. (2014), or Mukhopadhyay et al. (2018). While the former approaches require gamma priors, the procedure by Mukhopadhyay et al. (2018) is more flexible, allowing non-informative beta priors as well as prior distributions for the background rate.

Blinded, dynamic methods: For intermittent monitoring, group-sequential methods would need to be applied. For continuous monitoring, when data are processed after each exposure, the frequentist Sequential Probability Likelihood Ratio Test (SPRT) is a popular choice (Yao et al. 2013; Zhu et al. 2016). The SPRT requires the specification of the null and a simple alternative hypothesis. In scenarios where there is an effect but not near the specified alternative, the test behaves in undesirable ways. Kulldorff et al. (2010) pointed this out for single-event monitoring in a single-arm setting. They show how the so-called maxSPRT fixes the issue by letting the data determine the alternative hypothesis. If the null hypothesis is true, the maxSPRT tends not to stop the trial until the end. Sequential Generalized Likelihood Ratio (SGLR) tests also have a boundary to reject H1, and as such they tend to stop early to reject H1, under the H0 scenario. Shih et al. (2010) showed through simulations the considerable efficacy gains of the SGLR tests in terms of expected number of events before stopping.

On the Bayesian side, all the methods mentioned in "Blinded, static methods" are also used for the repeated analysis of accumulating data.

Unblinded, specific events, static: For this scenario, Wittes et al. (2015) gave a pragmatic rule that uses an unadjusted confidence interval as a "flagging device"; other information is pulled in to validate signals thus obtained. One interesting approach is to examine whether the signal disappears or strengthens when similar events are included in the definition.

Versions of Bayesian beta/binomial and gamma/Poisson methods exist that can also be used in the two-group unblinded setting.

Unblinded, specific events, dynamic: Specific events can be monitored by traditional group sequential designs (Jennison and Turnbull 1999). For continuous monitoring, the SPRT and its offsprings are available in the unblinded setting as well. A frequent application would be monitoring for mortality.

Unblinded, static, holistic: There are two seminal articles with methods that look at all AE data and consider multiplicity and correlation structures: One frequentist and one Bayesian.

The Bayesian one is by Berry and Berry (2004); they proposed a Bayesian hierarchical model that tries to incorporate biological correlations between events by considering the MedDRA structure. Decisions are based on the posterior distribution of differences between treatments. Similar events – those in the same SOC (System Organ Class) – will strengthen each other when all show a signal; if only one event in a group shows a signal but other events are "silent," there will be a "shrinkage" or downward correction toward the null hypothesis. Therefore, one can say that the model addresses multiplicity to some extent as it reels back possibly random, extreme differences.

This work spawned numerous follow-up articles to extend the model to more scenarios. The model, its extensions, and its limitations are described in detail in Chapter 7 about Bayesian safety surveillance methods.

The Frequentist work that tries to adjust for multiplicity by looking at all events is based on FDR: The "double-FDR" method by Mehrotra and Heyse (2004), Mehrotra, and Adewale (2012) proposes the natural clustering of AEs into SOCs and adjusts p values by using the FDR approach by Benjamini and Hochberg that was mentioned above. They perform this adjustment in two ways: One way is to pick the minimum p-value in an SOC as this reflects the strongest safety signal in that SOC. The other way is to adjust within each SOC. An AE is considered a safety signal if flagged in both of these ways.

Both approaches rely on the clustering of adverse events imposed by the MedDRA dictionary. One common criticism is that the MedDRA structure does not necessarily reflect biological reality. For instance, medically closely related terms may show up in different MedDRA SOCs. Of course, the methods would be easily applicable based on alternative clusterings that are more plausible biologically.

Unblinded, holistic, dynamic: Chen et al. (2013) applied the Bayesian hierarchical model in a group sequential manner.

Meta-Analysis: To evaluate a safety signal, we need to consider the totality of the evidence, not just the piece of evidence observed in ongoing, possibly blinded trials. Crowe et al. (2009) already emphasized the need for meta-analysis to integrate evidence for safety signals from completed studies. The huge body of literature for meta-analysis is not reviewed here. However, an important development is that new techniques are emerging which allow formal combination of evidence from disparate sources, such as randomized and non-randomized data sources (Röver and Friede 2020) or unblinded completed and blinded ongoing studies (Lin et al. 2020).

Links to other book chapters: What is offered here is only a top-level walk through of the considerable body of statistical literature applicable to pre-marketing safety surveillance. A more in-depth exploration is given in the methodological chapters of this book: Chapter 6 covers methods for the blinded review of safety data, Chapter 7 covers Bayesian methods, and Chapter 8 is an in-depth dive into likelihood-ratio-based tests. Chapter 9 is devoted to meta-analysis in the context of safety data.

2.2.6 Section 6: Communication of Safety Information

In addition to regulators that require regular updates of the accumulating safety data (DSUR, PBRER, "for cause" IND safety reports), there is increasing pressure on the industry to post clinical research information in the public domain (e.g., EMA 2018 about the implementation of Policy 070). Other stakeholders, including the investment and scientific communities, also demand updates on important findings. All of these pressures, taken together, lead to a bewildering landscape of documents, postings, and communications that are prone to disseminating conflicting messages regarding safety, if not held together by a common base that we call the "safety storyboard." The safety storyboard will evolve and therefore should be updated with the cadence of OASE / ASAP updates. The vision is that the safety storyboard should elucidate the key messages about risk in the context of the condition's epidemiology, unmet medical need, and benefit-risk considerations, including risk management.

2.3 The ASAP as a Modular Template-Driven Approach

In most companies, components of the procedures proposed here will already have been implemented via existing guidance and SOP documents. We propose a modular approach by which a company will draw from those ASAP concepts that are not yet addressed. The ASAP could function as a "master document"; it would contain pointers to relevant SOPs and guidance for those sections that those pre-existing documents already address.

While the ASAP gives us the structure of WHAT is needed for proactive planning of safety surveillance in a clinical program, the ASAP Template (Hendrickson et al. 2021) addresses the HOW by providing a concrete tool for implementation. This document expands on the condensed contents shown here, and it also provides a location and structure of a safety surveillance plan as proposed by the FDA guidance (2021).

2.4 The ASAP as a Safety Planning Tool to Respond to a Pandemic Such as COVID-19

The COVID-19 pandemic hit most nations unprepared and also revealed a lack of preparedness in clinical research. First, the crisis led to clinical study enrolment halts and gaps in data collection of subjects who were already enrolled, as people self-isolated at home and minimized visits to investigative sites. Second, the pandemic led to multiple reports of an unanticipated adverse events (COVID-19) among subjects in the clinical trials.

Let us dissect these effects separately and outline how the ASAP could govern a systematic response to the next pandemic that, based on expert views, is likely to occur within decades.

2.4.1 General Effect on Data Collection

Besides enrolment halts, one important effect of the COVID-19 pandemic was decreased intensity of follow-up and data collection of patients who were already enrolled in a trial. This was due to recommended or – depending on the country – enforced policies to self-isolate at home. Remote visits replaced many study visits, but all the tests mandated by the protocol could frequently not be conducted in such a setting.

Also, physical distancing and reluctance to seek in-person medical care during a pandemic led to decreased reporting of events or more severe outcomes of some events due to delays in treatment.

While these effects may affect all treatment arms equally, safety monitoring implications are further discussed below.

2.4.2 Unanticipated Infections of Subjects in the Clinical Trials Due to the Pandemic

Identification of infections: The MedDRA organization responded swiftly by making several new, "COVID-19 associated" preferred terms available. These included terms for "Asymptomatic COVID-19", "COVID-19", "Suspected COVID-19", and "SARS-COV-2 carrier", among others. A subsequent MedDRA update offered a new COVID-19 SMQ. Initially confirmed and suspected COVID-19 events were coded to the two relevant preferred terms of "coronavirus infection" and "coronavirus test positive." In some cases, events coded to these preferred terms were other types of coronavirus infection, particularly in pediatric patients. Also, as expected for a newly emerging disease, diagnostic testing was not widely available initially. Later, diagnostic results were generally only available for hospitalized patients. During this period, the diagnosis, particularly for non-hospitalized patients, was made primarily based on characteristic symptoms and the presence of relevant risk factors. However, there was no single pathognomonic symptom, and the understanding of the constellation of symptoms associated with COVID-19 evolved.

Consequently, early in the pandemic, both over-reporting and under-reporting of COVID-19 occurred. As a result, scrutiny of cases was needed to ensure the appropriate coding of adverse events. In addition, for trials ongoing at the start of the pandemic, ensuring proper recoding of previously reported events to the newly available COVID-19 preferred terms was needed with the MedDRA upgrade.

Pandemic infections as a dependent variable: The probability of becoming infected could be considered a dependent (outcome) variable, which may differ among the treatment arms. There could be a direct effect of the active drug, in the sense of increasing the risk of symptomatic infection or more severe outcomes. Also, there could be indirect effects of randomized treatment: For instance, in a trial of inflammatory disease, placebo patients may resort to an immunosuppressive rescue medication more frequently, and that medication might make them more susceptible to the new infection.

Pandemic infections as an independent variable: An infection event could also act as an independent variable that causes other downstream events, including withdrawal from the trial, hospitalizations, treatments for the infection, and other AEs that are complications of the illness or the medications given for treatment.

2.4.3 Using the ASAP To Respond to a Pandemic:

The ASAP is a dynamic, evolving document that can adapt to a clinical program's changing circumstances, such as the emergence of a pandemic. The following is a list of the most important considerations:

- Safety topics of interest: If the MoA or other prior information indicates possible immunosuppressive effects, the inclusion of pandemic infections – in this case of COVID-19 – should be considered a safety topic of interest.

- Data analysis approaches: The estimand approach could be used to answer pandemic-related safety questions, such as "percentage of patients infected" in both arms. That would give intercurrent events the proper attention; for instance, how would withdrawals before infection be handled?

- The ASAP section regarding the analysis of key gaps could elucidate novel gaps in data collection such as lack of a pre-specified mechanism to capture "withdrawal due to pandemic infection" and possibly establish conventions of how to deal with reporting pandemic-related adverse events. Identifying pandemic-related events for presentation in regulatory submission documents taking into account MedDRA version issues could also be addressed here.

- OASE: Due to the pandemic's various effects, background rates from external data sources may no longer be valid. For instance, in one ongoing cardiovascular outcome trial, a measurable dip in cardiovascular event rates was noted. Consequently, the availability of intra-study comparator arms becomes even more valuable.

  Undiagnosed infections may manifest themselves through their symptoms, some of which may not be initially recognized as part of the disease spectrum, e.g., the coagulopathy-related symptoms of COVID-19 or post-infection multi-system inflammatory syndrome. Such events would tend to disguise as "Type B" events, which would likely trigger IND safety reporting. In the case of COVID-19 again, the virus rolled over the world geographically in time, which is to say that COVID-19 initially was not everywhere at every timepoint. Therefore, apparent clusters of "Type B" events would need to be assessed, considering the locations where and times when the infections spiked. During periods of high general infection activity in a geographical area, a direct causal relationship to experimental treatment is less likely.

- Communication of safety information should include an evaluation of uncertainties introduced by the pandemic regarding the rate of adverse event reporting and the assessment of drug-relatedness of adverse events.

2.5 Anticipated Hurdles toward Implementation of the ASAP

The original publication (Hendrickson et al. 2021) seeks to facilitate the ASAP implementation by providing a template for the ASAP. Nevertheless, active change management will be required in organizations to address some anticipated objections against the concept:

Yet another document: The most obvious hurdle is the "Yet another document" reaction. One response would be that project-level documents and standardization make sense, facilitate health authority requests, and are already happening. The ASAP will assist this trend by providing a common foundation for collecting and analyzing project-wide safety data. The most powerful argument may be that the ASAP may make other documents of similar nature obsolete, such as safety surveillance plans or safety review plans.

We do not know yet where this drug is going: Attrition (projects dying away) is a good reason for not putting too much work into some project activities too early. The ASAP should be fit for purpose. In Phase 1, *don't produce a Cadillac version quite yet*! We are aware of a case example from a major pharmaceutical company where a lot of thinking and effort went into planning a Phase 3 meta-analysis, but then the drug died early without a single Phase 3 study being started. Certain items from the ASAP, such as an agreed list of safety topics of interest, will always make sense. The ASAP's modular structure allows for only those modules to be used that contribute value at a given point in time.

We are already doing it (informally). During the last ten years, interdisciplinary SMTs have become common, but they tend to perform aggregate reviews without a plan. Gaining alignment on aggregate analysis via a tool like the ASAP will result in project efficiencies and benefits related to more forward-thinking regarding safety planning.

We are already doing it, and it is governed by other documents, such as our ISS SAP. Whereas different parts of the ASAP may repeat information already described in other documents, the ASAP is the only document that can serve as a *single reference point* for all the essential safety analyses for a given project. Proactive safety strategies and thoughtful approaches for safety data assessment/presentation are critical at all levels (medical monitoring/CSR/CDT Section 2.7.4/DSUR/ISS etc.) as the project progresses through the development stages.

We do not have the resources. We need to emphasize that enabling the company to respond to unexpected safety issues effectively may save a program, which will easily justify resourcing. Providing examples of how the creation of an ASAP significantly benefited a development program is important.

Our program is small / the ASAP does not apply to orphan diseases.
Again, a "Cadillac plan" may not be justified for a rare disease with
only one or two studies. But just because a project is small does not
mean a sponsor should skip safety planning altogether. Certainly, the
"Final Rule" does not distinguish between different sizes of companies
or study development programs. Also, small programs for an orphan
disease could later balloon by adding on new and sometimes larger
indications.

An organization needs to have a certain maturity before considering
implementing the ASAP. In the following section, we elucidate a couple of
foundational requirements that indicate that state of maturity.

2.6 Foundation: Intelligent Data Architecture

The FDA IND annual safety report (INDAR) required *"a narrative or tabular
summary showing the most frequent and most serious adverse experiences by body
system."* (FDA 2015), which was a strong motivator for companies to integrate their accumulating clinical data on an ongoing basis. There is no such
requirement for the DSUR, which may be submitted to the FDA in place of
an INDAR. In a way, this focus on the uniform submission of DSURs across
regions was a "curse in disguise" because now a more intentional effort is
typically needed to structure ongoing data for easy integration.

In the introduction to this book, we described the four pillars of Safety
Analysis: Cross-disciplinary scientific engagement, Methods and analytic
tools, Intelligent data architecture, and Effective and efficient process. While
the ASAP contributes to the first and fourth pillar, the third pillar could be
considered a prerequisite, so let us spend some consideration on "intelligent
data architecture" that enables easy integration across closed and ongoing
trials. Elements of such an architecture include:

(1) Consistent implementation of data derivation rules

(2) Consistent MedDRA coding

(3) Data traceability and Single Source of Truth.

2.6.1 Consistent Implementation of Derivation Rules

Health authorities worldwide expect patient-level data from clinical trials
to be available in a well-defined standard called "Study Data Tabulation
Model" or SDTM (www.cdisc.org). Moreover, analysis-ready data containing

all important derived variables also need to be provided in a standard format called "Analysis Data Model" or ADaM (www.cdisc.org). Therefore, for each study, data should be available in those compatible SDTM and ADaM formats, facilitating data integration by simply "stacking" the ADaM or SDTM data.

Let us consider the first method: Stacking ADaM datasets. The danger is that such an approach does not consider the fact that derivations may have been implemented slightly differently for each study. Even if the same variable names are used in different analysis datasets (ADaM), they may not mean exactly the same. The advantage, of course, is that an analysis based on such a combined dataset, by study, will reproduce the study results precisely, and analyses across studies will "add up" to the single-study results. The big downside is that data could be analyzed inconsistently across studies which may not be acceptable in a formal integrated analysis.

A second method would be to stack together the SDTM datasets. Since there are fewer derived variables in the SDTM layer, the risk of inconsistent derivations would be smaller. However, it is not eliminated. For instance, the same treatment arm may have been called differently in two different studies, particularly when different organizations did both. This approach applies a consistent set of rules across the SDTM layer to create analysis datasets (in ADaM format) for integrated analysis. Now all data will be analyzed the same way, but there may be slight differences between the combined approach and the study-wise analyses that would need to be identified and explained in the Reviewer's Guide. Another potential issue with simply stacking SDTM datasets is that older studies may have been created using older SDTM versions meaning that the layout and purpose of variables are not aligned.

A third way to overcome this and some of the other issues mentioned above would be what could loosely be called *"Data Warehousing"*: A company defines its own standard of a core set of safety variables and always converts its internal data structure into that core set, on an ongoing basis for ongoing studies ensuring harmonization of the variable names and values. That approach facilitates creating safety SDTM datasets because they would be based on this data warehouse layer. The hope is that strict internal governance of the metadata for this core set, over time, would reduce inconsistencies among studies.

2.6.2 Consistent MedDRA Coding

Another reason for inconsistencies among studies is that they typically will have been coded using different MedDRA versions. Even if the same version was used, differences among studies could arise if the same coding conventions were not used. As an example, in a program where infusion events are important, a coding convention may be implemented to code both "Injection site reaction" and "Infusion site reaction" to the same lower level terms (LLT) of "Infusion site reaction" – but this approach would need to be done consistently for all studies.

These issues – inconsistent coding conventions and different MedDRA versions – create a situation where the same preferred term, used in different studies, may not mean exactly the same thing. For a proper pooled analysis, one would need to agree on a target MedDRA version and code all older studies into this version, based on the same MedDRA conventions. This is the process recommended in the FDA's Technical Conformance Guide (FDA 2020, Section 6.1.2.). Again, if this process is applied to older studies that are already closed, inconsistencies between the new "integrated" values and corresponding study level values will emerge; those should be identified and documented in the Analysis Data Reviewer's Guide.

We want to note here that MedDRA upgrades are not always innocuous or trivial; for instance, some past changes have significantly affected AE frequencies. A useful principle is *"up-versioning"* based on LLTs since this layer already reflects the implementation of coding rules. Also, LLTs typically will not change between MedDRA versions, except that some LLTs may be marked as *"non-current."* A notable exception to this rule is the occurrence of new or previously unrecognized diseases, which leads to the introduction of novel LLTs (such as the COVID-19 example discussed above).

One advantage of the *"Data warehouse"* approach described above is providing coding values for both the study-specific MedDRA version and the integrated MedDRA version. This way, both sets of values would be in the same dataset, making discrepancies easily identifiable.

2.6.3 Data Traceability and Single Source of Truth

The core principle of data integrity is that there must be a unidirectional flow of information from the source to the downstream converted datasets, such that:

- Upstream changes will trigger downstream data changes in an inevitable, automatic way.
- Downstream data cannot be edited or changed in isolation.

Generally speaking, if the links between the various data stages (raw, SDTM, ADaM, Tables/Listings/Graphs) are severed technologically or organizationally, data integrity will be at risk. One way to address this would be creating a Data Warehouse, as described above, that is closely connected technologically with the original data (typically EDC) and created and maintained by in-house resources.

2.7 Foundation: Background Rates

Based on current regulatory guidance, individual SAEs that are "anticipated events" for the study population may not necessarily need to be reported to the FDA (2021). Diseases are often associated with increased background rates of other co-morbid conditions, and sometimes these associations are not obvious. While the observation that depression rates are increased in patients with Alzheimer's disease may not be surprising, the well-established association between Parkinson's disease and melanoma is certainly not something one might have guessed. However, epidemiological studies have clearly shown that melanoma occurs more frequently among patients with Parkinson's disease. While the reason for this observation is not known, alterations in genes or proteins common to the pathogenesis of both Parkinson's disease and melanoma suggest a possible mechanistic connection between the two conditions (Bose et al. 2018). Unsurprisingly, when this relationship first emerged, the common treatment of Levodopa was blamed for the increase in melanoma cases, leading to warnings and precautions in the product label in several countries. Despite calls of removing such statements (Vermeij et al. 2009), at least the U.S. Levadopa label still has this warning.

Another example of the importance of disease-specific background rates is the clinical development program in Myelofibrosis of Fetratinib, a Jak2 selective inhibitor. In the pivotal trials, eight cases of Wernicke's encephalopathy were observed, which could not be explained through the drug's mechanism of action. Based on these events, the FDA put the program on clinical hold in 2013, and the sponsor later canceled the development of the drug. In 2017, one of the investigators re-analyzed the cases and found that the background rate of Wernicke's syndrome was elevated in Myelofibrosis, and that, with more thorough diagnostics, only some of the eight cases could be confirmed (Harrison et al. 2017). Both of those factors resulted in a re-analysis that showed that the incidence rate in the trials was *lower* than what would have been expected based on the background rate. We note that Fetratinib did obtain U.S. marketing approval for intermediate-2 or high-risk primary or secondary Myelofibrosis in 2019 after a checkered development history, but with encephalopathy as a boxed warning.

These examples should suffice to underline the importance of a thorough epidemiologic understanding of the target disease's epidemiology *and* all of its concomitant conditions. The goal is to establish a list of disease-specific background rates for unexposed patients. Since not all possible safety signals can be anticipated, having an appropriate "stand-by" database that could be interrogated for background rates once the need arises is desirable. These data would consist of a "standing cohort" of patients who would have qualified for the ongoing trial(s) (e.g., meet the study eligibility criteria and match the study population demographics). In the following, we list some data sources that are commonly used for those purposes.

(1) Dedicated natural history study

In recent draft guidance (FDA 2019), FDA advocates for the conduct of dedicated natural history studies for rare diseases. They define a natural history study as a *"preplanned observational study intended to track the course of the disease."* There are many potential benefits to such studies, including identifying and characterizing the target patient population, supporting the development of Clinical Outcome Assessments, and supporting biomarker development. Moreover, data from a natural history study could provide an external control arm to allow efficacy comparisons with a single-arm clinical trial.

While not specifically mentioned in the guidance, we believe that data from such a study – if properly sized - could also be used to provide reference background rates for safety events. Importantly, and similar to the *"standing cohort"* approach mentioned below, the data could be interrogated for occurrence, in the unexposed target population, of a new, unexpected suspected adverse reaction.

(2) Data from unexposed patients from prior clinical trials

One would think of clinical trials as an excellent source of background information. Still, there are obvious and less obvious issues with such data: The obvious one is that typical clinical trials are relatively small, yielding little power to detect differential rates for rare events. The Phase 3 program of Ruxolitinib in Myelofibrosis revealed a less obvious bias: This program consisted of one double-blind, placebo-controlled trial (Verstovsek et al. 2012), and one open-label of Ruxolitinib vs. best standard of treatment (BAT) (Harrison et al. 2012): While the safety profiles for the active drug were comparable in both studies, the safety profiles for the control patients were quite different: Nineteen percent of the placebo patients reported Grade 3/4 anemia, but only 13% of the BAT patients; 11% of placebo patients reported abdominal pain vs. 3% in the BAT arm, and 7% in placebo arm reported fatigue vs. zero in the BAT arm. Although this comparison admittedly involved two treatment arms in two different studies, the inclusion/exclusion criteria were similar. These data suggest that underreporting of adverse events went on when patients continued their usual treatments in an open-label manner and experienced their anticipated side effects.

(3) Literature review and Meta-analysis

An example of the results from literature research for the event "Syncope" is shown in the appendix of the original ASAP publication (Hendrickson et al. 2021). How much the rates vary among different sources is striking.

Such disparate results will have to be dealt with in some kind of integrated analysis, such as meta-analysis. It is now generally acknowledged that a Bayesian meta-analysis deals with the issue of heterogeneity better than older frequentist methods, in particular when there are only a few studies (Röver and Friede 2020): By "listening to" all other data, a Bayesian analysis will shrink extreme study-specific results to the middle, making them less extreme, and also produce a useful overall estimate whose spread incorporates both uncertainty and inconsistency.

(4) Registries

Registries span a wide range of different scenarios that have in common that they track patients in a non-interventional way, but data are collected purposefully for research. Disease registries enroll patients with the target disease, no matter what treatment they receive, while drug registries enroll patients that receive a certain drug for the condition. The former type is often sponsored and managed by non-pharmaceutical organizations, while the latter type is typically funded by the drug company.

Consequently, registries vary widely in scope and quality. Even in some highly regarded registries, collection of safety may be somewhat of an afterthought. For instance, verbatim terms may not be collected, no "seriousness" assessment may exist, and the safety data collection tool may simply be a series of checkboxes next to a list of system organ classes that may not correspond to MedDRA organ classes. However, other registries have become more sophisticated in safety data collection with detailed adverse event forms for safety topics of interest for the disease or relevant product classes used for treatment.

(5) Electronic Health Records (EHRs)

Real-world evidence (RWE) data sources include electronic claims databases and electronic medical records. When working with EHRs to establish background rates or to provide context information for events observed in an ongoing trial, several potential issues need to be considered:

- Coding is not done with MedDRA but with various dictionaries (ICD, CPT, SNOMWED, UK "read codes," etc.)

- Patient characteristics would need to be matched to the inclusion/exclusion criteria of the clinical trial, which may be more difficult than with some of the other data sources mentioned.

- Events captured in EHRs are usually those leading to seeking medical care. EHR will not capture all SAEs (for example, deaths occurring outside of the hospital).

2.8 Trends Likely to Influence the Further Direction of the ASAP

We would like to finish with consideration of some broad industry and regulatory trends that will likely influence the content and further evolution of the ASAP:

Data visualization: In the age of democratization of data analysis (Carmichael & Marron 2018), there will be an ever-increasing demand for data visualization and safety graphics to enable data customers and decision-makers to draw their own conclusions. Simple, static paper graphs will no longer meet an increasingly sophisticated client base. For instance, the notion that one would need to use listings to look up a far outlier to obtain the patient ID behind the plotting symbol needs to be relegated to a past century. Nowadays, the expectation is that the ID and possibly more patient data be revealed when the user hovers the cursor over a data point representing a study subject on a graphic. Another beneficial consequence of the democratization trend is that more people are making their programs publicly available in venues such as *Github*. Some useful examples of modern displays of safety data are given in Chapter 3 of this book (Safety Signal Evaluation).

RWE as the linchpin between pre-and post-marketing safety surveillance: Triggered by the 21st Century Cures act in the US and a similar trend in Europe, regulators are now open to accepting submissions based on RWE data for secondary indications. Examples of successful filings of this nature already exist, either fully based on RWE (Bartlett et al. 2019) or partially based, in which RWE contributed control data (Gökbuget et al. 2016). For such development programs, the ASAP will need to address the peculiarities of RWE data; some of those challenges are listed above in the discussion of the use of RWE data for background event rates. Such an ASAP could lead almost seamlessly into post-marketing surveillance as the same RWE data sources could be used for that purpose.

Transparency requirements: More and more pressure is being put on the industry to be fully transparent about their data and documents, including their publication and dissemination. We expect this trend to continue, leading to the release of more records to new audiences, including patient communities (e.g., layperson summaries of clinical trial outcomes). This expectation will lead to ever-increasing chances of communicating safety results in an inconsistent, contradictory way, which will make the concept of the ASAP "safety storyboard" increasingly important.

Benefit-risk: More and more, regulators look at risks not in isolation but in conjunction with benefits. Acceptable trade-offs between risks and benefits differ among various stakeholder groups (health authorities, insurers, and patients) and even within those groups. For instance, risk-averse patients look at benefit-risk trade-offs differently from early-adopting risk-takers (Ho et al. 2015). These complex issues will require careful planning that takes risk-benefit evaluations beyond the qualitative level. We anticipate that analogous to the ASAP, a Benefit-Risk Assessment Plan will emerge. Within the Safety Working Group, a task force to address benefit-risk planning has already been formed. We believe that a sound foundation of any benefit-risk assessment consists of thorough evaluations of both benefits and risks separately. More than a decade ago, the point was made that there is a gap in thoroughness between efficacy and safety, favoring efficacy (O'Neill 2008). The issues highlighted in that article have essentially remained the same since then. We believe now is the time to close that gap, and the ASAP is an important tool to this end.

Note

1 FDA Clinical Review Memo, 2017: https://www.fda.gov/media/105640/download

References

Ball G. 2011. Continuous safety monitoring for randomized controlled clinical trials with blinded treatment information. Part 4: One method. *Contemporary Clinical Trials* 32:S11–17.

Ball G, Kurek R, Hendrickson BA, Buchanan J, Wang WW, Duke SP, Bhattacharyya A, Li M, O'Brien D, Weigel J, Wang W, Jiang Q, Ahmad F, Seltzer JH, Herrero-Martinez E, and Tremmel L. 2020. Global regulatory landscape for aggregate safety assessments: Recent developments and future directions. *Therapeutic Innovation & Regulatory Science* 54(2):447–461.

Bartlett C, Mardekian J, Yu-Kite M, et al. 2019. Real-world evidence of male breast cancer (BC) patients treated with palbociclib (PAL) in combination with endocrine therapy (ET). *Journal of Clinical Oncology* 37:1055–1055.

Benjamini Y, and Hochberg Y. 1995. Controlling the false discovery rate: A practical and powerful approach to multiple testing. *Journal of the Royal Statistical Society. Series B (Methodological)* 57:289–300.

Berry SM, and Berry DA. 2004. Accounting for multiplicities in assessing drug safety: a three-level hierarchical mixture model. *Biometrics* 60(2):418–426.

Bose A, Petsko GA, and Eliezer D. 2018. Parkinson's disease and melanoma: Co-occurrence and mechanisms. *Journal of Parkinson's Disease* 8(3): 385–398.

Carmichael I, and Marron JS. 2018. Data science vs. statistics: Two cultures? *Japanese Journal of Statistics and Data Science* 1:117–138.

Chen W, Zhao N, Qin G, and Chen J. 2013. A Bayesian group sequential approach to safety signal detection. *J Biopharm Stat.* 23(1):213–230.

Chuang-Stein C, and Xia HA. 2013: The practice of pre-marketing safety assessment in drug development. *Journal of Biopharmaceutical Statistics* 23(1):3–25.

CIOMS 2005: Management of Safety Information from Clinical Trials. Report of CIOMS Working Group VI. https://cioms.ch/wp-content/uploads/2017/01/Mgment_Safety_Info.pdf

Colopy MW, Gordon R, Ahmad F, Wang WW, Duke SP, and Ball G. 2019: Statistical practices of safety monitoring: An industry survey. *Therapeutic Innovation & Regulatory Science* 53(3):293–300.

Crowe BJ, Xia HA, Berlin JA, Watson DJ, Shi H, Lin SL, Kuebler J, Schriver RC, Santanello NC, Rochester G, Porter JB, Oster M, Mehrotra DV, Li Z, King EC, Harpur ES, and Hall DB. 2009. Recommendations for safety planning, data collection, evaluation and reporting during drug, biologic and vaccine development: a report of the safety planning, evaluation, and reporting team. *Clin Trials* 6(5):430–440.

EMA 2018. External guidance on the implementation of the European Medicines Agency policy on the publication of clinical data for medicinal products for human use. Revision 4 – adopted Guidance. https://www.ema.europa.eu/en/documents/regulatory-procedural-guideline/external-guidance-implementation-european-medicines-agency-policy-publication-clinical-data_en-3.pdf

FDA 2005. Reviewer Guidance Conducting a Clinical Safety Review of a New Product Application and Preparing a Report on the Review https://www.fda.gov/media/71665/download

FDA 2010. Code of Federal Regulations Title 21 Food and Drugs; Chapter I Food and Drug Administration Department of Health and Human Services Subchapter D Drugs for Human Use Part 312 Investigational New Drug Application.

FDA 2012. Safety Reporting Requirements for INDs (Investigational New Drug Applications) and BA/BE (Bioavailability/Bioequivalence) Studies. Guidance for Industry and Investigators. https://www.fda.gov/regulatory-information/search-fda-guidance-documents/safety-reporting-requirements-inds-investigational-new-drug-applications-and-babe

FDA 2015. Safety Assessment for IND Safety Reporting. Draft Guidance for Industry, https://www.fda.gov/media/94879/download

FDA 2019. Rare Diseases: Natural History Studies for Drug Development. Draft Guidance for Industry. https://www.fda.gov/regulatory-information/search-fda-guidance-documents/rare-diseases-natural-history-studies-drug-development

FDA 2020. Study Data Technical Conformance Guide. https://www.fda.gov/media/136460/download

FDA 2021. Sponsor Responsibilities - Safety Reporting Requirements and Safety Assessment for IND and BA/BE Studies. https://www.fda.gov/regulatory-information/search-fda-guidance-documents/sponsor-responsibilities-safety-reporting-requirements-and-safety-assessment-ind-and

Gökbuget N, Kelsh M, Chia V et al. 2016. Blinatumomab vs. historical standard therapy of adult relapsed/refractory acute lymphoblastic leukemia. *Blood Cancer Journal* 6, e473. https://doi.org/10.1038/bcj.2016.84

Harrison C, Kiladjian JJ, Al-Ali HK, Gisslinger H, Waltzman R, Stalbovskaya V, McQuitty M, Hunter DS, Levy R, Knoops L, Cervantes F, Vannucchi AM, Barbui T, and Barosi G. 2012. JAK inhibition with ruxolitinib versus best available therapy for Myelofibrosis. *The New England Journal of Medicine* 366(9):787–798.

Harrison C, Mesa R, Jamieson C, Hood J, Bykowski J, Zuccoli G, and Brewer J. 2017. Case series of potential Wernicke's encephalopathy in patients treated with fedratinib. *Blood, 130*:4197–4197.

Hendrickson BA, Wang W, Ball G, Bennett D, Bhattacharyya A, Fries M, Kuebler J, Kurek R, McShea C, Tremmel L. 2021. Aggregate safety assessment planning for the drug development life-cycle. *Therapeutic Innovation & Regulatory Science*. doi: 10.1007/s43441-021-00271-2. Epub ahead of print. PMID: 33755928.

Ho MP, Gonzalez JM, Lerner HP, Neuland CY, Whang JM, McMurry-Heath M, Hauber AB, Irony T. 2015. Incorporating patient-preference evidence into regulatory decision making. *Surgical Endoscopy* 29(10):2984–2993.

ICH 2020. E9 (R1) addendum on estimands and sensitivity analysis in clinical trials to the guideline on statistical principles for clinical trials https://www.ema.europa.eu/en/documents/scientific-guideline/ich-e9-r1-addendum-estimands-sensitivity-analysis-clinical-trials-guideline-statistical-principles_en.pdf

Jennison C, and Turnbull BW. 1999. *Group sequential Methods with Applications to Clinical Trials*. Boca Raton: Chapman & Hall.

Kashiwabara K, Matsuyama Y, and Ohashi Y. 2014. A Bayesian stopping rule for sequential monitoring of serious adverse events. *Therapeutic Innovation & Regulatory Science* 48(4):444–452.

Kulldorff M, Davis RL, Kolczak M, Lewis E, Lieu T, and Platt R. 2010. A maximized sequential probability Ratio test for Drug and vaccine safety surveillance. *Sequential Analysis* 30:58–78.

Lin LA, Yuan SS, Li L, and Ball G. 2020. Meta-analysis of blinded and unblinded studies for ongoing aggregate safety monitoring and evaluation. *Contemporary Clinical Trials* 95:106068.

Mehrotra DV, and Heyse JF. 2004. Use of the false discovery rate for evaluating clinical safety data. *Statistical Methods in Medical Research* 13(3):227–238.

Mehrotra DV, and Adewale AJ. 2012. Flagging clinical adverse experiences: Reducing false discoveries without materially compromising power for detecting true signals. *Statistics in Medicine* 31(18):1918–1930.

Morton SC. 2005. Ephedra. *Statistical Science* 20:242–248.

Mukhopadhyay S, Waterhouse B, and Hartford A. 2018. Bayesian detection of potential risk using inference on blinded safety data. *Pharmaceutical Statistics* 17(6):823–834.

O'Neill RT. 2008. A perspective on characterizing benefits and risks derived from clinical trials: Can we do more? *Drug Information Journal* 42(3):235–245.

Röver C, and Friede T. 2020. Dynamically borrowing strength from another study through shrinkage estimation. *Statistical Methods in Medical Research* 29(1):293–308.

Schnell P, and Ball G. 2016. A Bayesian exposure-time method for clinical safety monitoring with blinded data. *Therapeutic Innovation & Regulatory Science* 50(6):833–838.

Shih MC, Lai TL, Heyse JF, and Chen J. 2010. Sequential generalized likelihood ratio tests for vaccine safety evaluation. *Statistics in Medicine* 29(26):2698–2708.

U.S. Department of Health and Human Services 2017. Common Terminology Criteria for Adverse Events (CTCAE). https://ctep.cancer.gov/protocoldevelopment/electronic_applications/docs/ctcae_v5_quick_reference_5x7.pdf

Vermeij JD, Winogrodzka A, Trip J, and Weber WE. 2009. Parkinson's disease, levodopa-use and the risk of melanoma. *Parkinsonism & Related Disorders* 15(8):551–553.

Verstovsek S, Mesa RA, Gotlib J, et al. 2012. A double-blind, placebo-controlled trial of ruxolitinib for Myelofibrosis. *The New England Journal of Medicine* 366(9):799–807

Wittes J, Crowe B, Chuang-Stein C, Guettner A, Hall D, Jiang Q, Odenheimer D, Xia HA, and Kramer J. 2015. The FDA's final rule on expedited safety reporting: Statistical considerations. *Statistics in Biopharmaceutical Research* 7(3):174–190.

Yao B, Zhu L, Jiang Q, and Xia HA. 2013. Safety monitoring in clinical trials. *Pharmaceutics* 5(1):94–106.

Zhu L, Yao B, Xia HA, and Jiang Q. 2016. Statistical monitoring of safety in clinical trial. *Statistics in Biopharmaceutical Research* 8(1):88–105.

3

Safety Signaling and Causal Evaluation

James Buchanan

Covilance, LLC, Belmont, CA, USA

Mengchun Li

TB Alliance Inc., New York, NY, USA

CONTENTS

3.1 How to Detect a Signal...46
    3.1.1 Preclinical Findings 46
    3.1.2 Single-Case Reports..46
    3.1.3 Aggregate Data – Clinical Development...47
        3.1.3.1 Adverse Events...48
        3.1.3.2 Laboratory Data 50
    3.1.4 Aggregate Data – Post-Marketing 51
3.2 How to Evaluate a Signal..52
    3.2.1 Confounding and Bias..52
    3.2.2 Causal Association...54
3.3 How to Determine When a Signal Represents a Product Risk..............59
    3.3.1 Strength of Association 60
    3.3.2 Consistency...61
    3.3.3 Specificity...61
    3.3.4 Temporality...61
    3.3.5 Biologic Gradient 62
    3.3.6 Plausibility 63
    3.3.7 Coherence..63
    3.3.8 Experimental Evidence 64
    3.3.9 Analogy 64
3.4 What Tools Are Available?...66
    3.4.1 Common Graphical Displays for the Adverse
        Event Review..66
    3.4.2 Common Graphical Displays for the Laboratory
        Data Review...70
    3.4.3 Other Tools, Platforms, and Resources..71
3.5 Summary..72
Notes..73
References..73

DOI: 10.1201/9780429488801-4

3.1 How to Detect a Signal

The term "signal detection" is commonly used but, even among safety professionals, can be subject to misunderstanding. The Council for International Organizations of Medical Sciences (CIOMS) Working Group VIII (2020) report succinctly defined a signal as "an event with an unknown causal relationship to treatment that is recognized as worthy of further exploration and continued surveillance". Thus, the process of detecting a signal is identifying a finding deserving further investigation because it raises the possibility of a causal relationship to the drug in question in the mind of the reviewer. Considering a large number of types of data sources in which such findings could be identified, and the methods for perusing these data, it is not surprising that the process of signal detection cannot be elegantly distilled into a single method. To add some clarity to the process, the following examples are offered to illustrate the variety of sources and methods by which signals come to light.

3.1.1 Preclinical Findings

Before a new drug is first administered to humans, the first signposts toward a safety signal can be gleaned from preclinical safety pharmacology and toxicology studies, often augmented by *in vitro* and *in silico* tools (Weaver & Valentin 2019, ICH M3). The elucidation of metabolic pathways and points of possible drug interactions, as well as the untoward effects in various animal models, can direct attention to specific outcomes of interest when human trials are initiated. Should the preclinical findings manifest in humans exposed to the drug, a signal can be raised and evaluated. The preclinical data can serve as a supportive material in the steps of evaluating evidence for a causal association as described later in this chapter.

3.1.2 Single-Case Reports

The types of data sources available for safety evaluation during clinical development and post-marketing can differ substantially, but some commonalities exist. Reports describing individual cases are classified as serious adverse event reports during clinical trials and as individual case safety reports in the post-marketing environment. Some of these individual reports are noteworthy enough to meet the definition of a "signal" in the mind of the reviewer. Such cases are characterized by an adverse event that rarely occurs spontaneously and usually only appears as a result of drug exposure, for example, Stevens-Johnson syndrome, toxic epidermal necrolysis, and agranulocytosis. To provide a consistent approach to determining the presence of such cases, the Designated Medical Events[1] list was created. The DME list is available at https://www.ema.europa.eu/documents/other/designated-medical-event-dme-list_en.xls. A construct of the European Medicines Agency (EMA), the DME list can be used to automate the identification of events that likely have a relationship to drug exposure and, hence, can be considered a "signal". In fact, companies who are charged with reviewing their post-marketing cases reported to EudraVigilance are instructed to consider the presence of DME terms as one of the criteria for considering a case report a "signal" deserving further evaluation. In clinical development, the list serves as an optional tool to assist with signal identification.

In addition to the DME list, the list of Important Medical Events (IME) can be used to augment the identification of signals from single-case reports either from clinical trials or post-marketing sources. Also a product of the EMA, the IME list includes terms that are considered medically significant and could have a substantial impact on the health of a patient. DME terms are a subset of the IME list of terms. Such events could arise by a variety of etiologies and, therefore, are not as closely related to the use of a drug as the events on the DME list. Nevertheless, some sponsors compare the reported safety events against the IME list in addition to the DME list to broaden the search for cases that may be worth further evaluation. Terms in the IME and DME lists span a wide variety of medical conditions and MedDRA System Organ Classes (SOCs). In this manner, they differ from Standardised MedDRA Query (SMQ) lists which facilitate searches for event terms relevant to a specific medical condition of interest. As is the case with SMQ lists, the IME list is specific to a given MedDRA version (The IME list for MedDRA version 22.1 is available at https://www.ema.europa.eu/en/documents/other/important-medical-event-terms-list-version-meddra-version-221_en.xls). As with the DME list, during clinical development, the IME list serves as an optional method to assist with signal identification.

3.1.3 Aggregate Data – Clinical Development

Beyond the review of individual case reports, the reviewer considers ways to evaluate aggregate data. Because the nature of these data sources can differ substantially between clinical development and post-marketing, the techniques of analyzing these datasets must be specific to the unique characteristics of each. This discussion will begin with a consideration of data sources typical during clinical development. The study protocol prospectively defines the type and frequency of data collected, which for the safety reviewer typically focuses on adverse events and laboratory results, augmented by vital signs, ECG, medical history, and concomitant medication use. These data elements are collected more comprehensively than is possible from post-marketing data sources and allow robust data analysis.

3.1.3.1 Adverse Events

Adverse events collected in the course of a clinical trial are accompanied by various attributes, such as the associated study drug dose (and other measures of exposure), severity (including changes in severity over time), time to onset and resolution, duration, outcome, actions taken concerning the study drug, and any required treatments as well as whether the event meets the regulatory definition of "serious"[2]. Linkage across other data domains allows the event to be associated with assigned study treatment and the subject's corresponding medical history and use of concomitant medications. With such a wealth of data, a variety of analyses can be performed to assist in identifying that finding worthy of further exploration, the signal.

The relationship between the extent of exposure and the occurrence of an adverse event can be explored in several ways which can highlight a signal. At its most simple level, event frequency can be assessed across a range of ascending doses. This is most readily accomplished during early single ascending dose (SAD) and multiple ascending dose (MAD) Phase 1 studies and Phase 2 studies that utilize more than one dose level. The usually small number of patients enrolled to each dose level precludes finding a dose relationship with less common adverse events, but it provides an early window into the safety profile of the drug.

To the extent that the treatment assignment is known in a multi-arm trial, the frequency of a given adverse event can be compared across treatment groups. A difference in the event frequency can draw the reviewer's attention and begin the determination of whether it should be considered a signal. However, little guidance is available to determine how large a frequency difference should be deemed a signal. Current US regulations stipulate that the sponsor should notify the Food and Drug Administration (FDA) in an expedited IND safety report "when an aggregate analysis of specific events observed in a clinical trial indicates that those events occur more frequently in the drug treatment group than in a concurrent control group"[3]. Absent from the requirement is guidance as to what constitutes the degree of difference that would trigger such a report. The FDA Reviewer Guidance "Conducting a Clinical Safety Review of a New Product Application and Preparing a Report on the Review" (2005) advises that "events occurring at an incidence of at least 5 percent and for which the incidence is at least twice, or some other percentage greater than, the placebo incidence would be considered common and drug-related". The guidance cautions the reviewer that such criteria are inevitably arbitrary and sensitive to the sample size. Few sponsors would interpret a two-fold frequency difference to represent sufficient evidence to conclude a causal relationship to the study drug; however, the example parameters could be practically applied as a tool to screen adverse event data for events that subsequently are considered safety signals. A focus solely on subjects experiencing the event demonstrating the frequency difference results in a derandomized dataset; hence, caution is warranted as confounding factors may be present, a concern further explored in a subsequent section.

The relative frequency of adverse events across treatment groups can be further explored by considering differences in the severity levels. Even when the frequency of a given event is comparable between treatment arms, there may be differences in the severity of their presentation which could flag the event to the attention of the reviewer. Such differences in the frequency of severity levels for an adverse event can be further explored to determine if subject demographics (e.g., age group, gender, and ethnicity) are associated with the severity differences. Similar to the analysis of the relative severity of an event across treatment groups is the assessment of the event duration across treatment groups. Continuous exposure to a drug that could give rise to an adverse event, an event that can have other non-drug etiologies as well, could lengthen the duration of the event beyond that anticipated for a non-drug cause.

Studies employing a single-arm design, or when the comparator arm is another active drug, make an evaluation of frequency differences challenging. However, separate datasets in a comparable population that has placebo treatment information can serve as a comparator. When the sponsor does not have internal clinical trial data available to serve as the comparator dataset, other sources are available. Transcelerate offers the Placebo and Standard of Care initiative that enables the sharing of de-identified data from the placebo or standard-of-care arms of clinical trials (https://www.transceleratebio-pharmainc.com/initiatives/historical-trial-data-sharing/#:~:text=The%20 Historical%20Trial%20Data%20Sharing,control%20arms%20of%20clinical%20trials.). When utilizing an external comparator group, it is important that the demographic characteristics match as closely as possible that of the study subjects. To avoid ascertainment bias, the frequency with which adverse events and laboratory tests are obtained should also be matched as closely as feasible to the target study. Alternatively, epidemiologic studies can provide information on background rates of targeted adverse events and medical conditions from a variety of databases. It is best to consult an expert epidemiologist for how best to obtain such information in a manner that informs the analysis.

Analogous to the dilemma of evaluating event frequencies in the absence of a control arm is the situation of the ongoing blinded clinical trial. Schnell & Ball (2016) proposed a Bayesian method for detecting higher than expected rates of adverse events in the drug treatment group of a blinded study. The method focuses on a limited number of adverse events of special interest (AESI) for which event rates can be estimated for a control group. For each AESI, a critical rate is defined for the treatment arm that would warrant investigation. Using blinded trial data, the software, available as the open-source R code, calculates the probability that the critical event rate is exceeded for each AESI. The tool allows for continuous safety monitoring during the course of the blinded clinical study. Additionally, Gould & Wang (2017) also described a Bayesian framework for monitoring the occurrence of adverse events using blinded data.

The actual dose administered is but one measure of drug exposure. The dose can be expressed based on patient weight, such as mg/kg. The reviewer is cautioned, though, that the results can be confounded depending on the distribution of gender if the weight of female subjects is generally less than that of male subjects. The analysis can also be confounded when there are differences in bioavailability or pharmacokinetic (PK) measures across doses. When the study drug administration extends over a period of time, and particularly when the duration of exposure is variable due to subjects coming off treatment for any number of reasons, the analysis should account for the time dimension. In such cases, event frequency is analyzed as a function of subject-time, e.g., subject-months or subject-years of exposure.

3.1.3.2 Laboratory Data

The search for adverse events that pique the interest of the reviewer can be extended to laboratory data. Laboratory data are more rigorously collected during clinical development; however, the same principles can be applied to post-marketing sources of laboratory data depending on the extent of availability. Signals may be evident in chemistry, hematology, vital sign, and even urinalysis data in the form of outlier values. Various data visualization techniques, explored in later sections, are available to facilitate the identification of such outliers. The astute reviewer will always include an examination of laboratory analytes of hepatic, renal, and bone marrow function, as recommended by the FDA Guidance for Industry on "Premarketing Risk Assessment" (2005). Laboratory findings can support the evaluation of clinical manifestations of an event when assembling a case series, as recommended by the FDA Guidance for Industry on "Good Pharmacovigilance Practices and Pharmacoepidemiological Assessment" (2005).

Elevations of hepatic transaminases and total bilirubin are early specific indicators of potential drug-induced liver injury. The FDA refined the work of Dr. Hyman Zimmerman (1978) into the concept of "Hy's Law" which continues to undergo further validation and refinement (CIOMS 2020). Drs Ted Guo and John Senior at the FDA were the first to develop a graphical tool to examine elevations of transaminases and bilirubin to identify possible Hy's Law cases (Senior 2014). This tool, evaluation of drug-induced serious hepatotoxicity or eDISH, is an approach that has been adopted by both industry and academia. The DIA-ASA Working Group has released an open-source tool (Hepatic Explorer) based on eDISH which affords an interactive interface to identify and explore potential cases of Hy's Law, as well as instances of isolated transaminase elevation (Temple's Corollary) and isolated hyperbilirubinemia. This interactive safety graphic is further described below.

Subjects who experience a decline in renal function show accompanying elevations in serum creatinine above the laboratory-defined upper range of normal. However, important declines in renal function may occur despite the creatinine values remaining within the normal range. For example, an increase in serum creatinine from 0.5 mg/dL to 1.0 mg/dL is still below the upper limit of normal in most assays yet represents a 50% loss of renal function. A review of serum creatinine values should always consider such changes within the normal range as well as looking for outliers above the normal range. Particularly important in the assessment of renal function is abrupt increases in serum creatinine (Thomas et al. 2015).

3.1.4 Aggregate Data – Post-Marketing

Signal detection in the post-marketing environment is hindered by the lack of detailed, high-quality datasets of adverse event reports and laboratory results. Instead, the reviewer is faced with datasets representing spontaneous reports from consumers and various healthcare providers. The quality of the information is typically less, and the extent of the information reported a fraction of what is occurring in the population of treated patients (Hazell & Shakir 2006). Clinical trial data allow a calculation of event frequency as both the number of subjects experiencing an adverse event (numerator) and the number of subjects exposed to the drug (denominator) are known. In the post-marketing space, adverse events are under-reported, hence the numerator is not clearly known, and the number of patients actually exposed to the drug cannot be known with certainty due to vagaries in the extent to which prescriptions are filled and patient non-compliance. Instead, the relative reporting rate of an event among drugs is the measure of interest. Signal detection methodologies in such spontaneous reporting databases rely on measures of reporting rates to discern differences suggestive of drug effects. The general hypothesis holds that if a drug is causally related to the occurrence of an adverse event, that event should be disproportionately reported for that drug compared to all other drugs that have no causal association to the event.

The two principal classes of methods to identify events that appear to be disproportionately reported are frequentist methods (e.g., proportional reporting ratio (PRR), relative reporting ratio (RRR), and reporting odds ratio (ROR) and Bayesian methods (e.g., multi-item gamma Poisson shrinker and the Bayesian confidence propagation neural network (BCPNN)). The reader is referred to any number of excellent reviews on these methodologies for more information (Hauben et al. 2005, Almenoff et al. 2007, Candore et al. 2015) including the Likelihood Ratio Test; this is further described in Chapter 8. The output of each of these methods is a statistical measure called a "signal of disproportionate reporting". It is exceedingly important to recognize that this signal of disproportionate reporting (SDR) is *not* a signal in the sense described earlier in this chapter; that is, this statistical measure is not by itself a finding of interest worth further exploration. Unfortunately, the appearance of an SDR score that exceeds a certain predefined threshold may be the result of factors other than an actual causal link with the drug of interest. Differences in reporting due to media attention, promotional efforts, and legal actions, for example, can preferentially increase the reporting rate for one drug compared to others. When a frequently co-prescribed concomitant medication embodies a risk for an adverse event, the SDR score for that event can appear artifactually inflated for the drug of interest, a phenomenon called signal leakage. A surprising number of duplicate reports are present in these systems and can produce elevated SDR scores. Thus, a careful medical review of each event with a "high" SDR score is necessary before concluding that the finding represents a signal in the clinical sense.

3.2 How to Evaluate a Signal

3.2.1 Confounding and Bias

When a signal arises, it is not commonly due solely to the drug of interest but rather is accompanied by a myriad of other factors that may have contributed to its appearance. The evaluation of confounders is important in the assessment of both clinical trial and post-marketing data. Consider, for example, the potential role of concomitant medications. In the case of a difference in adverse event frequency between two treatment groups, there may have also been a difference in the frequency of use of a concomitant medication that produces the observed adverse event. However, the astute clinician should not only be concerned with the independent contribution of the pharmacological effect of a concomitant medication but also consider the potential for an interaction between the concomitant medication and the drug of interest. Thus, confounding by concomitant medication can take two forms: 1) the concomitant medication as a direct independent contributor to the observed adverse event or 2) a drug–drug interaction between the concomitant medication and the drug of interest.

A frequent source of confounding is the underlying medical condition of the patient or comorbidities (confounding by indication). As in the example above of an imbalance in adverse event frequency, an imbalance in a medical condition that results in or predisposes to the appearance of the adverse event of interest may be present between the treatment groups. A difference in the distribution of patient characteristics, such as age, gender, and race, may play a role in the frequency of the adverse event between treatment groups to the extent that these characteristics influence the appearance of the event. Similarly, knowledge of the metabolic pathways for the drug of interest is important to be aware of the effect of genetic polymorphisms on the metabolic capacity of patients. An imbalance in the distribution of such genetic polymorphisms between treatment groups, resulting in different degrees of drug exposure, can contribute to the observed event frequency difference.

Post-randomization events can be a source of confounding, particularly when such events are not evenly distributed across treatment arms. The COVID-19 pandemic is an illustrative example. If infection with SARS-CoV-2 is not equally distributed across treatment arms, the signs and symptoms of COVID-19 will also be unevenly distributed and may appear as frequency imbalances in certain AEs. Unequal use of medications to treat COVID-19 can add additional AEs to the study arm with the greater use of these drugs. If there is an unequal distribution of COVID vaccine administration, their side effects may also appear more frequently in the group with a greater vaccination rate. To account for such post-randomization confounding, analytical strategies can include the use of the estimand framework (ICH 2009), regression analysis, and stratification.

When data are pooled across studies, either for ongoing aggregate data review or for marketing applications, care should be exercised so as not to give rise to unexpected differences in adverse event frequencies between treatment groups. Of note, among the association paradoxes, Simpson's paradox (aka the Yule–Simpson Effect) is relevant to safety analyses when an adverse event frequency difference appears between treatment groups in individual clinical studies or subgroups of aggregate data but disappears or reverses when these data sources are combined. This is the result of the existence of an influencing variable that is not accounted for, sometimes called a "lurking" variable. An illustration is provided by an analysis of the number of selective serotonin reuptake inhibitors (SSRIs) prescriptions and the rate of suicide in Japan (Nakagawa et al. 2007). In aggregate, the SSRI prescription and suicide rates were positively correlated. However, when evaluated by age groups, for both males and females, the correlation between SSRI prescription and suicide was negative. In this case, the "lurking" variable was age; the rate of suicide is not constant across age groups.

Simpson's paradox is important to consider whenever study data are aggregated, during ongoing aggregate analyses of clinical trial data as well as data pooling when preparing the Integrated Summary of Safety for a New Drug Application (NDA) submission. Care should be taken when data are pooled for these aggregate analyses so that safety trends are not obscured or untoward associations do not confuse the analysis. The bias introduced by the confounder can be removed by stratification, but only to the extent that the independent variable is known. The analysis plan should include an attempt to identify potential confounding variables so that stratification by those variables can be performed if required. Importantly, as illustrated by the SSRI example above, Simpson's paradox also can be relevant to the evaluation of post-marketing data.

Another example of the influence of an unidentified variable that produces an apparent association between a drug and an adverse event is the concept of protopathic bias. Protopathic bias occurs when a patient develops a medical condition that is unrecognized, the condition produces a corresponding symptom for which drug treatment is initiated, and, subsequently, the condition is diagnosed. The temporal sequence appears to associate the introduction of the drug with the appearance of the medical condition. Yet in actuality, the medical condition predated the use of the drug. Protopathic bias can complicate a variety of data analyses, from individual case assessment to epidemiological studies.

Protopathic bias was identified as a confounder in the assessment of the risk of colorectal cancer among users of proton-pump inhibitors (PPIs) (Yang et al. 2007). A nested case-control analysis of medical records in the UK found that the odds ratio for colorectal cancer in patients with PPI therapy within a year of the cancer diagnosis was 2.6 (95% CI 2.3–2.9, P < 0.001). While it had been noted that long-term acid suppression by PPIs increases the secretion of gastrin (Koop et al. 1990) which, in turn, has growth-promoting effects on the gastric mucosa (McWilliams et al. 1998), the association between PPI use and risk for colorectal cancer was no longer statistically significant when PPI use exceeded one year prior to the cancer diagnosis. The authors concluded that the apparent risk found with PPI use up to one year before the cancer diagnosis was an example of protopathic bias; specifically, PPIs had been prescribed to treat the symptoms of the undiagnosed colorectal cancer, the symptoms of which manifested within the year prior to diagnosis but after the initiation of PPI therapy.

3.2.2 Causal Association

Once a signal has been identified and having considered sources of bias and confounding, evidence is sought that supports or refutes a causal association between the event and the drug of interest. Little guidance has been provided by regulatory authorities concerning the suggested types of analyses to perform to evaluate signals. However, the FDA has issued several guidances that provide the Agency's thinking on this topic. The Premarket Risk Assessment Guidance for Industry (2005) speaks to evaluations of temporal association, the relationship to the dose level, PK parameters or cumulative dose, and the relationship to demographic subsets of patients. The Good Pharmacovigilance Practices and Pharmacoepidemiologic Assessment

Guidance for Industry (2005) similarly recommends evaluating a temporal relationship, the relationship to dose, presence of a positive dechallenge and rechallenge, and the demographic characteristics of the patients experiencing the adverse event. The guidance document intended for internal FDA medical reviewers, Conducting a Clinical Safety Review of a New Product Application and Preparing a Report on the Review (2005), describes recommended analyses including exploring the dose relationship, temporal relationship, demographic interactions, drug–drug interactions, and an interaction with the underlying disease state.

There is a multitude of factors that can be considered in the search for causal evidence, building upon the recommendations from regulatory guidances. One such factor is the relationship of the frequency of the event to measures of drug exposure. This can take the form of a simple dose-response relationship. Early in development when SAD and MAD studies are conducted, such a relationship to the dose level can be investigated. Later Phase 2 studies may utilize more than one dose level affording another opportunity to look for a relationship to the dose level. But other measures of drug exposure can be investigated. For example, the risk for developing an adverse event can be assessed as a function of cumulative exposure. This can identify if there is a threshold effect; that is, the risk for an event does not increase until a certain cumulative exposure has been achieved. An assessment of the relationship between an event and the extent of exposure should also consider PK variables. Given the variation in the maximal concentration (Cmax) and area-under-the-curve (AUC) values achieved for a given dose across subjects, which may obscure a relationship based solely on the dose level, a more sensitive assessment may be looking at the frequency of the event as a function of Cmax and AUC. Recognition of a relationship between event frequency and Cmax, for example, could lead to formulation changes that minimize the Cmax attained after administration. The ability to perform these analyses for PK parameters, though, is frequently limited to the early Phase 1 studies where these data are regularly collected. Thus, it is important to look for evidence of a relationship to the dose level and PK parameters in those studies that afford the opportunity, recognizing that the adverse event dataset will be limited.

As there are multiple measures of drug exposure, there are multiple ways to assess temporal relationships. The most direct measure is the time to onset. However, this may be time to onset from the initiation of therapy or time to onset from the most recent dose, particularly in the latter case when dosing is done on an interval basis (e.g., once a week, once every three weeks, etc.). If an adverse event is unrelated to the drug of interest, it should appear randomly over time, but a clustering of events with a similar time to onset is suggestive of a drug-related effect. Measures of the hazard rate over time can determine if the risk is constant over time or if there is evidence of tolerance or sensitization. The cumulative incidence function (a Kaplan–Meier plot) can visually display how events accumulate over time and can be applied to single-arm or multiple-arm trials as well as post-marketing data sources. Histograms can show clustering of events over time and also give an indication of the proportion of subjects so affected.

One of the most compelling measures of a relationship between an adverse event and a drug is the presence of a positive dechallenge and rechallenge; specifically, did the event resolve when dosing stopped (ideally in the absence of treatment given to resolve the event) and did the event reappear when the drug was reintroduced. However, the opportunity to rechallenge a study subject may be limited by the study protocol. The ability to temporarily withhold study drug treatment provides an opportunity to assess the impact of a rechallenge. However, if the protocol only allows study drug discontinuation, then a rechallenge would be effectively prohibited by the protocol. This is an example of how protocol design can impact the ability to collect safety information. There is a greater ability to acquire dechallenge and rechallenge information in the post-marketing environment.

The extent to which an adverse event appears to be consistent with the known pharmacology of the drug is replicated in animal models or is consistent with the known adverse effects of the class of drugs can be supportive of a causative role. It is important that theoretical risks identified based on the drug's pharmacology or class effects should simply serve as a guide for what to look for in the safety data. The FDA (2012) and the EMA (European Commission 2011) have specifically stated that class effects and those anticipated from the pharmacological properties of the drug should not be assumed to be part of the drug's safety profile until such events have been observed in treated subjects and confirmed to be due to the drug.

A less well-recognized method of assessment is the degree to which a subject experiences multiple instances of an adverse event. Adverse events that spontaneously arise independently of the use of the drug of interest are less likely to recur multiple times. In contrast, a subject is more likely to experience multiple recurrences of the adverse event if there actually is a causal relationship assuming that the disease or underlying medical condition is not responsible for the recurrences. The ability to detect recurrent adverse events is typically limited by the methods by which data are collected on case report forms and the types of data outputs commonly used for review purposes. Adverse event tables are generated which are subject-based; that is, a given adverse event is counted only once per subject. This provides a frequency estimate of the proportion of subjects that experience one or more instances of an event. It does not provide a measure of how many times the subject experienced that particular event[4]. The fact that a particular condition could be described by more than one MedDRA preferred term can also obscure the occurrence of repeat episodes. Consider a subject who experiences single instances of fatigue, malaise, lethargy, and malaise. All four terms are separate MedDRA preferred terms yet all four express a common medical condition. Taken individually, there would be no evidence of recurrent events in this subject, but once it is recognized that collectively they describe the same condition then it becomes apparent that the subject experienced four instances of this common event. The assessment of recurrent events is more relevant to clinical trials that follow a subject over time and less so to post-marketing reports that typically represent a single point in time.

Continuing on the topic of aggregating MedDRA preferred terms, this technique can be used to strengthen or discount a signal arising from an increase in the frequency of a given adverse term relative to a comparator or other baseline value. It is also a tool to identify signals not evident when the relative frequencies are evaluated based on single, unaggregated terms. Using the example above, if a greater frequency of fatigue was observed in the study drug group compared to a control, the next analysis should be to evaluate the relative frequency of the combination of fatigue, malaise, lethargy, and malaise. If the drug actually produces fatigue-type reactions, then the addition of the other similar terms should demonstrate a greater frequency imbalance when assessed as the aggregate of terms. On the other hand, if the aggregate term analysis finds no relative difference in frequency between groups, then this is taken as evidence of variation in how events are characterized by the investigator (or coded by the sponsor) between treatment groups and evidence refuting a drug effect. When there are low counts of any one of these terms, it will be difficult to discern a difference in frequency across treatment groups, in which case the only way to ascertain a difference is to consider the terms in aggregate. This can be accomplished by SMQs, the use of MedDRA high level term, or high level group term groupings, or even by constructing custom groupings of MedDRA preferred terms.

There is no specific guidance from regulators as to which terms should be considered in aggregate. In the FDA guidance to their medical reviewers (Reviewer Guidance 2005), the example of somnolence, drowsiness, sedation, and sleepiness was given as terms likely referring to the same event. Nevertheless, the Agency's approach to aggregate term analysis can be discerned from the medical reviewer reports within the drug approval packages FDA makes available on their website. By observing how various aggregations of preferred terms are commonly used in evaluations across submissions within a particular division, companies can get a sense of how their data will be analyzed upon making a submission to that division and can proactively integrate a similar aggregate review into their routine safety data review practices. The practice of aggregating preferred terms can also be applied to post-marketing data (Hill et al. 2012). SDRs are typically calculated for individual preferred terms, but there is value in calculating SDR scores for aggregated terms as well as for SMQs and higher MedDRA levels.

In addition to seeking support across other MedDRA preferred terms that describe a similar medical condition, the review should also look into the Investigations SOC for supportive terms. For example, while the clinical terms describing liver injury include acute hepatic failure, hepatic failure, hepatitis, liver disorder, and liver injury, the reviewer should also look into the Investigations SOC that describes adverse events pertaining to the corresponding relevant laboratory abnormalities, such as "alanine aminotransferase increased", "aspartate aminotransferase increased", "transaminases increased", and "liver function test abnormal". Using an SMQ or company-defined aggregate term list can facilitate this review. Beyond these MedDRA preferred terms, the actual laboratory data should be reviewed. Many study protocols stipulate that a laboratory abnormality not be considered as an adverse event unless it is deemed "clinically significant", typically meaning that the laboratory abnormality required a medical intervention and/or was accompanied by clinical signs or symptoms. Trends in laboratory abnormalities that support an adverse event finding can offer evidence in support of a relationship to the study drug. Alternatively, the addition of laboratory data into an assessment can discount a causal role for the drug. A frequency imbalance in a particular adverse event, e.g., anemia, may not be supported when the actual hemoglobin and hematocrit data across treatment groups show no imbalance in outliers. In this case, the apparent imbalance in "anemia" may solely be due to an imbalance in the propensity of investigators to label an out-of-range hemoglobin/hematocrit as "anemia".

The search for corroborating evidence in the Investigations SOC and laboratory data can be expanded into the non-clinical data. Clinical trial and post-marketing findings that are accompanied by toxicology support carry additional weight in support of a causal role of the drug. A finding that is replicated across clinical and non-clinical datasets is less likely to be a product of chance alone.

Support for a causal relationship between an adverse event and the study drug may be provided when there is evidence that the event segregates into one or more subgroups based on demographics or medical history. A particular demographic feature may constitute a risk factor, such as a racial predisposition based on a polymorphism in a metabolizing enzyme, or an age group may represent a higher risk group based on changes in metabolic capacity or presence of comorbid conditions that interact with the drug in a negative manner, such as reduced renal clearance in the elderly for a drug predominantly subject to renal excretion. Subgroup analyses are also valuable to perform as a signal detection exercise as the imbalance in event frequency may only be observed at the level of a subgroup, whereas the difference is masked when the data are evaluated overall. Such subgroup analysis should also be considered for post-marketing data.

The aforementioned methods of evaluating data for evidence of a causal relationship between an event and drug can be difficult to apply in the post-marketing setting where the spontaneously reported information is more limited in quality. However, evidence of a temporal relationship, positive dechallenge/rechallenge information, consistency with pharmacological and toxicological effects as well as class effects, and consistency with clinical trial data can be sought using well-constructed case series. The approach to evaluating a case series involves first carefully establishing a case definition; that is, formal criteria for including or excluding a case for analysis. For example, the case definition for pancreatitis is 1) abdominal pain consistent with acute pancreatitis (acute onset of a persistent, severe, epigastric pain often radiating to the back), 2) serum lipase or amylase at least three times the upper limit of normal, and 3) characteristic findings on contrast-enhanced computed tomography, magnetic resonance imaging, or transabdominal ultrasonography (Banks et al. 2013). The reports that contain the event of interest that meets the case definition can then be assembled into a case series for further detailed evaluation.

The review of the case series should identify which are confounded; that is, have other possible etiologies than the drug of interest (Good Pharmacovigilance Practices and Pharmacoepidemiologic Assessment Guidance for Industry 2005). Confounded cases should not be excluded from evaluation but assessing unconfounded cases initially can be illuminating. The cases in the series are examined for similarities in terms of clinical and laboratory manifestations, the time course of the event, demographic characteristics, doses used, and presence of concomitant medications and comorbid conditions. These features allow an evaluation as described above to determine the extent to which there is evidence in support of a causal role for the drug.

3.3 How to Determine When a Signal Represents a Product Risk

While the aforementioned regulatory guidances recommend approaches to signal detection, they are silent on how to assess whether the evidence as a whole supports or refutes a causal role of the drug. Wittes et al. (2015) described a statistical methodology to assist in determining when a signal represents a reportable event in keeping with the FDA's Final Rule (2012). However, it is important to point out that this method does not confirm a causal association, rather it can be considered evidence for the strength of association described below. The determination of when a body of evidence is sufficient to conclude that an adverse event is causally associated with a drug is inherently a subjective one. Unfortunately, this results in considerable variation between reviewers as to what constitutes a convincing body of evidence upon which to make this decision. Multiple attempts have been made to reduce such variability by constructing an algorithmic approach. Examples include the French imputability method (Theophile et al. 2012), the Naranjo method (Naranjo et al. 1981), and several others (Hire et al. 2013). The challenge has been that fundamentally such scoring systems are still based on placing subjective values to various criteria despite their appearance of having a quantitative framework. An alternative approach is provided by the seminal work of Sir Austin Bradford Hill, an epidemiologist, who proposed criteria to be considered in establishing a causal relationship (Hill 1965). The criteria span nine aspects of causal inference: 1) strength of association, 2) consistency, 3) specificity, 4) temporality, 5) biological gradient, 6) plausibility, 7) coherence, 8) experimental evidence, and 9) analogy. As Bradford Hill first presented his criteria in the setting of occupational medicine, it is useful to reframe his tenets in the context of drug safety evaluations. This framework is equally applicable to clinical trial and post-marketing environments.

3.3.1 Strength of Association

The underlying assumption is that strong associations are more likely to be causal than weak associations which are more likely to be the result of confounding or bias. The strength of association is expressed as a statistic, often a relative risk, but also other measures such as the odds ratio or hazard ratio. In the post-marketing environment, the statistic is a disproportionality score, either from a frequentist method such as a PRR, RRR, and ROR or a Bayesian method such as the Empiric Bayes geometric mean or information component. During clinical development when the most common adverse events due to the drug are being discovered, the relative risk can be greater than 2, but later when searching for more infrequent drug-induced events it is uncommon to find relative risks greater than 2. Thus, the finding of a weak strength of association for an uncommon event should not completely discount the possibility of a causal association.

An example of a strength of association assessment is provided by an evaluation of the association between saxagliptin and heart failure (Cebrian Cuenca et al. 2017). One randomized trial reported a hazard ratio (HR) of 1.27 (95% CI 1.07–1.51). A subsequent analysis that included all hospitalizations showed an HR of 1.26, with the observation that the risk was only observed within the first 314 days. A sensitivity analysis excluding first hospitalizations for heart failure did not find an elevated risk (HR 1.06; 95% CI 0.75–1.50), illustrating that different models can yield different results and underscoring the importance of statistical expertise in the interpretation of the findings.

3.3.2 Consistency

Consistency of an association is present when the finding is apparent in different populations, different circumstances, and different times. Consistency can be sought across different clinical trials, patient populations, indications, or geographic regions depending on the stage of the product life cycle. The more often the signal is replicated across studies, populations, indications, and regions the more evidence there is of a causal association instead of random chance. That being said, some product risks are limited to particular patient subgroups, perhaps because of interacting concomitant medications frequently used in this group or comorbidities or other medical conditions that place them at higher risk, for example. If such a subgroup is not represented across trials, patient populations, indications, or regions, then consistency will not be observed.

An example of consistency is afforded by the analysis of cisapride as a cause of arrhythmias (Perrio et al. 2007). Spontaneous reports of cisapride and QT prolongation-associated arrhythmias came from several patient populations with different levels of risk and susceptibility. These included interacting medications, cardiovascular conditions, and renal failure. Reports in adults spanned nine countries, and reports in children came from three countries.

3.3.3 Specificity

If a drug causes a narrowly defined adverse event that is unlikely to have other etiologies, then the specificity criterion is met. When the drug is associated with many adverse events, some of which could arise from other etiologies, yet the frequency of a particular adverse event is greatly increased then there is specificity in the magnitude of the association. An example is the appearance of sclerosing peritonitis with practolol. Nevertheless, specificity is perhaps one of the weakest of the criteria since drugs can have a multitude of off-target effects. The lack of finding specificity between the drug and an event should not be taken as evidence of a lack of causal association.

The association of zinc with anosmia illustrates the concept of specificity (Davidson & Smith 2010). Intranasal zinc gluconate has been observed to produce hyposmia and anosmia in some patients. This nasal spray does not appear to result in other adverse events and is apparently well-tolerated otherwise. The hyposmia/anosmia is a unique specific effect of the product.

3.3.4 Temporality

At its most basic, temporality means that the drug exposure must have preceded the appearance of the adverse event. However, as discussed previously, protopathic bias can artifactually produce an apparent temporal relationship.

Temporality is more commonly assessed in terms of time to onset and is frequently cited as a recommended evaluation in several guidances as previously discussed. When there is a temporal relationship, the onset of the event is not randomly distributed in time but rather shows evidence of clustering. Although such clustering may occur relatively soon after treatment initiation if there is a cumulative threshold effect or delayed sensitization, the onset may be delayed. Within the concept of temporality is the assessment of dechallenge and rechallenge: when the drug is withdrawn does the adverse event resolve and does it recur upon reexposure to the drug. It is generally assumed that resolution upon dechallenge has close temporal proximity to the withdrawal of the drug and a recurrence upon rechallenge is also has a close temporal association with its reintroduction.

The rapid time to the onset of hearing loss following treatment with hydrochlorothiazide demonstrates the principle of temporality (Belai et al. 2018). Among 34 post-marketing cases of patients treated with hydrochlorothiazide who experienced hearing loss that included the time to reaction onset, the median time to onset was three days. In six reports, the hearing loss developed on the second day following the initiation of treatment with only hydrochlorothiazide. Nevertheless, the time to the onset of an event must be weighed in the context of biologic plausibility as is illustrated by the finding of fractures in patients treated with SSRIs (Warden & Fuchs 2016). Non-clinical studies had suggested that SSRIs could reduce skeletal mass leading to the hypothesis that SSRIs could contribute to secondary osteoporosis and fractures. Self-controlled case series, cohort, and case-control studies that reported risk across different time windows indicated a rise in the risk of fractures within the first two–four weeks of treatment with SSRIs. While this demonstrated a temporal association, it is unlikely that even if SSRIs negatively impacted bone density, the effect could manifest within one month and to an extent that would increase fracture risk. Indeed, the same studies indicated a 2.6-fold increase in fall risk in the first four weeks of SSRI use. Thus, the apparent relationship of SSRIs and risk of fractures appeared to be due to the heightened fall risk, both as a direct effect of SSRIs on gait and also as an example of confounding by indication – depression is independently associated with an increased risk of falls.

3.3.5 Biologic Gradient

The dose-response relationship is well known in epidemiology but is equally applicable to drug safety. Recommendation for a dose-response evaluation is highlighted in several guidances and was discussed previously in this chapter. It is useful to bear in mind that a dose response need not be limited to an assessment of the dose level, but also other measures of exposure including weight-adjusted exposure, cumulative exposure, and exposure as a function of PK parameters such as Cmax and AUC. The finding of a dose-response relationship can be a compelling piece of evidence.

Drawing again from the cisapride experience is an example of the biologic gradient (Perrio et al. 2007). Individual reports and clinical studies highlighted that QTc prolongation was associated with high doses. A dose-response relationship was also evident from drug interaction studies where medications that inhibit CYP3A4, the main enzyme responsible for cisapride metabolism, increased exposure to cisapride, and the frequency of QTc prolongation. Additionally, *in vitro* studies found that the inhibition of the human Ether-à-go-go-Related Gene (HERG) potassium channel by cisapride, a measure of the risk of clinical QTc prolongation, is dose-dependent.

3.3.6 Plausibility

The plausibility criterion can be met to the extent that a credible mechanism of action or pathophysiologic process can be posited as the reason a drug may cause a particular adverse event. However, particularly early in development, all of the pharmacologic actions of the drug may not be known, and a plausible mechanism may not be apparent. The lack of a plausible mechanism does not preclude the possibility of a causal relationship, but if one is present it can strengthen the body of evidence.

The relationship between hormone replacement therapy and breast cancer provides a useful example of plausibility (Colditz 1998). Evidence at the time supported that cell proliferation was the process by which DNA damage accumulated attendants with the risk of breast cancer. Research indicated that estrogens caused not only nongenotoxic cell proliferation but also a genotoxic effect. It was observed that women who used replacement hormones showed a greater increase in breast density by mammography than women who did not receive replacement hormones, which provided clinical evidence of the cell proliferation effect of estrogens.

3.3.7 Coherence

Whereas "plausibility" focuses on the effects of the drug that might reasonably result in the adverse event, "coherence" starts with the known etiologies of the adverse event and asks if the drug's action is consistent with any of those known etiologies. Bradford Hill specifically stated that the "data should not seriously conflict with the generally known facts of the natural history and biology of the disease". Admittedly, this criterion may be difficult to apply when the natural etiology of the event is either multi-factorial or poorly understood. This criterion appears to be the most difficult to apply as there has been considerable lack of consistency in the interpretation of its meaning across multiple causality evaluations. Some authors have considered the coherence criterion to have been met when multiple lines of evidence (such as other Bradford Hill criteria) are supportive (Spitzer 1998) when epidemiological/observational data are consistent with laboratory findings (Warden & Fuchs 2016) or when there is no alternative cause (Davidson & Smith 2010) instead of evaluating the data in terms of the known causes of the events. Besides misinterpreting the intent of the coherence evaluation, caution is also warranted in assuming that the etiologic mechanisms leading to the event are completely known. The mechanism by which the drug produces the adverse event may reflect a new, unrecognized path that is entirely valid. Perhaps the best example of coherence is the circumstance when an event principally only arises following drug exposure, such as hemolytic anemia and progressive multi-focal leukoencephalopathy. In the case of hemolytic anemia, the development of autoimmune antibodies that target erythrocytes can also be elicited by certain drugs. In the case of progressive multi-focal leukoencephalopathy, the causative JC virus is facilitated by certain immunosuppressant drugs. Examples of such events are presented in the previously mentioned EMA's Designated Medical Events list.

3.3.8 Experimental Evidence

As much as "consistency" sought for the signal across different clinical experiences, "experimental evidence" is data arising from *in vitro* and *in vivo* studies that support the clinical finding. This may take the form of animal pharmacology and toxicology studies, studies of receptor binding and transporter inhibition, and even metabolic studies using hepatocytes or liver homogenates. One could argue that clinical trials contribute experimental evidence to this criterion; however, the contribution of these data sources was previously described under "consistency". Data that support a mechanism of action or pathophysiological process linking a drug to an adverse outcome meet the criterion of experimental evidence. However, caution is also warranted to avoid circular reasoning wherein, for example, *in vitro* data inform a putative mechanism of biological plausibility, and the same *in vitro* data are then applied as experimental evidence in support of the same putative mechanism.

Several *in vitro* studies provide experimental evidence useful in the analysis of the association between ciprofloxacin and the risk for hypoglycemia in non-diabetic patients (Berhe et al. 2019). Two studies in rat islet cells showed that exposure to quinolone antibiotics increased insulin secretion via blockade of adenosine triphosphate-dependent potassium channels. Other studies in rat islet cells suggested that fluoroquinolones could also augment the stimulated release of insulin, which was consistent with the observation that the majority of the clinical reports occurred postprandially.

3.3.9 Analogy

The most commonly cited analogy is a reference to class effects. When a particular adverse event is common to the class of drugs to which the drug of interest belongs, then it is reasonable to suspect a causal link when that event is observed with the drug of interest. Of particular note, and as previously mentioned, is the caution expressed by both the FDA and the EMA that class effects are not *a priori* assumed to be risks associated with a new product in the class until 1) that event has occurred in subjects treated with the drug and 2) the evidence is supportive of a causal relationship. To avoid succumbing to circular reasoning in the sole reliance on the analogy criterion, there must be additional supportive evidence of a causal association.

The review of causation criteria previously cited for cisapride offers a useful example of analogy (Perrio et al. 2007). As articulated in ICH E14 (2005, 2016), drugs that delay cardiac repolarization, as measured by the prolongation of the QT interval, create an electrophysiological environment that favors the development of cardiac arrhythmias, of most concern torsade de pointes. Drugs such as terfenadine, dofetilide, astemizole, and moxifloxacin produce QT prolongation mediated by inhibition of the HERG channel, similar to the effect of cisapride.

As can be appreciated from the preceding discourse, the application of the Bradford Hill criteria to the determination of causality is an exercise in accumulating a body of evidence that in aggregate convinces the reviewer that a causal association exists. The nature of the evidence differs, and there is no set threshold above which causation can be defined. Furthermore, it can be seen that some criteria exist mostly independently, such as strength of association, temporality, and biologic gradient, whereas other criteria may appear to have some degree of overlap with other criteria, such as consistency, analogy, and experimental evidence. The criterion of coherence appears to be the least well understood in its application to drug safety causation determination. Before applying these criteria, the reader is urged to understand the intent of each to make the best use of them. In the end, the number of data elements and how compelling each contribute to the body of evidence that is either convincing of a causal association or is not. If not, the reviewer understands where the knowledge gaps are and can address them with further research, which in turn can strengthen or weaken causal association.

The assessment of causality may appear complicated when the data sources span both preclinical, clinical, and post-marketing environments. Which data source should be given more weight is a valid question the reviewer may pose. In the context of the Bradford Hill approach, it is perhaps relevant to consider that it is not so much the source of the data but rather the strength of the evidence of each criterion that should be the focus. As we have seen from the examples above, some criteria may be given more weight due to their independence from other criteria (e.g., strength of association, temporality, and biologic gradient) followed by those criteria that exhibit a degree of overlap (e.g., consistency, analogy, and experimental evidence).

However, a company that wishes to establish a process for determining the causal association between an adverse event and a drug should link into other internal procedures for safety data review, signal detection, and evaluation which should be carefully planned and documented as part of a quality system. The approach to aggregate safety assessment is further described in Chapter 2. How identified risks factor into benefit-risk assessment planning is discussed in Chapter 13.

3.4 What Tools Are Available?

Safety clinicians frequently review tabular summaries of clinical trial adverse event and/or laboratory data information, looking either for notable differences in event frequency between treatment groups or when frequencies depart from an expected background rate for single-arm studies. This can be tedious when there are many adverse events to review and difficult to prioritize when frequency differences are discovered, or too much laboratory data to compare longitudinally or transversely, through absolute value, normalized value, etc. In the following section, we will introduce some available tools to facilitate the AE data review and some others for laboratory data review.

3.4.1 Common Graphical Displays for the Adverse Event Review

Several graphical displays facilitate this review and prioritization, of particular note the dot plot with relative risk and volcano plot.

A graphic that displays a dot plot showing the relative frequency of treatment-emergent adverse events for each treatment arm paired with a display of the associated relative risk (or other statistical measures) affords the ability to sort the adverse event terms in the descending order of risk. The user can readily focus attention on those adverse events with the greatest risk differences. An example is provided in Figure 3.1 (Amit et al. 2008).

The volcano plot is an alternative to the dot plot display. The frequency of adverse events between two treatment groups can be assessed by relative risk, or other statistical measures, on the x-axis and a p-value threshold on the y-axis. This type of display draws the user to the adverse events with the greatest differences in risk with the greatest statistical significance. The point size can be adjusted to display a size proportional to the number of adverse events. An example is provided in Figure 3.2 (Zink et al. 2013).

The analysis of adverse events is complicated by the fact that multiple comparisons are performed. When multiple statistical tests are performed, such as done in the volcano plot example, adjustments must be made to the alpha level to maintain a 95% confidence in concluding the result is statistically significant. Zink et al. (2013) explored various approaches to the multiplicity issue as it applies to the volcano plot.

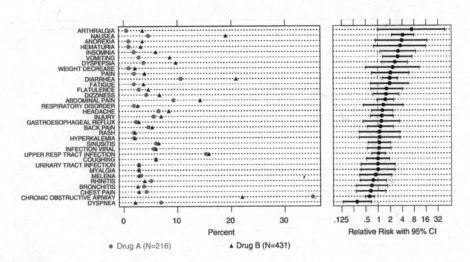

FIGURE 3.1
Adverse event dot plot with relative risk

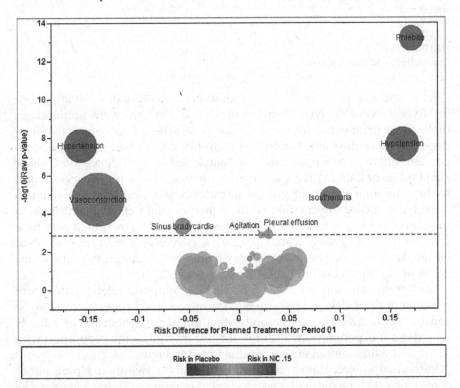

FIGURE 3.2
Volcano plot

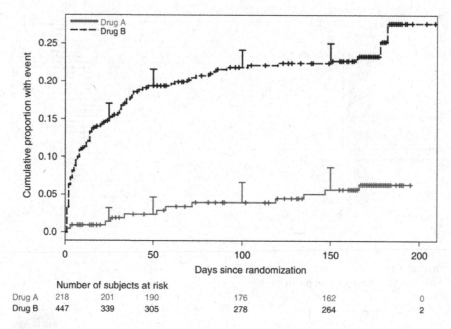

FIGURE 3.3
Cumulative incidence function

As discussed previously, an exploration of temporality is important in adverse event analysis. Time to onset and how an event accumulates throughout drug exposure are two such parameters. This cumulative incidence function illustrates periods of greatest risk and when risk may no longer accumulate. The curves can be accompanied by confidence intervals or standard error bars to give a sense of the degree to which the curves representing different treatment groups statistically separate. It is important to consider including information on the number of subjects that contribute to the curves at various time points as a loss of subjects on treatment over time can contribute to greater variability in the curves and indicate periods of elevated risk that may be an artifact due to the loss of subjects from the study. An example is provided in Figure 3.3 (Amit et al. 2008).

Histograms are another method to display temporal relationships from which periods of elevated risk can be determined. The histogram can also be constructed to give the user an understanding of the proportion of subjects who did not experience the event of interest as well as the proportion who did develop the event over time, as illustrated in Figure 3.4.

Individual subject data can be displayed via a swimmer's plot, either across subjects for a particular adverse event of interest (Figure 3.5) or across adverse events within a single subject (Figure 3.6). In the former case, the relative time to the onset of the event can be compared across subjects in

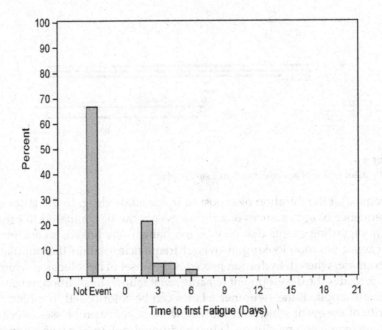

FIGURE 3.4
Histogram

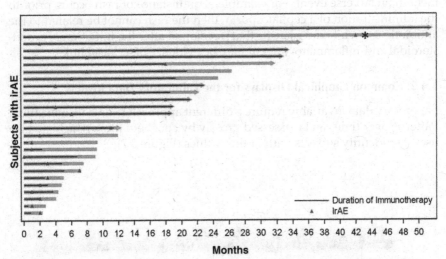

Each bar represents one subject from start of immunotherapy to death/last follow-up. Right arrow cap indicates that patient is still alive.

FIGURE 3.5
Swimmer's plot of time to immune-related adverse event (Owen et al. 2018)

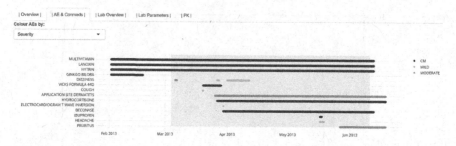

FIGURE 3.6
Visual patient profile of adverse events in a single subject

the context of the duration of exposure to the study drug. In the latter case, the sequence of appearances of adverse events can be compared to provide insight regarding events that may be mechanistically linked, an assessment that cannot be done looking at overall frequencies within the population. For example, when dehydration precedes the onset of hypotension which, in turn, precedes the development of acute renal failure. In addition, when step grading is applied, the swimmer's lanes can be color-coded to indicate the severity of the event which can show how the severity progresses over time, both worsening and resolution. When additional swimmer's lanes depict the time course of concomitant medication administration, an assessment can be made of the possible role of the concomitant medication in the development of an adverse event. For example, if an instance of rash occurs prior to the administration of a cephalosporin, then the rash cannot be related to the antibiotic use while the gastritis that developed after the initiation of a non-steroidal anti-inflammatory agent may be evidence of a causal link.

3.4.2 Common Graphical Displays for the Laboratory Data Review

Laboratory data invariably require a different approach to data visualization. Patterns over time can be assessed grossly by spaghetti plots which are also useful to identify subjects with outlier values (Figure 3.7).

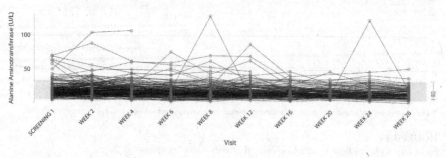

FIGURE 3.7
Spaghetti plot of alanine aminotransferase over time[5]

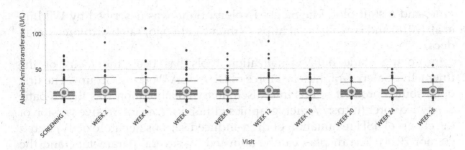

FIGURE 3.8
Summary of alanine aminotransferase values over time[6]

The search for outlier values in laboratory data can also be performed using tools that display data aggregated by the time interval as well as illustrating values outside of the normal range. In the example in Figure 3.8, each time point displays the mean, median, first and third quartiles, 5% and 95% range, and maximum and minimum values. Outlier values can be readily identified and, in this case, the corresponding subject information is accessed by hovering the cursor over the interactive display.

3.4.3 Other Tools, Platforms, and Resources

This chapter has described several techniques for signal detection and signal evaluation that could be facilitated by data visualization techniques. Commercially available tools can greatly facilitate the data review process, such as offerings from Spotfire®, JMP® Clinical, Tableau®, and Qlic® to name a few. There are also open-source tools. CTSPedia is a website created by the Clinical and Translational Science Awards program's Biostatistics/Epidemiology/Research/Design Online Resources and Education taskforce to share resources among researchers worldwide. Within this website is a section pertaining to drug safety that offers a variety of graphical displays for various safety data. Each graph is accompanied by a description of its use, example displays, and the SAS code to run the graph.

A website by the name of Gersonides offers a variety of statistical tools, including graphics for clinical trials, some of which are relevant for safety analyses such as dot charts and tree maps for adverse events and laboratory tests over time. The open-source R code is available for each of these displays.

Rho® has developed an open-source suite of interactive data visualization tools applicable to safety monitoring. Adverse events can be evaluated using the Adverse Events Explorer and Adverse Events Timeline. Laboratory data can be analyzed using a histogram, outlier explorer, test results over time, and a shift plot. The Safety Explorer Suite was described by Wildfire et al. (2018) and is available at https://rhoinc.github.io/safety-explorer-suite/docs/.

Improving static data visualization tools has been the focus of the Interactive Safety Graphics taskforce of DIA -ASA Working Groups. The first open-source product of this interdisciplinary collaboration was the Hepatic Safety Explorer (https://safetygraphics.github.io/), an interactive version of the classic eDISH (evaluation of drug-induced serious hepatotoxicity) graph (Senior 2014). The display can be adjusted by several parameters, and the user can "drill down" into individual patients to visualize the course of laboratory analytes, changes in the R value[7] over time, and the calculation of P_{ALT}, an estimate of the percent hepatocyte loss. The display can be changed from the standard plot of fold-change of ALT to total bilirubin to the nR modified Hy's Law approach (Robles-Diaz et al. 2014). Unique to the tool is an accompanying clinical workflow that provides a diagram of recommended analyses with a discussion of why the analysis is conducted and how to interpret the findings, all referenced to the medical literature.

Disproportionality analysis is applied to large post-marketing datasets of spontaneous adverse event reports where neither the numerator (number of patients experiencing an adverse event) nor denominator (number of patients exposed to the drug of interest) is known. Instead, reporting rates are evaluated by a number of frequentist and Bayesian methods. Commercial tools to calculate SDRs using these methodologies are offered by Oracle® Empirica Signal, LifeSphere® Signal and Risk Management, DrugLogic® Qscan, and Sciformix® Scignal Plus, for example. The SDR results can be visually displayed by third-party tools that format the data into a heat map or tree map, for example.

The reader is also referred to Chapter 15 for additional considerations on visual analytics for safety assessment.

3.5 Summary

This chapter has explored the processes by which safety signals can be identified from clinical trials and post-marketing sources and then assessed for evidence of a causal association to the drug of interest. The variety of data sources and methods for interrogation precludes a single methodology for identifying a safety signal. Rather it is important to bear in mind that the signal is "an event with an unknown causal relationship to treatment that is recognized as worthy of further exploration and continued surveillance". Several examples of signal detection in the setting of single-case reports, aggregate data, and laboratory data, across the clinical trial and post-marketing environments were described.

All safety signals should be evaluated for the potential presence of confounding factors or sources of bias in the finding. This is an important consideration as it represents a common source for frequency differences between groups or other instances in which an event distinguishes itself from the background.

Just as there is no one method to reliably identify a signal, there is no single method to assess the signal for evidence of a causal association to the drug of interest. Several algorithms have been proposed to assist with the evaluation of causal association, but none are robust enough to replace good clinical judgment. It bears stressing, though, that a thorough causal evaluation must not rely solely on clinical judgment but also be supported by quantitative and visualization tools. Such an amalgamation was described in the context of the Bradford Hill framework for the assessment of causal association. The strength of this process is dependent on a multi-disciplinary collaboration of clinicians, statisticians, and data scientists.

Notes

1 Terms on the Designated Medical Events list are those adverse events that represent serious medical concepts often causally associated with drugs across multiple pharmacological/therapeutic classes. The list is used by the European Medicines Agency, as well as EEA Member States, to identify reports of suspected ADRs that deserve special attention, irrespective of statistical criteria used to prioritize safety reviews.
2 An adverse event is "serious" when it results in death, is life-threatening, requires hospitalization (or prolongation of an existing hospitalization), a significant or persistent disability, congenital anomaly, or is deemed an important medical event
3 21CFR312.32(c)(1)(i)(C)
4 This refers to multiple independent occurrences of an adverse event rather than the practice of "step grading" where entries are made for each change in severity grade within a single instance of an AE.
5 Safety Explorer Suite, https://becca-krouse.shinyapps.io/safetyGraphicsApp/
6 Safety Explorer Suite, https://becca-krouse.shinyapps.io/safetyGraphicsApp/
7 The R value, calculated as (ALT/ULN)/(alkaline phosphatase/ULN), is a measure of the extent to which ALT elevations may be related to a hepatocellular process, a cholestatic one, or a mixed hepatocellular/cholestatic picture.

References

Almenoff JS, Pattishall EN, Gibbs TG, et al. Novel statistical tools for monitoring the safety of marketed drugs. *Clin Pharmacol Ther.* 2007;82: 157–166.

Amit O, Heiberger,RM, Lane PW. Graphical approaches to the analysis of safety data from clinical trials. *Pharmaceut Statist*. 2008;7:20–35.

Banks PA, Bollen TL, Dervenis C, et al. Classification of acute pancreatitis – 2012: revision of the Atlanta classification and definitions by international consensus. *Gut*. 2013;62:102–111.

Belai N, Gebrehiwet S, Fitsum Y, Russom M. Hydrochlorothiazide and risk of hearing disorder: a case series. *J Med Case Rep*. 2018;12:135–138.

Bradford Hill A. The environment and disease: association or causation. *Proc R Soc Med*. 1965;58:295–300.

Buchanan JF, Li M, Hendrickson BA, Roychoudhury S, Wang W. Assessing adverse events in clinical trials during the era of the COVID-19 pandemic. 2021, in press.

Candore G, Juhlin K, Manlik K, et al. Comparison of statistical signal detection methods within and across spontaneous reporting databases. *Drug Safety* 2015;38:577–587.

CIOMS Working Group. Drug-induced liver injury (DILI): Current status and future directions for drug development and the post-market setting. A consensus by a CIOMS Working Group. Geneva. 2020.

Colditz GA. Relationship between estrogen levels, use of hormone replacement therapy, and breast cancer. *J Natl Cancer Inst*. 1998;90:814–823.

Davidson TM, Smith WM. The Bradford Hill criteria and zinc-induced anosmia: a causality analysis. *Arch Otolaryngol Head Neck Surg*. 2010;136:673–676.

Gould AL, Wang WB. Monitoring potential adverse event rate differences using data from blinded trials: the canary in the coal mine. *Statist Med*. 2017;36:92–104.

Guidance for Industry, Good Pharmacovigilance Practices and Pharmacoepidemiologic Assessment. U.S. Department of Health and Human Services, Food and Drug Administration, Center for Drug Evaluation and Research (CDER), Center for Biologics Evaluation and Research (CBER), March 2005.

Guidance for Industry and Investigators, Safety Reporting Requirements for INDs and BA/BE Studies. U.S. Department of Health and Human Services, Food and Drug Administration, Center for Drug Evaluation and Research (CDER), Center for Biologics Evaluation and Research (CBER), December 2012.

Guidance for Industry, Premarketing Risk Assessment. U.S. Department of Health and Human Services, Food and Drug Administration, Center for Drug Evaluation and Research (CDER), Center for Biologics Evaluation and Research (CBER), March 2005.

Hauben M, Madigan D, Gerrits CM, et al. The role of data mining in pharmacovigilance. *Expert Opin Drug Safety* 2005;4:929–948.

Hazell L, Shakir SAW. Under-reporting of adverse drug reactions: a systematic review. *Drug Safety* 2006;29:385–396.

Hill AB. The environment and disease: association or causation? *Proc R Soc Med*. 1965;58:295–300.

Hill R, Hopstadius J, Lerch M, et al. An attempt to expedite signal detection by grouping related adverse reaction terms. *Drug Safety* 2012;35:1194–1198.

Hire RC, Kinage PJ, Gaikwad NN. Causality assessment in pharmacovigilance: a step towards quality care. *Sch J App Med Sci*. 2013;1:386–392.

ICH Topic E 14, The Clinical Evaluation of QT/QTc interval prolongation and proarrhythmic potential for non-antiarrhythmic drugs. CHMP/ICH/2/04, 2005.

ICH guideline E14: the clinical evaluation of QT/QTcinterval prolongation and pro-arrhythmic potential for nonantiarrhythmic drugs (R3) - questions and answers. EMA/CHMP/ICH/310133/2008, 2016.

ICH guideline M3(R2): on non-clinical safety studies for the conduct of human clinical trials and marketing authorisation for pharmaceuticals. EMA/CPMP/ICH/286/1995, 2009.

Information from European Union Institutions, Bodies, Offices and Agencies. Detailed guidance on the collection, verification and presentation of adverse event/reaction reports arising from clinical trials on medicinal products for human use ('CT-3'), (2011/C 172/01).

Koop H, Klein M, Arnold R. Serum gastrin levels during long-term omeprazole treatment. *Aliment Phrmacol Ther.* 1990;4:131–138.

McWilliams DF, Watson SA, Crosbee DM, et al. Coexpression of gastrin and gastrin receptors (CCK-B and delta CCK-B) in gastrointestinal tumor cell lines. *Gut.* 1998;42:795–798.

Nakagawa A, Grunebaum MF, Ellis SP, et al. Association of suicide and antidepressant prescription rates in Japan, 1999–2003. *J Clin Psych.* 2007;68:908–916.

Naranjo CA, Busto U, Sellers EM, et al. A method for estimating the probability of adverse drug reactions. *Clin Pharmacol Ther.* 1981;30:239–245.

Owen DH, Wei L, Bertino EM. Incidence, risk factors, and effect on survival of immune-related adverse events in patients with non-small-cell lung cancer. *Clin Lung Cancer* 2018;19:e893–900.

Practical Aspects of Signal Detection in Pharmacovigilance: Report of CIOMS Working Group VIII. Geneva, 2010.

Reviewer Guidance, Conducting a Clinical Safety Review of a New Product Application and Preparing a Report on the Review. U.S. Department of Health and Human Services, Food and Drug Administration, Center for Drug Evaluation and Research (CDER), February 2005.

Robles-Diaz M, Lucena MI, Kaplowitz N, et al. Use of Hy's Law and a new composite algorithm to predict acute liver failure in patients with drug-induced liver injury. *Gastroenterology* 2014;147:109–118.

Schnell PM, Ball G. A Bayesian exposure-time method for clinical trial safety monitoring with blinded data. *Ther Innov Reg Sci.* 2016;50:833–838.

Senior JR. Evolution of the Food and Drug Administration approach to liver safety assessment for new drugs: current status and challenges. *Drug Saf.* 2014;37 (Suppl 1):S9–17.

Spitzer WO. Bias versus causality: interpreting recent evidence of oral contraceptive studies. *Am J Obstet Gynecol.* 1998:179:S43–S50.

Theophile H, Andre M, Arimone Y, et al. An updated method improved the assessment of adverse drug reaction in routine pharmacovigilance. *J Clin Epidemiol.* 2012;65:1069–1077.

Thomas ME, Blaine C, Dawnay A, et al. The definition of acute kidney injury and its use in practice. *Kidney Int.* 2015;87:62–73.

Warden SJ, Fuchs RK. Do selective serotonin reuptake inhibitors (SSRIs) cause fractures? *Curr Osteoporosis Rep.* 2016;14:211–218.

Weaver RJ, Valentin JP. Today's challenges to de-risk and predict drug safety in human "mind-the-gap". *Toxicol Sci.* 2019;167:307–321.

Wildfire J, Bailey R, Krouse RZ. The Safety Explorer Suite: Interactive safety monitoring for clinical trials. *Ther Innov Regul Sci.* 2018;52:696–700.

Wittes J, Crowe B, Chuang-Stein C, et al. The FDA's Final Rule on expedited safety reporting: statistical considerations. *Stat Biopharm Res.* 2015;7:174–190.

Yang YX, Hennessy S, Propert K, et al. Chronic proton pump inhibitor therapy and the risk of colorectal cancer. *Gastroenterol.* 2007;133:748–754.

Zink RC, Wolfinger RD, Mann G. Summarizing the incidence of adverse events using volcano plots and time intervals. *Clin Trial.* 2013;10:398–406.

4

Safety Monitoring through External Committees

Amit Bhattacharyya
Alexion Pharmaceuticals, Boston, MA, USA

William Wang
Merck & Co, Inc, Kenilworth, NJ, USA

James Buchanan
Covilance, LLC, Belmont, CA, USA

CONTENTS

4.1 Introduction – Safety Monitoring and Assessment during Clinical Development..78
4.2 Role of DMCs 78
4.3 Evaluating the Need for a DMC...80
4.4 Composition of the DMC...81
4.5 DMC Operations...83
4.6 DMC's Scope for Monitoring Trial and Safety of Participating Patients...83
    4.6.1 Investigational New Drug (IND) Safety Expedited Reporting to Health Agencies...84
    4.6.2 Expanded DMC to Serve as a SAC (or Safety Assessment Entity) 89
    4.6.3 Quantitative Analysis of ADRs to Support the SAC (or Safety Assessment Entity)...90
4.7 Safety Monitoring in Studies with Multiple Drugs or Diseases – Complex Designs...92
4.8 Safety Monitoring in Pragmatic Trials...93
4.9 Graphical Assessment of Safety Endpoints 94
4.10 Summary..95
References..96

DOI: 10.1201/9780429488801-5

4.1 Introduction – Safety Monitoring and Assessment during Clinical Development

Patient safety is of paramount interest in clinical trials with many stake-holders, e.g., sponsors, patients, regulators, and trial subjects to ensure that the treatment for a disease is efficacious and safe for its intended use. During the conduct of the trials, there are many functions, internal and external, for the sponsors who have the obligation to monitor and report patient safety to the appropriate authorities. Data Monitoring Committees (DMCs) play a significant role in monitoring the safety of an investigational drug during development. The author hopes that this chapter on Safety Monitoring and Assessment will provide the reader an overview of the roles of DMCs, their roles during a trial, the composition, and operations. In addition, the recent (2015) suggestions from the US regulatory agency about establishing a Safety Assessment Committee (SAC) are discussed in relation to or supplemental to the DMC in certain cases. The role of the SAC is to help guide the sponsors to decide on the causality of the serious unexpected adverse event (AE) that needs expeditious reporting to the regulatory agencies.

4.2 Role of DMCs

One of the key components of continuing safety monitoring during trials is the initiation of an independent Data Monitoring Committee (also called IDMC) for a trial when the level of risk in a trial warrants more detailed and independent analysis than the sponsor can provide by internal safety monitoring processes. The DMC is sometimes called the Data Safety Monitoring Board or DSMB. Setting up such a committee has been considered a key tool in monitoring accruing results during ongoing clinical trials during the past half-century (Greenberg, 1988).

The concept of a DMC was introduced in 1967 by the Greenberg Report with an intent to monitor the conduct and safety of single trials. It was mentioned at the time that trials might be terminated because of either the clear establishment of intervention's benefit, or because of sufficient evidence of harm, or because the trial no longer was viable or of interest, or other compelling reasons. The usage of DMCs has increased steadily in recent times as they are utilized very commonly in current clinical trial practice. Initially, DMCs were used primarily for large late-development trials targeting major mortality–morbidity endpoints; over time, their usage has expanded in smaller trials, in studies with less serious outcomes (particularly in vulnerable populations), in earlier stages of clinical development, and in simultaneously monitoring multiple trials within a development program.

Regulatory guidance documents (EMA 2005, FDA 2006) have helped to establish a common landscape regarding the role and operating processes of DMCs. The International Conference on Harmonization (ICH) Guidance E6 (R2) also emphasized that DMCs should assess the progress of a trial, both in terms of safety and efficacy, and should make recommendations to the sponsor as to whether to continue, modify, or stop a trial. It is recommended that the reader should read the regulatory guidance on DMCs (EMA 2005, FDA 2006) for a detailed scope, responsibility, and monitoring process as recommended by the health agencies. This chapter covers some of the important concepts from the guidance and illustrates a few examples.

The following definition of a DMC is taken from the Food and Drug Administration (FDA) guidance (2006):

> A clinical trial DMC is a group of independent experts with pertinent expertise that reviews on a regular basis accumulating data from one or more ongoing clinical trials. The DMC advises the sponsor regarding the continuing safety of trial subjects and those yet to be recruited to the trial, as well as the continuing validity and scientific merit of the trial.

The European Medicines Agency (EMA) Guidance (2005) also has similar definitions and recommendations, especially appropriate for multinational studies.

DMCs are primarily in place to ensure that patient safety is not compromised in an ongoing trial, and these committees consider safety from several perspectives. The most straightforward aspect is monitoring for the emergence of serious adverse events (SAEs) or toxicities and stopping a trial for evidence of harm that cannot be effectively mitigated. While the primary focus of DMCs is patient safety and ethical conduct of the trials, efficacy results are also monitored as appropriate to enable benefit-to-risk assessments by the committee (see Chapter 13 for further discussion of benefit-risk assessments). DMCs are hence charged with protecting subjects from assuming unnecessary risks of clinical trial participation when a trial appears to be futile (i.e., no chance for participating patients to benefit). The scope of a DMC might not be limited to the monitoring of a single trial; some DMCs review a set of trials that are part of a drug development program or a clinical research network. DMCs are often consulted before the sponsor finalizes any protocol amendment reflecting trial management changes. The DMC members often, with extensive experience in their clinical practice, share their insight into the current trial or bring knowledge from other trial experiences in such discussions on protocol amendments.

More complex clinical trial designs in recent times have broadened the scope of DMC deliberations and potential recommendations, often accompanied by pre-specified interim analysis to certain pre-specified time points of interest or after a certain number of events have occurred. These include stopping a trial early due to overwhelming efficacy or due to futility based on pre-specified stopping criteria. Stopping for efficacy may include situations where there is overwhelming evidence (i.e., beyond a reasonable doubt) of a mortality or morbidity benefit such that placebo-treated patients are protected from the risk of no treatment and the trial can be brought to rapid completion to expedite the availability of effective therapy to the broader patient population. Stopping for futility includes situations when stopping a clinical trial in which the interim results suggest that it is unlikely to achieve statistical significance and limited resources can be saved which could be used on more promising research, as well as prevent patients from receiving a potentially non-efficacious drug. However, stopping early for benefit must be balanced against the risk of stopping too early due to the need for sufficient safety data for a new drug to assess the proper benefit-risk balance. Similarly, declaring futility also assumes risks, such as the potential for missing a delayed treatment effect, an effect on important secondary endpoints. This also adds ethical, scientific, and logistical challenges to DMC operations. Addressing analytical complexities in a clinical trial with complex design often requires an experienced DMC biostatistician member well-versed with the statistical methodologies utilized in such innovative design.

4.3 Evaluating the Need for a DMC

All clinical trials require safety monitoring, but not all trials require formal monitoring by an independent committee that is external to the sponsor, its vendors, and the study investigators. DMCs are typically established for randomized, multi-site studies that evaluate treatments intended to prolong life or slow the progression of disease. They are generally not needed for trials addressing lesser outcomes, such as relief of symptoms unless the trial population is at an elevated risk for more severe outcomes (FDA, 2006).

Regulatory guidance (FDA, 2006; EMA, 2005) across the globe recommend establishing a DMC to enhance the safety of trial participants in situations in which safety concerns are unusually high, which includes, but not limited to:

- The trial endpoint is such that a highly favorable or unfavorable result or even a finding of futility at an interim analysis might ethically require termination of the trial before its planned completion;
- There are *a priori* reasons for a particular safety concern, as, for example, if the procedure for administering the treatment is particularly invasive;

- There is prior information suggesting the possibility of serious toxicity with the trial treatment;
- The trial is being performed in a potentially fragile population, such as children, pregnant women, or the very elderly, or other vulnerable populations, such as those who are terminally ill or of diminished mental capacity;
- The trial is being performed in a population at elevated risk of death or other serious outcomes, even when the trial objective addresses a lesser endpoint;
- The trial is large, or long duration, and multi-center.

In studies with one or more of these characteristics, the additional oversight provided by a DMC can further protect trial participants. This applies to both the blinded and open-label studies.

DMCs are commonly employed for a single large clinical trial and may not always be charged with providing oversight of an entire clinical program. However, CIOMS IV (1998) recommends that sponsors ensure that important new safety information is communicated to the DMC even if the information did not originate from the trial the DMC is monitoring. To address this, and for an overall safety monitoring across a clinical program, the same DMC may be employed to monitor multiple studies within a clinical program.

Since sponsors are responsible for monitoring patient safety, there are several other mechanisms to ensure patient safety even if a formal DMC is not constituted for a trial. These may include the sponsor's internal safety review committees or safety monitoring committees with oversight from their safety/pharmacovigilance functions.

4.4 Composition of the DMC

The regulatory guidance documents provide detailed recommendations on the composition of a DMC, which should be composed of medical experts (clinicians), biostatisticians, ethicists, and others with experience in the conduct of clinical trials and the therapeutic area of the trial. The ability of DMCs to provide the anticipated additional assurance of patient safety and trial integrity depends on the quality and proper selection of DMC members.

The clinician members of DMCs have expertise in relevant clinical specialties, and at least one biostatistician DMC member should be knowledgeable about statistical methods for clinical trials. Prior DMC experience is important when considering the committee as a whole, and so, it is highly desirable that at least some members have had prior DMC service. Prior DMC experience is particularly important for the biostatistician if there is only one biostatistician serving on the DMC. The DMC Chair should have prior DMC experience, preferably in a prior leadership role in the profession. The DMC Chair officiates all DMC meetings and is responsible for the communication with the sponsor representative about the DMC recommendations after the DMC meetings. The recommendations from the DMC are non-binding as the sponsor is ultimately responsible for making a final decision on whether to continue or stop the trial, modify the protocol, or continue the trial, after receiving the DMC recommendation.

Unlike other individuals and committees responsible for the execution of a trial, members of the DMC are typically unblinded to aggregate or individual data by the treatment arm during the conduct of the trial. The DMC provides independent advice, usually to the sponsor or its designee, on the continuing scientific validity and safety of the trial and the efficacy of the therapy under investigation. In studies where endpoints often need adjudication of the clinical endpoints, DMCs also make sure that the endpoints are being reported in a timely fashion. If the endpoints in a trial are adjudicated by a separate adjudication committee, the DMC will request timely access to the adjudicated events so that they could be incorporated appropriately during the DMC safety review meetings. In some DMC charters, the DMC likes to review the data in a partially blinded fashion, i.e., when the DMC reviews the summary data for the multiple treatment groups without identifying the actual doses of the treatments unless it is necessary. This provides an unbiased interpretation of the results by the DMC, especially when identifying the multiple doses that may influence the DMC's decision and its impact.

On some occasions, especially in an open-label trial, or in the very early stages of a trial, the safety monitoring activity could be conducted by an internal (to sponsor) committee for safety evaluation, often called the internal Safety Board. However, in most trials in Phase 2 and Phase 3, an external independent DMC is constituted to evaluate safety based on unblinded aggregated data, while the internal Safety Board reviews trial safety issues in a blinded fashion.

Because of the sensitivity to the exposure of unblinded treatment information during the conduct of a trial, DMC members are mandated to keep strict confidentiality of the trial and are free of serious conflicts of interest during and after the trial that they were entrusted with monitoring. ICH E9 (1998) guidance stressed that "The independence of the DMC is intended to control the sharing of important comparative information and to protect the integrity of the clinical trial from adverse impact resulting from access to trial information", even though the concept of independence for a DMC can be complex. DMC members are generally selected by the sponsor and remuneration for DMC services could be perceived as a conflict, although it is important to note that DMCs are in place to protect patient safety and not the business interests of the sponsor. Involving highly knowledgeable individuals on a DMC is desirable, but these individuals may be more likely than non-experts to have conflicts that need to be managed. Although some conflicts may exist, DMC members should not have relationships that would result in financial, academic, career, or other gains based on the trial outcome.

4.5 DMC Operations

ICH E9 states that "The DMC should have written operating procedures and maintain records of all its meetings, including interim results; these should be available for review when the trial is complete". To satisfy this requirement, a DMC charter should be agreed upon between the sponsor, DMC members, and involved third-party vendors before the start of the DMC activities. The charter will describe in detail the responsibilities of the DMC, its structure, format for reports, statistical guidelines for recommending trial termination, contractual and indemnification information, and processes for conducting meetings, ensuring confidentiality, rules concerning the avoidance of serious conflicts of interests, and communication with the sponsor, and any relevant information. Having a well-detailed charter is vital for the DMC, similar to having a detailed protocol and an analysis plan for a clinical trial. A charter with pre-specified procedures reduces any concern about the DMC being inappropriately influenced by interim data that could bias the trial results and interpretation. Health Authorities, such as the FDA and EMA, may request that the sponsor submit the charter in advance of the performance of any interim analyses, and before the initiation of the trial.

4.6 DMC's Scope for Monitoring Trial and Safety of Participating Patients

The DMC's oversight of patient safety is focused on the incidence of AEs, SAEs, and adverse events of special interest (AESI) on a continual basis at their scheduled meetings over the course of the trial falling into the following broad categories:

1. Anticipated risks: These are known due to the population under study, their disease state, and the treatment class. These are probably the simplest for the DMC because the role of safety monitoring for these outcomes is to ensure that the balance of risk and benefit continues to favor benefit for the patients. While some of these known

risks could be serious, the DMC may still recommend continuation if DMC feels that the patients are being well cared for and that they have been adequately informed of the risks.

2. Unexpected risks: In these situations, the DMC may recommend the continuation of the trial without changes because the benefits may still outweigh the risks. The DMC may continue to monitor more occurrences of these AEs before making any conclusive assessment of the perceived risks.

3. Unexpected Serious risks: These are the real challenges for the DMC to evaluate whether changes would be needed for the trial. If the AEs are considered to have a causal relationship to the investigational drug, the DMC will further evaluate if the benefit still outweighs the risks, especially in the situation of unexpected SAEs with dire consequences. If the risk is unacceptably high compared to the benefit, the DMC may recommend stopping the trial. For some medically devastating events, e.g., a single fatal liver failure, the DMC will recommend stopping the trial, if the event is suspected to be associated with the experimental drug.

DMCs also review data related to the conduct of the trial to satisfy themselves of the quality of the trial and its ultimate ability to address the scientific questions of interest, in addition to effectiveness and safety outcomes. These may include recruitment of the patients, their planned follow-ups, adherence to the protocol, discontinuation of patients, and potential for imbalance across patient sub-populations of interest or demographics. Often DMCs could identify early indications of any potential deviation from the planned treatment allocations and randomization issues that can affect trial validity. The DMC will issue recommendations to the sponsor regarding trial conduct when concerns arise that some aspects of trial conduct may threaten the safety of participants or the integrity of the trial.

4.6.1 Investigational New Drug (IND) Safety Expedited Reporting to Health Agencies

While the DMC, with the knowledge of the treatment allocation, is positioned to make recommendations for the safety of patients in a trial or a program, the committee is not generally expected to monitor individual patient-level safety on a regular basis. The investigator, who is the closest to the participant of a clinical trial, observes the participant at the bedside or at the protocol-defined clinic visits. On noting a SAE, the investigator has a regulatory obligation to report that event to the sponsor. Most sponsors also employ a "pharmacovigilance" (PV) team, either internal or external, who is responsible for reviewing AEs, especially those deemed to be "serious", to assess the safety of the experimental treatment. The local Institutional Review Boards (IRBs) or Ethics Committees (ECs) review information about the events that occur at their sites. The PV team may be in the best position to determine if the AE may be attributable to the study drug (with their knowledge of potential "identified risks") if they had the knowledge of the treatment allocation. However, in a blinded trial, the investigator, the PV group, or the IRB/EC may observe a surprising and worrisome AE, but they do not know whether the event occurred in a treated or a control participant. Furthermore, the investigator and the IRB/ECs do not know how many of these events have occurred in the trial as a whole at other sites.

The US FDA as well as other regulatory agencies requires the sponsor to report all unexpected SAEs attributed to the investigational treatment in an expedited manner to the agency. In the last decade, the FDA issued several guidelines (2010, 2012) on expedited safety reporting. The 2010 Revised IND Safety Rule first introduced the term suspected "adverse drug reaction" (ADR) and made clear that a "reasonable possibility" means that there is evidence to suggest a causal relationship between the drug and the AE. It also mandated expedited (7- and 15-day) reporting of potential serious risks, including serious and unexpected suspected adverse reactions. However, the regulation calls for the sponsor to determine whether the event has a "reasonable possibility" of relatedness and subsequently needs to be reported to the agency.

The FDA Guidance on implementing the Final Rule states,

> The sponsor must report in an IND safety report any suspected adverse reaction to study treatment (i.e., including active comparators) that is both serious and unexpected [FDA Guidance 2012, 21 CFR 312.32(c)(1)(i)]. Before submitting an IND safety report, the sponsor needs to ensure that the AE meets all three of the definitions: (i) suspected adverse reaction; (ii) serious; and (iii) unexpected. If the AE does not meet all three of the definitions, it should not be submitted as an IND safety report.
>
> **(US Department of Health and Human Services and Food and Drug Administration 2012, p. 8–9)**

The intra-industry CTTI-IND Safety Reporting working group (Wittes et al., 2015) provided practical interpretations of the final rule in terms of identifying and the reportable actions for these suspected, serious, and unexpected AEs, reportable as suspected ADRs, into three broader groups as described below:

Category A. An event that is uncommon and known to be strongly associated with drug exposure (e.g., angioedema, hepatic injury, Stevens–Johnson syndrome, and agranulocytosis) even though these events can happen in rare situations. Such events are informative as single cases when fully investigated and characterized. The occurrence of even one of these events would meet the definition of a *suspected adverse reaction*. It is expected that sponsors should unblind themselves in each of these events and, if it occurs in association with the investigational drug, they are obligated to report it to the agency. There could be other plausible non-drug-related causes for the event, too. For example, not all hepatic injury is drug-induced, particularly in patients with certain underlying diseases such as heart failure.

Category B. An event that is not commonly associated with drug exposure but is otherwise uncommon in the population exposed to the drug (e.g., tendon rupture in any recipient of a product or a myocardial infarction in a study of young women), such that a few occurrences are sufficient to attribute causality. If the event occurs in association with other factors strongly suggesting causation (e.g., clear temporal association or reoccurrence on rechallenge), a single case may be sufficiently persuasive to meet the definition of a *suspected adverse reaction*. Often, more than one occurrence of a specific event is necessary before the sponsor can judge that there is a reasonable possibility that the drug caused the event. Determining the threshold for the number of occurrences, at which it is reasonable to assume causality, is often debated and will need medical and statistical considerations.

Category C. Events that an aggregate analysis of data from a clinical development program indicates occur more frequently in the drug treatment group than in a concurrent or historical control. These events may be known consequences of the underlying disease or condition under investigation or other events that commonly occur in the study population independent of drug therapy. Events in this category are the most problematic for determining whether they need expedited reporting because of the difficulty in identifying whether an individual event is causally related solely to the drug. These occurrences could be considered anticipated, so that interpretation of individual events may not be reportable as such, but collectively, they might be considered reportable due to a higher frequency in the drug treatment group.

Under the FDA guidance describing the final rule (2012), instead of reporting each category C event individually (because it will not be possible to conclude that the event is a suspected adverse reaction), the sponsor should compare "at appropriate intervals" the number of such events in each arm of a trial and report the events to the FDA expeditiously as IND safety reports if an observed excess rate in the treatment arm relative to control suggests a reasonable possibility that the drug caused the AE. If the trial is part of a larger development program, the sponsor should evaluate the entire clinical trial database (both blinded and unblinded studies) to determine whether to submit the events as expedited reports.

The boundaries between categories B and C are sometimes challenging, especially if some events in a particular patient population based on demographics or other comorbidities may be considered "Anticipated", while for others, they may be not. For example, a heart attack in a 25-year-old woman would likely fall into category B while one in a 75-year- old man would fall into category C, making it challenging to categorize "heart attack" into category B or C solely on the basis of the event term.

The FDA draft guidance in December 2015 titled "Safety Assessment for IND Safety Reporting - Guidance for Industry" also advocated a systematic approach to improve the safety reporting requirement for SAEs for human and biological products developed under the IND application. The recommendation provides clarity on how to conduct safety decisions on categories C above and may often be applied to category B. The recommendations, among other things in this guidance, included the creation of a SAC for a drug development program, consisting of a group of medical and scientific experts. It is important to note that the SAC for such IND safety reporting is recommended for overseeing the safety reporting for the entire program and not for an individual trial within the program. The guidance suggests that the SAC will be charged with reviewing safety data at the program level of the investigational drug and other relevant safety information outside of the program to make a judgment about the likelihood that the investigational drug could cause any SAE. Specifically, it recommended to focus on the following: (1) "an aggregate analysis of specific events observed in a clinical trial indicates that those events occur more frequently in the drug treatment group than in concurrent or historical control groups" and (2) "clinically important increase in the rate of a serious suspected adverse reaction". Once implemented, the SAC streamlines the safety reporting to the FDA for AEs that occurred more frequently in the treated group compared to the untreated group across a program, and also of expected serious adverse reactions that appeared above a defined threshold based on the expected background rate of such events. Additionally, whereas sponsors traditionally were expected to review blinded safety data, the new guidance intends for the review of unblinded safety information both for the establishment of "relatedness" and for the detection of increased incidence of expected events. Through the implementation of a SAC and the Safety Surveillance Plan, the 2015 guidance recommends that sponsors develop a consistent and documented process across all compounds by which SAEs be evaluated on a periodic basis based on their incidence rates on the program level against the expected rates in the population of interest, the disease severity, potential effects due to the treatment class, and all other available information.

It is to be noted that although SACs have been referred to by the FDA 2015 draft guidance on IND safety reporting, unlike DMCs, there is no uniform agreement about their role in drug development or has there been any new action on their regulatory status since the draft guidance at the time of this writing. In other words, even if the usefulness for this SAC has been suggested, unlike DMCs there has been a lack of effort in making this implementation quite harmonized globally. The FDA further clarified their position regarding SACs and DMCs in revised draft guidance in June 2021 titled "Sponsor Responsibilities - Safety Reporting Requirements and Safety Assessment for IND and Bioavailability/Bioequivalence Studies". The new 2021 draft guidance was based on response to industry feedback on the 2015 draft guidance, specifically, (1) trial integrity challenges related to repeated unblinding, (2) trial complexity issues related to the interaction of the SAC and the DMC, (3) separating true safety signals from the background noise and the risk of over-reporting, and (4) the ability of the DMC to function as the SAC. The 2021 guidance on the sponsor's responsibility, together with upcoming guidance on the investigator's responsibility, will replace the 2015 guidance and the 2010 final rule for IND safety reporting.

This new draft guidance, while keeping in place many of the proposals from the 2015 draft guidance in principle, has, however, softened the mandate through some practical considerations on how these reviews may be accomplished without the perceived burden and complexity noted by sponsors. The new 2021 draft guidance no longer specifically recommends the use of the SAC. Instead, it indicates that a sponsor may choose to designate an entity to review the accumulating safety information to make a recommendation to the sponsor whether the safety information meets the reporting requirements. If a DMC is in place, it may be used to conduct aggregate analyses to help the sponsor assess whether the reporting criteria have been met. The entity should include individuals with knowledge about the investigational drug, the disease being treated, including the epidemiology of the disease, and the characteristics of the study population. These members should be qualified by training, and experience to make clinical judgments about the safety of the drug identification of a new type of clinical safety concern may warrant adding additional expertise to the entity reviewing safety data the roles and responsibilities of each individual or group of individuals in the entity should be clearly defined in the plan for safety surveillance. But unlike the 2015 draft guidance, the current draft guidance does not require the sponsor to submit a Safety Surveillance Plan (aka safety monitoring plan) to the IND but it should be available upon inspection. The plan is expected to document the process for how the data will be analyzed and how reporting decisions are made. The plan should be updated as appropriate as knowledge of the safety profile expands.

4.6.2 Expanded DMC to Serve as a SAC (or Safety Assessment Entity)

Although a DMC may share the same trial-specific unblinded information with the SAC and some overlapping scopes, these two independent committees have different roles. In general, the DMC's focus is on monitoring the trial or program by assessing the ongoing safety and efficacy for the active treatment against the placebo or standard of care comparator. Their recommendations center on the conduct of the trial based on aggregate safety data at regular intervals. DMCs do not have a reporting responsibility to the regulatory agency. Traditionally, DMCs are not constituted for all Phase 2 and 3 industry-sponsored trials, but a SAC may be required for all development programs. Also, DMCs meet on a set schedule, either quarterly or bi-yearly, or based on enrollment status in a trial, and not on the basis of the timing of safety events. Additionally, rapid evaluations and decisions must be made by a SAC as opposed to a more deliberative approach often used in DMCs.

However, it may be advantageous to extend the scope of the DMC to include SAC roles, by expanding the charter to include evaluation of individual SAEs in relation to the aggregated analysis, and also including additional DMC members to support epidemiology of populations and disease areas, and other potential experts as required. DMC members will likely need additional training to understand the reporting requirements and their implications. Given that the DMC reviews unblinded data, the committee is well placed to review the patient-level ADRs, and along with additional aggregate information about the safety of the patients, could potentially flag potential signals for reporting to the agencies, especially for the Type C events mentioned earlier. However, to have DMC members review these AEs in an expedited fashion, mechanisms of *ad hoc* reviews using proper technology and principles need to be developed and agreed upon in the charter. From a scientific and safety perspective, DMCs also need to fully understand the prevalence (background) rates for reported safety events in order to properly identify meaningful potential risks vs "noise" before recommending reporting the ADR to the regulatory agencies. Also, this will make the DMCs remit into a program-level safety monitoring, a natural enhancement from traditional trial-level monitoring roles.

In view of such an expanded DMC role, there will be a need for a stage-gate developed by the sponsor before an SAE is reported to the DMC for a review for it being a potential ADR to be part of IND safety reporting. Given the extensive knowledge about the investigational drug, the disease population and the epidemiology of the disease, and the background rates for any observed events in the population and the treatment class, the sponsor could develop a quantitative assessment of safety data from the current and other ongoing studies based on the blinded data and develop an algorithm to determine the initial threshold for DMC review for IND reporting. This will help the committee only review a handful of specific events that gets flagged by the blinded review by the sponsor. Such an expanded role for DMC still needs deliberate planning and establishes scope and processes on when they act as SAC evaluating such ADRs, and when they meet for regular DMC roles for assessing overall safety for the treatment in an aggregate manner. Once these two distinctly different roles are established and processes defined and agreed by the charter, the expanded DMC could be the most efficient way of handling the IND safety reporting as mandated by regulatory agencies. In order to expand the DMC roles to also include the responsibilities of SAC, therefore significant paradigm changes will be needed, even if this seems to be a reasonable choice.

4.6.3 Quantitative Analysis of ADRs to Support the SAC (or Safety Assessment Entity)

Monitoring for safety presents statistical challenges. For safety signals, the SAC usually focuses on the rare, unexpected events (i.e., category B events as described before). However, the DMC and SAC also evaluate for not so rare AEs (i.e., category C, as described before) and evaluate if the frequency of occurrence is higher in the treatment group, compared to the control. Often statistical tests, even if used to test the differences in the occurrences of many AEs, are seen as more of a "hypothesis generation" exercise rather than "hypothesis confirmatory tests" because the statistical power for such signal detection will be very small.

Moreover, with so many tests, there are issues of multiplicities and hence the inflation of the False positive rates, the Type-I error in statistical terminology. However, one cannot specify the number of hypotheses relevant to safety upfront, and hence the multiplicity adjustments, in the traditional sense, are impossible. On the other hand, traditional multiplicity adjustments may lead to a high rate of false negatives, which are of particular concern when evaluating safety. Therefore, for rare events in an ongoing development program, the interpretation will usually be based more on clinical judgment than on the statistical significance of p-values. Bayesian approaches (Berry and Berry 2004; Xia et al. 2011) in Chapter 7 discuss these scenarios, especially when a reasonable amount of prior information is available (e.g., for the event rate in the control arm), while most often the prior information is not reliable in the literature.

Wittes et al. (2015) provided some decision rules for reporting category B and category C events, based on statistical thinking, as follows:

Possible Decision Tree for Reporting Category B Events

If the AE of category B is considered definitely not caused by the investigational product, these should not be reported to the agency expeditiously. Similarly, if there is no imbalance based on summary data from a category C AE, these should not be reported expeditiously to the agency either.

However, the causality assessment may not be clear, in which case ....

If there is an increase (Type B or Type C), evaluate if the increase in rate satisfies the following conditions:

- Does a one-sided 80% confidence interval of the difference between the observed and anticipated include 0? (or, does the one-sided confidence interval of a relative measure include 1?) The anticipated rate comes from epidemiological or natural history data. The observed rate may come from a single trial or from a combined analysis of all trials of the product.
- Is the relative risk less than 3 for category B; or less than 2 in category C AEs? (For studies without controls, the risk is relative to expectation in a relevant historical population; for studies with controls, the risk is relative to the controls or to the historical population, or both.)
- Does lumping similar events make the signal disappear?

If the answers to all three are "yes", the data do not show sufficient evidence of imbalance to report.

If the answers to all three are "no", the evidence of causality is clear enough to report the event.

Otherwise, the imbalance is unclear, in which case, the following may be useful

No increase in rate	Increase in rate Unclear	Clear increase in rate
Do Not Send Report to Agency	Consider sending the report to the agency if at least one of the following conditions hold • Other safety outcomes (e.g., AEs and laboratory data) support causality. • The mechanism of action supports causality. • Investigators need information about this event to manage patients. • Information could influence either patients' willingness to participate in the trial or behavior of patients outside the trial. • Pooling similar endpoints strengthens the signal. • Event is potentially fatal or disabling	Send to Agency
	Otherwise, do not report but reanalyze as new events	

The choice of 3 (for category B) or 2 (for category C) for relative risk cut-off is a suggestion and subject to the study teams' decision, given the disease, patient populations, and other factors which may affect the safety profile for an investigational drug. Similarly, the use of a one-sided 80% confidence interval is a suggestion; some teams would opt not to use confidence intervals at all, and some would use 90% or a different confidence interval. Finally, while statistical analysis and such rules are supporting evidence, the medical professional and safety physicians should have the final say in these safety issues impacted by many patients' profiles and their confounding medical factors.

As noted before, category C AEs are traditionally monitored when the DMC is engaged for a trial. To consider DMC to act as the SAC and help the sponsors in the evaluation of safety signals for expeditious reporting to the agency, the DMC needs to review and evaluate these events in an aggregated way much more frequently than a traditional DMC meeting frequency setup and will have a program-wise safety oversight responsibility. If a DMC is not engaged for a trial, the SAC needs to be formed, ideally external to the sponsor, to have such safety oversight of the trial, and the program as a whole, as proposed by the draft guidance from the FDA (2015).

Please note that the new 2021 FDA draft guidance no longer specifically recommends the use of the SAC per se. Instead, it indicates that a sponsor may choose to designate a Safety Assessment Entity. The discussion in Section 4.6.3 can also be applied to safety assessment entities.

4.7 Safety Monitoring in Studies with Multiple Drugs or Diseases – Complex Designs

In recent times, especially in oncology and rare-disease studies, there is increased interest in expediting late-stage drug development through developing trial designs that test multiple drugs and/or multiple disease sub-populations in parallel under a single protocol, without a need to develop new protocols for every trial. Designs of such trials, often referred to as umbrella, basket, or platform trials, are examples of such a framework often called "master protocol". In contrast to traditional trial designs, where a single drug is tested in a single disease population in one clinical trial, master protocols use a single infrastructure, trial design, and protocol to simultaneously evaluate multiple drugs and/or disease sub-populations in multiple sub-studies, allowing for efficient and accelerated drug development. However, such complex designs bring significant challenges in identifying causality of safety issues due to multiple diseases or multiple treatments, which has attracted attention of health agencies. FDA draft guidance (2018a) on the master protocol provides an expectation from regulatory agencies on more frequent safety monitoring and reporting of the safety events in order to ensure quality data in support of drug approval based on these complex designs. Examples of potential challenges in drug safety in such trials include (i) difficulty in the attribution of AEs to one or more investigational drugs if multiple drugs are administered within various arms and the trial lacks a single internal control for those drugs or (ii) assessing the safety profile of any given investigational drug (still under IND) if multiple drugs are studied across multiple protocols.

The FDA Guidance recommends having a DMC to monitor efficacy results as well as the safety endpoints. The DMC charter should authorize the committee to conduct pre-specified and ad hoc assessments of efficacy and futility and recommend protocol modifications or other actions, including sample size adjustment and discontinuation or modification of a sub-study based on futility or overwhelming evidence of efficacy. It also proposes an independent committee responsible for the assessment of safety who should complete the real-time review of all SAEs as defined in FDA regulations and periodically assess the totality of safety information in the development program. For studies with pediatric populations, appropriate pediatric expertise is also recommended along with an ethicist in this committee reviewing safety. The guidance further recommended that the sponsor should establish a systematic approach that ensures rapid communication of serious safety issues to clinical investigators and regulatory authorities under IND safety reporting regulations. In addition, the approach should describe the process for rapid implementation of protocol amendments in existing or new substudies to address serious safety issues. In such cases of amendments, the current and future trial subjects should be informed of newly identified risks via changes in the informed consent document and, if appropriate, recommending re-consent of current patients to continue trial participation.

Similar suggestions for safety monitoring by DMC and SAC have also been recommended in the FDA guidance (2018b) on "Expansion Cohorts: Use in First-In-Human Clinical Trials to Expedite Development of Oncology Drugs and Biologics"

4.8 Safety Monitoring in Pragmatic Trials

The DMC may have a slightly different take when monitoring pragmatic clinical trials (PCTs). The PCT trials are conducted within the real-healthcare-practice framework where it is common to consider: (a) clinically relevant alternative interventions to compare, (b) a diverse population of trial participants, (c) heterogeneous practice settings, and (d) a broad range of health outcomes. Ellenberg et al. (2015) identified four different areas within the standard DMC framework where special care needs to be considered for any DMC deliberations. They are (a) outcomes, (b) protocol adherence, (c) patient eligibility, and (d) intensity of follow-up.

For outcomes in pragmatic trials, DMC needs to be comfortable monitoring more subjective outcomes like patient reported outcomes (PROs), which are prevalent more than the traditional clinical trials. Moreover, these PRO measures are often found in the treating clinician's notes or derived from electronic health records (EHRs) or claims databases, rather than by a central adjudication group. First, the sites for PCTs may be more heterogeneous than in a traditional trial in terms of patient characteristics and treatment approaches of practitioners, as well as for lab data assessment, special considerations with standardized lab data may have to be utilized by the DMC for assessing safety using the lab parameters. Second, DMCs need to give special considerations to protocol adherence in a PCT. Otherwise, a failure to find a difference in treatment effects will be difficult to interpret, as this may be due to the true equivalence of treatments or to widespread failure to adhere to the assigned interventions as described in the protocol. Third, DMCs need to give special attention to the patient eligibility criteria as a substantial number of randomized participants not meeting the trial's eligibility criteria would raise concerns about the quality of trial conduct. Because many PCTs are open-label studies (to mimic "real world" set-up), it may be particularly important for DMCs to monitor ineligibility rates by treatment arm to assess any inconsistency. More frequent follow-up from DMCs is warranted, especially, during the early stages of a PCT, to ensure the quality of data generated, especially given the heterogeneity of patients, clinical practice, and other factors.

From the analytical aspect, an experienced statistician is also needed to help DMC evaluate complex interim data and its analysis to make informed recommendations. Complexities of PCTs, such as the use of cluster designs and the reliance on EHR data, make statistical expertise especially important for DMC decision-making. DMCs monitoring PCTs may also be better served by experts in the practical medical setting and often a patient representative.

4.9 Graphical Assessment of Safety Endpoints

Traditionally, the DMC reviews thousands of pages of output from demographics, a list of occurrences of many AEs including some considered as SAEs, in a short time, and assesses the safety aspects of the patients enrolled in the trial. This has been a tremendous burden in the process on the DMC which could be significantly improved through additional visualization tools that could answer more questions about patient safety through various pre-specified graphics, than the hundreds of tables generated at each instance. Additionally, utilizing interactive graphics, as discussed in Chapter 2, would also help in visualizing patient data interactively and exploring relationships within correlated safety events during the DMC meetings to have a comprehensive assessment of safety. Statisticians, working closely with the safety physicians, could develop such key safety tables, listings, and appropriate graphical displays to quantitatively assess these occurrences of AEs. Together with the sponsor, and the processes developed for DMC, SAC, and similar safety committees, patients, investigators, and regulatory agencies, one can ensure the scientific rigor and integrity that is required to develop a safe drug for the patients in need of treatment for a disease with unmet need.

4.10 Summary

Monitoring safety in the lifecycle of clinical development is difficult due to its uncertainties. While some of the safety issues can be anticipated (due to the disease population) *a priori* or are expected (due to prior knowledge of the treatment effect on safety endpoints), the timing of such AEs is unknown. Moreover, it is impossible to ascertain which patient in active treatment will experience the event and who will not. The AE may also be experienced by a patient under a placebo due to their disease state or in certain unknown circumstances. Because of this unplanned nature of occurrences of the multiple potential safety events, unlike a few key efficacy endpoints, one cannot power a trial based on safety events, nor will it serve any purpose in the development of drugs for assessing its safety. For the same reason, there is rarely a request from an agency for multiplicity adjustments for safety endpoints, because of the nature of the events and the importance of erring on the conservative side. Safety is part and parcel of clinical development and even if some adverse events may be anticipated, but not all of the safety concerns can be anticipated a priori. Studies with active controls may even pose additional challenges with respect to safety issues. Because of these uncertainties, adequate safety monitoring tools and processes need to be established for each trial or a suite of trials in a clinical development program.

The DMC and potentially a SAC, if established, along with organizational internal safety groups, provide such key support to trial monitoring, from the patient safety perspective. Their critical review of the potential imbalances in the aggregated rate of AEs and ADRs across treatment groups, as well as reviewing individual patient safety, are important to identify potential new unexpected safety issues for an investigational treatment in a specific disease population. These periodic reviews are the backbone of conducting clinical trials with patient safety as the guiding star.

References

Berry SM, Berry DA. Accounting for Multiplicities in Assessing Drug Safety: A Three-Level Hierarchical Mixture Model. *Biometrics*, (2004) 60, 2, 418–426.

CIOMS Working Group IV. (1998) *Benefit-Risk Balance for Marketed Drugs: Evaluating Safety Signals*. Geneva.

Ellenberg SS, Fleming TL, DeMets DL. (2002) *Data Monitoring Committees in Clinical Trials: A Practical Perspective*. John Wiley & Sons Ltd., ISBN: 0-471-48986-7.

Ellenberg SS, Culbertson R, Gillen DL, Goodman S, Schrandt S, Zirkle M. Data monitoring committees for pragmatic clinical trials. *Clin Trials*. 2015 October;12(5):530–536.

Greenberg Report. Organization, Review, and Administration of Cooperative Studies (Greenberg Report): A Report from the Heart Special Project Committee to the National Advisory Heart Council, May 1967. *Controlled Clinical Trials 9* (1988): 137–148.

European Medicines Agency, Committee for Medicinal Products for Human Use (CHMP). Guidelines on Data Monitoring Committees (2005). https://www.ema.europa.eu/en/documents/scientific-guideline/guideline-data-monitoring-committees_en.pdf, (accessed June 30, 2021).

International Council for Harmonisation of Technical Requirements for Pharmaceuticals for Human Use (ICH). ICH Harmonised Guideline E6(R2): Integrated Addendum to ICH E6 (R1): Guideline for Good Clinical Practice (2016). https://database.ich.org/sites/default/files/E6_R2_Addendum.pdf

International Council for Harmonisation of Technical Requirements for Pharmaceuticals for Human Use (ICH). ICH Harmonised Tripartite Guideline E9: Statistical Principles for Clinical Trials (1998). https://database.ich.org/sites/default/files/E9_Guideline.pdf, (accessed June 30, 2021).

US Food and Drug Administration. Guidance for Industry: Safety Assessment for IND Safety Reporting (draft, December 2015). https://www.fda.gov/media/94879/download, available at the safety book folder under https://community.amstat.org/biop/workinggroups/safety/safety-home

US Food and Drug Administration. Guidance for Industry: Sponsor Responsibilities—Safety Reporting Requirements and Safety Assessment for IND and Bioavailability/Bioequivalence Studies (draft, June 2021) https://www.fda.gov/media/150356/download (accessed June 30, 2021).

US Food and Drug Administration FDA CDER's Small Business and Industry Assistance (SBIA) Webinar: Overview of FDA draft guidance on "Sponsor Responsibilities—Safety Reporting Requirements and Safety Assessment for IND and Bioavailability/Bioequivalence Studies", https://www.youtube.com/watch?v=qwYVh4czK0w June 2021 (accessed June 30, 2021).

US Food and Drug Administration. Guidance for Clinical Trial Sponsors: Establishment and Operation of Clinical Trial Data Monitoring Committees (2006). https://www.fda.gov/media/75398/download, (accessed June 30, 2021).

US Food and Drug Administration. Final Rule: Investigational New Drug Safety Reporting Requirements for Human Drug and Biological Products and Safety Reporting Requirements for Bioavailability and Bioequivalence Studies in Humans (2010). https://www.fda.gov/drugs/investigational-new-drug-ind-application/final-rule-investigational-new-drug-safety-reporting-requirements-human-drug-and-biological-products

US Food and Drug Administration. Guidance for Industry and Investigators Safety Reporting Requirements for INDs and BA/BE Studies (2012). https://www.fda.gov/files/drugs/published/Safety-Reporting-Requirements-for-INDs-%28Investigational-New-Drug-Applications%29-and-BA-BE-%28Bioavailability-Bioequivalence%29-Studies.pdf, (accessed June 30, 2021).

US Food and Drug Administration. Guidance for Industry: Expansion Cohorts: Use in First-In-Human Clinical Trials to Expedite Development of Oncology Drugs and Biologics Guidance for Industry (DRAFT, August 2018a). https://www.fda.gov/media/115172/download, (accessed June 30, 2021).

US Food and Drug Administration. Guidance for Industry: Master Protocols: Efficient Clinical Trial Design Strategies to Expedite Development of Oncology Drugs and Biologics (draft, September 2018b). https://www.fda.gov/media/120721/download, (accessed June 30, 2021).

Wittes J, Crowe B, Chuang-Stein C, Guettner A, Hall D, Jiang Q, Odenheimer D, Xia HA, Kramer J. The FDA's Final Rule on Expedited Safety Reporting: Statistical Considerations. *Statistics in Biopharmaceutical Research* 2015 July 3;7(3):174–190.

Xia HA, Ma H, Carlin BP. Bayesian Hierarchical Modeling for Detecting Safety Signals in Clinical Trials. *J Biopharm Stat.*, 2011;21:1006–1029.

Part B

Statistical Methodologies for Safety Monitoring

5

An Overview of Statistical Methodologies for Safety Monitoring and Benefit-Risk Assessment

William Wang
Merck & Co, Inc, Kenilworth, NJ, USA

Ed Whalen
Pzifer Inc., New York, NY, USA

Melvin Munsaka
AbbVie, Inc., North Chicago, IL, USA

Judy X. Li
Bristol-Myers Squibb, San Diego, CA, USA

CONTENTS

5.1   Introduction 102
5.2   A Perspective of Statistical Methods in Safety Monitoring 103
    5.2.1   Blinded versus Unblinded Safety Assessments 105
    5.2.2   Bayesian and Frequentist Approaches..106
    5.2.3   Post-Marketing PV versus Pre-Marketing Evaluation.............108
    5.2.4   Static versus Dynamic Evaluation..109
    5.2.5   Analyses from Patient-Level Data to Integrated Safety
        Evaluation..110
5.3   Benefit-Risk Assessment Methodologies...110
5.4   Concluding Remarks: Safety Monitoring and Benefit-Risk
    Methodologies...113
Bibliography 114

DOI: 10.1201/9780429488801-7

5.1 Introduction

In June 1997, the Food and Drug Administration (FDA) approved a calcium channel blocker (mibefradil) for hypertension and angina, but with some concerns about drug–drug interactions. Safety monitoring continued after drug approval in the form of monitoring spontaneous reports and further drug–drug interaction studies with the FDA requiring the addition of more drug–drug interactions to the product label. See, for instance, historical documents at http://www.fda.gov/ohrms/dockets/ac/98/briefingbook/1998-3454B1_03_WL32.pdf. In June 1998, only one year following approval, the FDA asked the sponsor to withdraw the drug as a result of the accumulating evidence on drug–drug interactions. This extreme case illustrates the value of continual safety monitoring of biopharmaceutical products throughout the drug development life cycle.

The passage, in 1962, of the Kefauver-Harris amendment led to an emphasis on the proper study of efficacy claims in the drug industry. The U.S. Kefauver Harris Amendment or "Drug Efficacy Amendment" is a 1962 amendment to the Federal Food, Drug, and Cosmetic Act. Events of the last 20 years, such as the previous example with mibefradil, have exposed the importance of the proper study and interpretation of drug safety. In recent years, the FDA has issued guidance regarding safety monitoring and reporting for an investigational new drug (IND) to encourage fuller development of safety profiles, as seen in their 2012 guidance and 2015 draft guidance, respectively. As O'Neill pointed out, statistical methodology for safety monitoring has not been well developed to match that for efficacy, and pre-marketing safety evaluation must balance statistics with clinical discernment (O'Neill, 2002, 2008). Clinical discernment or medical judgment puts safety monitoring findings in the context of benefit-risk assessment. Quantitative evaluation of safety needs to connect with benefit-risk evaluation in developing/implementing ways to minimize the drug risks while preserving its benefits (FDA guidance on pre-marketing risk assessment, 2005).

As described in previous chapters (see Chapters 1 to 3), safety monitoring reporting (SMR) has several defining characteristics:

- Safety monitoring is a process and involves a wide range of stakeholders.
- It assesses side effects across a spectrum of frequencies and magnitudes, ranging from easily detected side effects to those that rarely occur but have potentially high impact on patient well-being.
- Monitoring can be carried out dynamically via pharmacovigilance (PV) as outlined in Chapter 2 (aggregate safety assessment planning) and Chapter 3 (safety signaling).

- Safety monitoring and its risk management is an iterative process of assessing a product's benefit-risk balance (FDA guidance on pre-market risk assessment, 2005).
- It serves to lay the foundation for an integrated analysis of safety and benefit-risk analysis in regulatory submissions, such as an NDA, or for a possible advisory committee meeting.

The contents of this book chapter look at SMR from the statistical perspective while keeping the larger SMR landscape in mind. The focus is on the strategies to identify issues during the pre-marketing development phases but will include descriptions of methods typically used in post-approval surveillance. Section 5.2 describes various aspects of the statistical methods used in SMR. In an effort to present a unique view in our summary, we chose to summarize the literature using the following five perspectives on data and analysis: blinded versus unblinded assessments; Bayesian versus frequentist approaches; post-marketing methods in pre-marketing evaluations; static versus dynamic evaluations; and analyses from patient-level data to meta-analytical evaluations. Section 5.3 connects SMR with benefit-risk evaluation and provides an overview of benefit-risk assessment methodologies.

5.2 A Perspective of Statistical Methods in Safety Monitoring

A webster definition of "perspective" is "the capacity to view things in their true relations or relative importance" or "the interrelation in which a subject or its parts are mentally viewed" (http://www.merriam-webster.com/dictionary/perspective assessed on April 4 2021). Statistical thinking and methodologies can apply to drug safety data in many ways. To provide some useful perspectives on the statistical methodologies, we would want to examine relevant methods in a meaningful relationship. The perspectives of different stakeholders in safety monitoring will interact with decisions made to frame the safety goals, methods for getting to them, and decision processes based on the derived information. See Figure 5.1 for an outline of these components.

The following points outline one, but not the only, way to break out some of those statistical aspects to provide a coherent framework for discussing drug safety statistics, especially in the development (pre-marketing) phases. The following points align in part with Figure 5.2 which illustrates the various dimensions of safety monitoring.

Breaking down statistical methods and thinking for safety monitoring and reporting

- Blinded and unblinded assessment
- Bayesian and frequentist approaches

- Post-marketing PV and pre-marketing evaluation
- Static and dynamic safety evaluation
- Analyses from patient-level data to meta-analytical evaluation

Unlike a typical statistical application, such as confirming a primary endpoint, safety monitoring requires multiple statistical views of safety data.

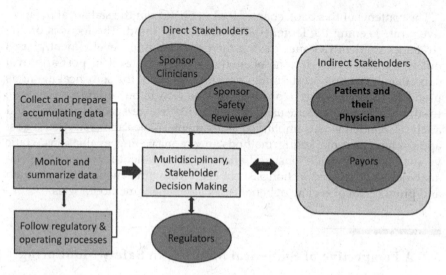

FIGURE 5.1
Stakeholders and data in safety monitoring

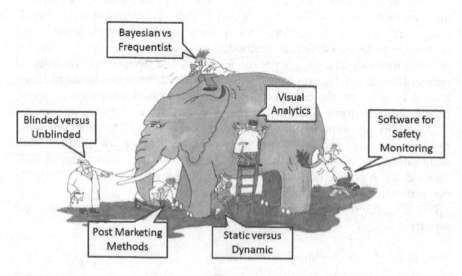

FIGURE 5.2
Multidimensional nature of safety monitoring

These features form part of the development of an overall safety picture as the sponsors and regulators work to understand, track, and fill out the safety profile of a drug.

The following subsections cover some of the key specifics regarding each of these six general areas.

5.2.1 Blinded versus Unblinded Safety Assessments

Blinded safety monitoring applies to the setting in which a clinical trial is ongoing and there is some interest in assessing safety outcome(s) of interest in a blinded fashion in the sense that no treatment information is made available. An immediate point to note is that blinded looks at safety data will not be as informative and as efficient as in an unblinded look at the data where the treatments are known. But clearly, during drug development many ongoing studies may hold important clues to the safety profile. Upon study completion, the data typically get unblinded for the assessment of efficacy and safety. Any unblinding prior to study completion is either part of a planned or unplanned set of interim analyses or the result of unblinding specific patients who have experienced serious adverse events that require knowledge of the blinded treatment for deciding on countermeasures to the event. A safety data monitoring committee can make much better judgments about a drug program if they have some knowledge of the actual treatments received, but it comes at the expense of no longer having a fully blinded study.

Assessing safety data in a blinded manner comes with various challenges and issues, including ethical considerations, statistical considerations, and design considerations. Ball (2011), Ball and Piller (2011), Ball and Silverman (2011), Gould and Wang (2015), Mukhopadhyay, Waterhouse, and Hartford (2018) have discussed these issues and challenges as they pertain to blinded safety monitoring. For example, proactive safety signal monitoring using blinded data in an on-going clinical trial will inevitably raise logistical questions regarding monitoring patient safety while at the same time maintaining the study blind. Another consideration arises regarding available information on the safety concern as well as how to address blinded safety monitoring in the case where some studies are still ongoing. There is the question of how one harnesses information from the completed studies and those that are ongoing?

The above considerations raise the question: how much information can blinded safety monitoring provide? As it turns out, a good deal of safety information can be gained from blinded data as suggested in the definition itself and some consequences of defining it this way. Another question that needs to be addressed in blinded safety monitoring is that of the statistical method that can be leveraged for blinded safety analysis along with linking blinded and unblinded analyses of safety data. Various statistical methods and approaches, including frequentist, Bayesian, and likelihood-based methods that can be used in applications of blinded safety monitoring, are highlighted in Chapter 6.

5.2.2 Bayesian and Frequentist Approaches

We focus our review on safety signal detection and evaluation in clinical development. In practice, the usual approach for safety data analysis involves the comparison of the proportion of subjects who experience an adverse event (AE) between treatment groups for each type of AE. This involves a large number of analyses with inadequate statistical power and no meaningful control of type 1 error. Unadjusted analyses may result in a large number of false positive results while using simple adjustments (e.g. Bonferroni procedures) are generally too conservative and counterproductive for considerations of safety. Thus, it is important to develop an approach to analyze AE data that addresses concerns about multiplicity and the imprecision of data inherent in the analysis of a small number of events.

The Bayesian approach for safety monitoring incorporates prior knowledge about the safety profile of the treatments and updates knowledge based on accumulating data. The potential advantages of the Bayesian approach relative to classical frequentist statistical methods also include the flexibility of incorporating the current knowledge of the safety profile, originating from multiple sources, into the decision-making process. However, it is sometimes difficult to specify the prior distributions reliably.

In early development, most information will come from the literature or early-phase sponsor trials where the events to be followed come from knowledge of mechanisms of action or small numbers of trial data. This leads naturally to consideration of Bayesian methods. At later stages, frequentist methods may be more helpful especially for events of the Tier 1 type, based on the hierarchy reported in the study by Crowe et al. (2009).

Weaver et al. (2016) introduced three approaches using simple Bayesian methods to assess pre-specified AE: Single conjugate prior; meta-analytic predictive (MAP) prior, which comprises a mixture of conjugate priors; and a robust mixture prior that incorporates a robust parameter and a weakly informative component to the MAP prior so that the prior-data conflict could be handled more flexibly. Berry and Berry (2004) proposed a three-level hierarchical mixed model to account for multiplicities in AE assessment. The basic level is the type of AE. The second level refers to the body system which contains a number of types of possibly related AEs. The highest level is the collection of all body systems. The proposed three-level hierarchical mixed model provides an explicit method for borrowing information across various types of AEs. The probability that a drug caused a type of AE is greater if its rate is elevated for several types of AEs within the same body system than if the AEs with elevated rates were in different body systems. Xia et al. (2011) expanded Berry and Berry's method into a hierarchical Poisson mixture model which accounts for the length of the observation of subjects and improves the characteristics of the analysis for rare events. This makes the applicability of this method reach beyond Tier 2 events (see Crowe et al., 2009, for the tier definitions).

DuMouchel (2012) proposed a multivariate Bayesian logistic regression (MBLR) method to analyze safety data when there are rare events and sparse data from a pool of clinical studies. It is designed to be a compromise between performing separate analyses of each event and a single analysis of a pooled event. It requires the selection of a set of medically related issues, potentially exchangeable with respect to their dependence on treatment and covariates. As with the Berry and Berry method, MBLR assumes that the events are classified into similar medical groupings in order to use a shrinkage model to allow borrowing strength across similar events. However, Berry and Berry do not consider covariates or the use of logistic regression.

Gould (2008a, 2008b, 2013) proposed an alternative Bayesian screening approach to detect potential safety issues when event counts arise from binomial or Poisson distributions. The method assumes that the AE incidences are realizations from a mixture of distributions and seeks to identify the element of the mixture corresponding to each AE. One of these distributions applies to events where there is no true difference between the treatments and another one applies to events where there is a treatment effect. The components of the mixture are specified a priori, and the calculations then determine the posterior probability that the incidences for each AE are generated from one or the other of the mixture components. It directly incorporates clinical judgment in the determination of the criteria for treatment association.

The Bayesian approaches can be contrasted with the new double false discovery rate (FDR) method from Mehrotra and Adewale (2012), which is a frequentist method for screening frequent (tier 2) AEs, controlling for multiplicity. The double FDR method (Mehrotra and Heyse, 2004) involves a two-step application of adjusted P values based on the Benjamini and Hochberg FDR (Benjamini and Hochberg, 1995), available in SAS proc multtest. In the new double FDR method, they proposed a new flagging mechanism that significantly lowers the FDR without materially compromising the power for detecting true signals, relative to the common no-adjustment approach. The R Package *c212* is available to apply the Bayesian hierarchical model and the new double FDR method along with other methods for error control when testing multiple hypotheses: http://personal.strath.ac.uk/raymond.carragher/files/c212/c212-manual.pdf

Comparing the frequentist and Bayesian methods, one key advantage of the Bayesian methods is the ability to interpret results while taking consideration of the nature of the AE. It is possible to even pre-specify probability criteria for interpreting the posterior distributions, while the frequentist methods do not adjust for the nature of an AE – all events are treated equally. The Bayesian methods can also incorporate historical data and provide posterior-based credibility intervals for each AE. The ability to continuously monitor a trial under the Bayesian philosophy is also a clear advantage. On the other hand, the frequentist methods are easy to implement, while in the Bayesian methodology estimation uses MCMC methods, requiring consideration of the convergence of the chains across each parameter of the model. This can be daunting for the novice user of these techniques. Frequentist methods fit well with situations involving a fixed number of looks at the data and where well-defined notions of type I error control are needed.

5.2.3 Post-Marketing PV versus Pre-Marketing Evaluation

Much progress has been made to develop statistical methods that handle spontaneous report safety data.

Among the more common methods for analyzing spontaneous reports are the proportional reporting ratio (PRR), see Hauben and Zhou (2003); Bayesian confidence neural network (BCPNN), see Bate et al. (1998); and the multi-item Gamma Poisson shrinker (MGPS), see DuMouchel (1999) or Szarfman, Machado, and O'Neill (2002). The use of these methods raises awareness of the need for PV in the developing countries where data sources may not be as rich and varied as developed regions; their work also illus-trates that BCPNN and PRR can work with RCT data and provide similar 'signals' relative to what is known for a drug. The use of these methods in clinical trials has not seen much activity, even though they provide another lens through which RCT data may be viewed. This raises the prospect for a thorough investigation of their strengths and weaknesses in the RCT setting relative to more conventional, denominator-based methods.

In general, there are many methods developed for spontaneous report and observational data such as electronic health record (EHR) data or claims data. Figure 5.3 provides an outline of the level of data from early-devel-opment experiments to later drug lifecycle data sources. The long, straight arrow indicates that more data sources become available as a drug moves

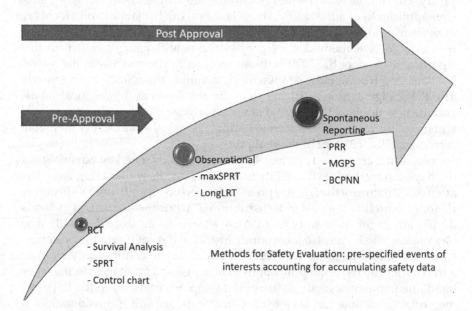

FIGURE 5.3
Safety evaluation during the life cycle

into the post-approval phase. Certainly, RCTs still get done and can provide critical safety data, but questions that require much larger exposures or a longer treatment duration may not be doable in an RCT. Therefore, claims data, EHRs, and spontaneous report data fill an information gap that RCTs cannot. The shorter arrow refers to the limits of pre-approval data to primarily RCTs and other clinical trial data. Some epidemiological data may help describe general safety concerns for the indicated patient population, or for drugs in the same class as the drug under study but will not extend to an investigational drug's specific concern.

Figure 5.3 will help in the next subsection, as well here, to understand the history and current use of methods for the dynamic SMR strategy such as the sequential probability ratio test; PRR; reporting odds ratio; Bayesian confidence neural network (BCPNN); and longitudinal likelihood ratio test (LongLRT). The particular methods and a few others are discussed with an emphasis on their one-time analyses (static) uses and ongoing (dynamic) reviews of accumulating safety data.

5.2.4 Static versus Dynamic Evaluation

With Figure 5.3 in mind, SMR activities can be thought of as an ongoing set of reviews of the safety data that get updated as new data become available either within a given study, across a growing set of studies, or over the passage of time as in the case of many observational data sources and spontaneous report sources. Some of the methods in the preceding section are better suited for a dynamic evaluation of data, while others work nicely for a one-time review as in the production of summaries of clinical safety at the filing of an NDA.

In general, the focus here is on the dynamic approach because it better fulfills the needs for detecting side effect issues as early as possible in a drug development program. Crowe et al. (2009) have proposed such a paradigm in the setup for repeated analyses of clinical trial data in development. Cumulative meta-analyses also aid in the pre-approval setting with a dynamic framework in which studies are pooled for a meta-analysis after the completion of each new study. These ideas were further expanded as aggregate safety assessment planning by the ASA safety working group (see Chapter 3).

Chapter 8 provides comprehensive descriptions of the static/dynamic methods based on whether they typically get used in a static or dynamic SMR setting. The distinction between methods better suited for one strategy or the other is in part to make the discussion of the various methods easier. Some methods, e.g. Cochran-Mantel-Haenszel, were designed for a single analysis with specific operating characteristics in mind to control type I error and maintain a prescribed level of power. Methods such as the LongLRT are more general and can accommodate a series of analyses as data accrue – making them better suited for a dynamic data setting. See, for instance, Hu et al. (2015) and Huang et al. (2011), Huang et al. (2013), and Huang et al. (2014).

The descriptions point to a tendency for LRT-based methods to be used in dynamic settings. Most methods in the 'static' category can be used in a dynamic setting but would require some false positive adjustment such as an FDR method or the method outlined in Hu et al. (2015).

5.2.5 Analyses from Patient-Level Data to Integrated Safety Evaluation

Integrated or meta-analytical approaches play an important role from safety monitoring to NDA submissions. In the past, and to an extent today, most involved patient-level data that were pooled across several studies with no adjustments for study and producing simple summary statistics, such as the proportion of patients with a given AE or a similar rate using treatment exposure as the denominator. In recent years, important safety concerns have emerged – e.g. cardiovascular side effects in the development of anti-diabetic treatments – in which sponsors and regulators need to look deeper at certain side effect questions. Some questions can be approached prospectively, while others may arise from the SMR process. In either case, simple statistics will not suffice, and more sophisticated statistical methods may come into play, see Crowe et al. (2016), Chuang-Stein and Beltangady (2011).

Prior to a drug application filing, cumulative meta-analysis can be used to update data every time there is a new trial available, for the evaluation of benefit or harm. Crowe et al. (2016) provided a broad description of the value and strategies for incorporating these meta-analyses into a drug's development process. In addition, a proactive aggregate safety assessment planning framework has been developed to aid the safety monitoring process throughout drug development life cycle (Hendrickson et al. 2021). That framework includes a careful decision to choose the right safety estimand in order to integrate the safety information across multiple studies (See Chapter 3 and Chapter 14 for a more detailed discussion).

5.3 Benefit-Risk Assessment Methodologies

Safety monitoring, safety assessment, and risk minimization form what the FDA calls *risk management (FDA guidance on premarketing risk assessment 2005)*. Specifically, risk management is an iterative process of (1) assessing a product's benefit-risk balance, (2) developing and implementing tools to minimize its risks while preserving its benefits, (3) evaluating tool effectiveness and reassessing the benefit-risk balance, and (4) making adjustments, as appropriate, to the risk minimization tools to further improve the benefit-risk balance. Such processes should be continuous throughout a product's lifecycle.

Consequently, the qualitative and quantitative characterization of safety profiles is essential for benefit-risk assessment throughout product life cycle. In the last 20 years, structured benefit-risk assessment has become an important area of methodological innovation. Mt-Isa et al. (2014) have provided a comprehensive review of the statistical methodologies for benefit-risk assessment.

Benefit-risk methods largely can be divided as qualitative versus quantitative methods. Qualitative methods are often referred to the descriptive approach that utilizes the summary statistics of key efficacy endpoints and key safety endpoints. The descriptive approach typically includes a structured approach for how this summary information weighs together into decision making. In contrast, the quantitative methods refer to an approach that integrates endpoints by quantifying their clinical relevance and trade-offs, usually requiring judgment in assigning numerical weights to benefit and risk.

As discussed in the study by Mt-Isa (2014), there are multiple qualitative frameworks that can be used to describe and evaluate benefit-risk in a structured way. Among the most popular frameworks, these include the FDA benefit-risk framework, the benefit-risk action team framework, and PROACT-URL framework that has been promoted by the EMA. All these frameworks share the same general stepwise process of (1) planning, (2) evidence gathering, (3) analysis, (4) exploration, and (5) conclusion and dissemination. In particular, for the PROACT-URL, it is built on top of the decision-making theory by Hammond et al. (1999) with the following key components: Problem framing, Objectives, Action, Consequence and Trade-off (PROACT); PROACT-URL further supplements the PROACT theory with consideration of uncertainty, risk tolerance, and linked decisions (URL). The IMI PROTECT has reviewed and provided case studies on this framework (http://protectbenefitrisk.eu/index.html).

As a foundation concept to these frameworks is a product benefit-risk value tree. Value trees are a visual, hierarchical depiction of key benefits (i.e efficacy) and key risk through an explicit visual map of their attributes that are of value to the benefit-risk decision-makers. From these value trees, quantitative methods include population level methods such as multi-criteria decision analysis (MCDA) and stochastic multi-criteria acceptability analysis (SMAA) as well as patient-level methods. MCDA is a popular quantitative method for synthesizing multiple benefits and risk criteria. For an overview of MCDA in healthcare from the International Society for Pharmacoeconomics and Outcomes Research (ISPOR), see Thokala et al. (2016) and Marsh et al. (2016). It is recommended when desiring to combine benefit and risk results to form an overall view. MCDA allows superimposing value judgments on the key efficacy/safety data by assigning weights and utility scores so an overall benefit-risk value can be derived. This overall benefit-risk value can be used to evaluate multiple treatment options. It can be useful in circumstances where complex or difficult decisions need to be made and allows synthesizing multiple benefits and risks defined using different scales (e.g. binary and continuous). SMAA further extends MCDA by bringing in analysis of the sampling variation and preference uncertainty, which are almost inevitable in real practices (Pelissari et al. 2020). It integrates benefits and risks and ranks the different options in terms of being the best option (second best, third best, etc.). Utility functions are still required, but typically defaulted to be linear. The implementation of SMAA requires more mathematical and computational knowledge, but recent applications have made it easier with R packages such as "smaa" and "hitandrun", (Tervonen and Lahdelma 2007; Tervonen 2014). An alternative to SMAA when only accounting for uncertainty in the preference weight is the predicted choice probability analysis (PCP); for an example see Obse et al. (2016) where they calculate the probabilities of uptake (the PCP in their context) for different possible health insurance plans in a discrete choice experiment (DCE).

In all three quantitative methods, two main sources of information are needed to conduct the evaluation. First, the important clinical benefits and risks need to be agreed upon with the regulator or health technology assessment organization. Second, the value (positive or negative) to patients of each clinical feature should be estimated through various types of patient preference studies, again with agreement between the sponsor and the evaluating organization on which preferences to evaluate. The best-worst scaling and the DCE mentioned earlier are two examples for assessing patients' utilities relative to the agreed benefits and risks. See, for example, Hauber et al. (2016), Johnson et al. (2013), Kuhfeld (2010), or Hauber et al. (2017) for further reading.

These methods have seen limited use in regulatory settings, but this is changing notably and sponsors will plan as rigorously as with any other components in the development program. In such a setting, it is reasonable to expect that a regulator will have or will want from the sponsor, a list of pertinent candidate efficacy and safety parameters as the basis of the quantitative benefit-risk assessment (QBRA). Secondly, developers need to consider both any additional data collection in RCTs and design features needed in any patient preference studies to ensure that they provide data acceptable to the regulator. Finally, analysis plans for the QBRA would likely need regulatory input as with any other study work. Early planning will be essential to meet these ends.

In abstract, the QBRA should follow a simple path of framing the QBRA questions, planning for data collection and analyses, and the interpretation of results. For example, a sponsor may frame the QBRA question to differentiate between the candidate treatments and existing treatments. They could plan to do an MCDA to evaluate a new antidepressant versus current therapies. The new candidate may have both superior efficacy and safety (in which case no QBRA is needed) or have a disadvantage in either benefit (efficacy) or risk (safety). To further the example, we may take the clinical inputs from Tervonen 2014, where we would have Hamilton depression scale responder rates as the clinical efficacy input, and three clinical safety inputs in the AEs of 1) nausea, 2) insomnia, and 3) anxiety. Assuming that the regulator(s) agree with the list of clinical inputs and preference data can be obtained, then the next step is to perform the MCDA, its sensitivity analyses, and any stochastic analyses such as SMAA.

Benefit-risk evaluation depends on understanding the preferences of patients, usually, relative to the clinical inputs. Preference studies can inform regulators with greater and more detailed information on the patient burden associated with disease and treatment. These studies provide decision-makers with a systematic approach to assessing patient needs and a complementary perspective to that from patient advocacy groups. Thus, their use will likely increase in future drug development and may even one day integrate with phase 3 studies when possible.

At each step of the process, surprises may arise. In planning, the list of safety events may need to change as more safety data on the drug are compiled. If so, then the sponsor may need to re-evaluate the list of inputs and seek regulatory feedback or agreement. This may also change the plans for any patient preference studies. Because the QBRA is a relatively new addition to drug development, in practice it may be wise to assume that the unexpected will occur, plans will need to change, and prepare with alternatives to satisfy regulators' critiques of the QBRA.

Furthermore, the patient-level analysis method refers to the evaluation of benefits and risks using patient-level information. It can provide important insight into the interaction of benefits and risks at the subgroup or patient level. Two such examples are the method of the composite benefit-less-risk score by Chung-Stein et al. (1991) and the method of desirability of outcome ranking by Evans et al. (2015). These methods use patient's experiences on efficacy and safety to derive an individualized benefit-risk measurement, which can be used to evaluate the overall benefit-risk comparisons.

5.4 Concluding Remarks: Safety Monitoring and Benefit-Risk Methodologies

This chapter provides an overview of statistical methods and strategies for answering drug safety questions in the pre-approval setting. The statistics community has developed many methods to address such questions in the post-approval setting and made significant headway in the development phases as well. However, much remains to be done. The importance of safety monitoring and reporting continues to increase as evidenced by the 2010 FDA IND safety reporting final rule and its subsequent 2012, 2015 guidances.

There are many methods for safety monitoring in the literature and we have highlighted only a sampling of the methods from a few unique angles. Within the field of statistics, the multifaceted nature of SMR can be expressed by the analogy with the blind trying to understand the elephant. Bearing this in mind we

- can proactively articulate the need to address different perspectives and stakeholders
- hope that these different views presented in this paper provide some guidance in our own work on safety monitoring and reporting
- recognize that the ultimate selection of a statistical method will depend on the question at hand or safety issue(s) focus and the context

The work in this manuscript represents a first phase of the ASA safety monitoring working group toward the goal of developing a systematic approach to safety monitoring and reporting. In addition, there is a strong need to pair up safety monitoring methodologies with the benefit-risk methodologies. It should be noted that, in the spectrum from qualitative/descriptive methods to the fully integrated quantitative benefit-risk methods, there is not a clear binary division of qualitative versus quantitative methodologies, but rather a continuum of the level of quantification. That's why in the comprehensive review by Mt-Isa et al. (2014) and by Jiang and He (2016), there are large numbers of methods that fall into the categories of benefit-risk metrics indices, survey utility techniques, and estimation techniques. These include the well-known number-needed-to-treat and number-needed-to-harm. These also include many innovative ways to visualize the safety and benefit-risk information both statically and interactively, for which we have a dedicated chapter to cover.

In the years ahead, the biopharmaceutical industry will see an exponential increase in the need for a clear and deep understanding of the benefit-risk assessment. Statisticians must play a key role in the growth of knowledge development but will do so only if fully committed and motivated to engage in the research and application of our field to the safety of our products.

Bibliography

Amit, O., Heiberger, R. M., and Lane, P. W., Graphical approaches to the analysis of safety data from clinical trials. *Pharmaceutical Statistics*, 2008, 7, 20–35.

Ball, G., Continuous safety monitoring for randomized controlled clinical trials with blinded treatment information. Part 4: One method. *Contemporary Clinical Trials*, 2011, 32, S11–S17.

Ball, G., and Piller, L. B., Continuous safety monitoring for randomized controlled clinical trials with blinded treatment information - Part 2: Statistical considerations. *Contemporary Clinical Trials*, 2011, 32, S5–S7.

Ball, G., Piller, L. B., and Silverman, M. H., Continuous safety monitoring for randomized controlled clinical trials with blinded treatment information - Part 1: Ethical considerations. *Contemporary Clinical Trials*, 2011, 32, S2–S4.

Ball, G., and Silverman, M. H., Continuous safety monitoring for randomized controlled clinical trials with blinded treatment information - Part 3: Design considerations. *Contemporary Clinical Trials*, 2011, 32, S8–S10.

Bate, A., Lindquist, M., Edwards, I. R., Olsson S., Orre, R., Lansner, A., and De Freitas, R. M., A Bayesian neural network method for adverse drug reaction signal generation. *European Journal of Clinical Pharmacology*, 1998, 54, 315–321.

Benjamini, Y., and Hochberg, Y., Controlling the false discovery rate: a practical and powerful approach to multiple testing. *Journal of the Royal Statistical Society: Series B*, 1995, 289–300.

Berry, S., and Berry, D., Accounting for multiplicities in assessing drug safety: a three-level hierarchical mixture model, *Biometrics*, 2004, 60, 418–426.

Boland, K., and Whitehead, J., Formal approaches to safety monitoring of clinical trials in life-threatening conditions. *Statistics in Medicine*, 2000, 19, 2899–2917.

Chen, W., Zhao, N, Qin, G., Chen, J., A Bayesian group sequential approach to safety signal detection, *Journal of Biopharmaceutical Statistics*, 2013, 23, 213–230.

Chuang-Stein, C., Beltangady, M., Reporting cumulative proportion of subjects with an adverse event based on data from multiple studies. *Pharmaceutical Statistics*, 2011, 10(1), 3–7.

Chuang-Stein, C., Mohberg, N.R., and Sinkula, M.S., Three measures for simultaneously evaluating benefits and risks using categorical data from clinical trials. *Statistics in Medicine*, 1991, 10(9): 1349–1359.

Crowe, B., Chuang-Stein, C., Lettis, S., Brueckner, A., Reporting adverse drug reactions in product labels. *Therapeutic Innovation & Regulatory Science*, 2016, 50(4): 455–463.

Crowe BJ, Xia HA, Berlin JA, Watson DJ, Shi H, Lin SL, et al. Recommendations for safety planning, data collection, evaluation and reporting during drug, biologic and vaccine development: a report of the safety planning, evaluation, and reporting team. *Clinical Trials*, 2009, 6:430–440.

Duke, S. P., 2000. Decision making and safety in clinical trials – graphs make a Difference!, https://www.ctspedia.org/wiki/pub/CTSpedia/GraphicsPresentationArchive/Duke_Susan_DIA11_v3.pdf

Duke, S. P., Bancken, F., Crowe, B., Soukup, M., Botsis, T., and Forshee, R., 2000a. Translating complex clinical trial and post-market safety data into visual stories: Seeing is believing: Good graphic design principles for medical research, http://www.fda.gov/downloads/BiologicsBloodVaccines/ScienceResearch/UCM490777.pdf

Duke, S. P., Bancken, F., Crowe, B., Soukup, M., Botsis, T., and Forshee, R., Seeing is believing: good graphic design principles for medical research. *Statistics in Medicine*, 2015, 34, 3040–3059.

DuMouchel, W., Bayesian data mining in large frequency tables, with an application to the FDA spontaneous reporting system. *The American Statistician*, 1999, 53, 177–190.

DuMouchel, W., Multivariate Bayesian logistic regression for analysis of clinical study safety issues. *Statistical Science*, 2012, 27, 319–339.

Evans S., Rubin, F., et al., Desirability of Outcome Ranking (DOOR) and Response Adjusted Days of Antibiotic Risks (RADAR). Clinical Infectious Disease, 2015, advance access published online July 2015.

Fayers, P., Ashby, D., and Parmar, M. Tutorial in biostatistics Bayesian data monitoring in clinical trials. *Statistics in Medicine*, 1997, 16, 1413–1430.

FDA, Guidance for Industry, Premarketing Risk Assessment. March 2005. https://www.fda.gov/media/71650/download (accessed April 4, 2021).

Gould, L., Detecting potential safety issues in clinical trials by Bayesian screening. *Biometrical Journal*, 2008a, 50(5), 837–851.

Gould, A. L., detecting potential safety issues in clinical trials by Bayesian screening. *Biometrical Journal*, 2008b, 50, 1521–4036.

Gould, L., Detecting potential safety issues in large clinical or observational trials by Bayesian screening when event counts arise from poisson distributions. *Journal of Biopharmaceutical Statistics*, 2013, 23(4), 829–847.

Gould, A. L., and Wang, W., Monitoring potential adverse event rate differences using data from blinded trials: the canary in the coal mine. *Statistics in Medicine*, 2017, 36, 92–104.

Gruber S., and van Der Laan M.J. An application of targeted maximum likelihood estimation to the meta-analysis of safety data. *Biometrics*, 2013, 69(1), 254–262.

Hammond, J.S., Keeney, R.L., and Raiffa, H., *Smart Choices: A Practical Guide to Making Better Decisions*. Broadway Books, New York, NY, 1999.

Harrell, F., Exploratory analysis of clinical safety data to detect safety signals, 2005. http://biostat.mc.vanderbilt.edu/wiki/pub/Main/FHHandouts/gsksafety.pdf

Hauben, M., and Zhou X., Quantitative methods in pharmacovigilance: focus on signal detection. *Drug Safety*, 2003, 26, 159–186.

Hauber, A.B., Gonzalez, J.M., Groothuis-Oudshoorn, C.G.M., Prior, T., Marshall, D.A., Cunningham, C., et al. Statistical methods for the analysis of discrete choice experiments: a report of the ISPOR conjoint analysis good research practices task force. *Value in Health*, 2016 July, 19, 300–315.

Hauber, A.B., Obi, E.N., Price, M.A., Whalley, D., and Chang, C.L., Quantifying the relative importance to patients of avoiding symptoms and outcomes of heart failure. *Current Medical Research and Opinion*, 2017 September, 33(11), 2027–2038.

Hendrickson, B.A., Wang, W., Ball, G. et al. Aggregate safety assessment planning for the drug development life-cycle. *Therapeutic Innovation & Regulatory Science*, 2021. https://doi.org/10.1007/s43441-021-00271-2

Herson, J., 2000. Safety monitoring, Chapter 11. In *Statistical Methods for Evaluating Safety in Medial Product Development*, by A. Lawrence Gould. Wiley, New York.

Hu, M., Cappelleri, J.C., Lan, K.K., Applying the law of iterated logarithm to control type I error in cumulative meta-analysis of binary outcomes. *Clinical Trials*, 2007, 4(4), 329–340.

Hu, N., Huang, L., and Tiwari, R. Signal detection in FDA AERS database using Dirichlet process. *Statistics in Medicine*, 2015, 34, 2725–2742.

Huang, L., Zalkikar, J., and Tiwari, R.C., A likelihood ratio test based method for signal detection with application to FDA's drug safety data. *JASA*, 2011, 106, 1230–1241.

Huang, L., Zalkikar, J., and Tiwari, R. C., Likelihood ratio test-based method for signal detection in drug classes using FDA's AERS database. *Journal of Biopharmaceutical Statistics*, 2013, 23, 178–200.

Huang, L., Zalkikar, J., and Tiwari, R. T., Likelihood ratio based tests for longitudinal drug safety data. *Statistics in Medicine*, 2014, 33(14), 2408–2424.

Inovative Medicine Initiative (IMI) 2000. Pharmacoepidemiological Research on Outcomes of Therapeutics (PROACT) http://protectbenefitrisk.eu/index.html accessed on December 5, 2020.

Jennison, C., and Turnbull, B., Group sequential tests for bivariate response: interim analyses of clinical trials with both efficacy and safety endpoints. *Biometrics*, 1993, 49(3), 741–752.

Jiang, Q., and He, W., *Benefit-Risk Assessment Methods in Medical Product Development: Bridging Qualitative and Quantitative Assessments*, Chapman and Hall/CRC, Boca Raton, 2016.

Johnson, F.R., Lancsar, E., Marshall, D., Kilambi, V., Mühlbacher, A., Regier, D.A., et al. Constructing experimental designs for discrete-choice experiments: report of the ISPOR Conjoint Analysis Discrete-Choice Experiment Experimental Design Good Research Practices Task Force. *Value in Health*, 2013, 16, 3–13.

Kuhfeld, W. *Marketing Research Methods in SAS: Experimental Design, Choice, Conjoint, and Graphical Techniques*. SAS Institute Inc., Cary, NC, 2010.

Kulldorff, M., Davis, R. L., Kolczak, M., Lewis, E., Lieu, T., and Platt, R., A maximized sequential probability ratio test for drug and vaccine safety surveillance. *Sequential Analysis*, 2011, 30, 58–78.

Li, L., and Kulldorff, M., A conditional maximized sequential probability ratio test for pharmacovigilance. *Statisticis in Medicine*, 2010, 29, 284–295.

Mehrotra, D.V., and Adewale, A.J., Flagging clinical adverse experiences: reducing false discoveries without materially compromising power for detecting true signals. *Statistics in Medicine*, 2012, 31, 1918–1930.

Mehrotra, D.V., and Heyse, J.F., Use of the false discovery rate for evaluating clinical safety data. *Statistical Methods in Medical Research*, 2004, 13, 227–238.

Mt-Isa, S., Hallgreen, C.E., Wang, N. et al. (2014), Balancing benefit and risk of medicines: a systematic review and classification of available methodologies. *Pharmacoepidemiology and Drug Safety*, 2014. DOI:10.1002/pds.3636.

Mukhopadhyay, S., Waterhouse, B., and Hartford, A. (2018). Bayesian detection of potential risk using inference on blinded safety data. *Pharmaceutical Statistics*, 17(6), 823–834.

O'Neill, R.T., Regulatory perspectives on data monitoring. *Statistics in Medicine*, 2002, 21, 2831–2842.

O'Neill, R.T., Pre-Market Safety Must Balance Statistics With Clinical Discernment, the FDA pink sheet, Nov 10, 2008, Vol 70, No 045.

Pelissari, R., Oliveira, M., Amor, S., Kandakoglu, A., and Helleno, A., SMAA methods and their applications: a literature review and future research directions. *Annals of Operations Research*, 2020, 293, 433–493.

Resnic, F. S., Zou, K. H., Do, D. V., Apostolakis, G., and Ohno-Machado, Exploration of a Bayesian updating methodology to monitor the safety of interventional cardiovascular procedures, *Medical Decision Making*, 2004, 24, 399–407.

Sarker, A. Ginn, R. Nikfarjam, A., O'Connor, K, Smith, K., Jayaraman, J., Upadhaya, T., Gonzalez, G. Utilizing social media data for pharmacovigilance: a review. *Journal of Biomedical Informatics*, 2015, 54, 202–212.

Schnell, P. M., and Ball, G. A., 2000. Bayesian Exposure-Time Method for Clinical Trial Safety Monitoring With Blinded Data. To appear in Therapeutic Innovation and Regulatory Science.

Schuemie, M.J. Methods for drug safety signal detection in longitudinal observational databases: LGPS and LEOPARD. *Pharmacoepidemiology and Drug Safety*, 2011, 20, 292–299.

Shih, M. L., Lai, T. L., Heyse, J. F., and Chen, J., Sequential generalized likelihood ratio tests for vaccine safety evaluation. *Journal of Statistics in Medicine*, 2010, 29, 2698–2708.

Szarfman, A., Machado, S. G., and O'Neill, R. T., Use of screening algorithms and computer systems to efficiently signal higher-than-expected combinations of drugs and events in the US FDA's spontaneous reports database. *Drug Safety*, 2002, 25, 381–392.

Tervonen, T., JSMAA: Open source software for SMAA computations. *International Journal of Systems Science*, 2014, 45(1), 69–81.

Tervonen, T., and Lahdelma, R., Implementing stochastic multicriteria acceptability analysis. *European Journal of Operational Research*, 2007, 178, 500–513.

Thall, P., and Simon, R., Practical Bayesian Guidelines for Phase IIB Clinical Trials. *Biometrics*, 1994, 50, 337–349.

Thokala, P., et al. Multiple criteria decision analysis for health care decision making – emerging good practices: report 1 of the ISPOR MCDA emerging good practices task force. *Value in Health*, 2016, 19, 1–13.

Van Der Laan, M., and Rose, S., (2011), *Targeted Learning: Causal Inference for Observational and Experimental Data*, Springer, New York, ISBN 978-1-4419-9781-4.

Wen, S., Ball, G., and Dey, J., Bayesian monitoring of safety signals in blinded CTs. *Annals of Public Health and Research*, 2015, 2(2), 1019.

Wittes, J., A statistical perspetive on adverse event reporting in cinical trials. Biopharm. Report, 1996.

Xia, A. H., Ma, H., and Carlin, B. P., Bayesian hierarchical modeling for detecting safety signals in clinical trials. *Journal of Biopharmaceutical Statistics*, 2011, 21, 1006–1029.

Zink, R. C., Huang, Q., Zhang, L. Y., Bao, W. J., Statistical and graphical approaches for disproportionality analysis of spontaneously-reported adverse events in pharmacovigilance, *Chinese Journal of Natural Medicines*, 11, 314–320.

Zink, R. C., Wolfinger, R. D., Mann, G., Summarizing the incidence of adverse events using volcano plots and time intervals, *Clinical Trials*, 2013, 10, 398–406.

6

Quantitative Methods for Blinded Safety Monitoring

Melvin Munsaka
AbbVie, Inc., North Chicago, IL, USA

Kefei Zhou
Bristol Myers Squibb, New York, NY, USA

CONTENTS

6.1 Introduction 119
6.2 Defining Blinded Safety Monitoring 120
6.3 Challenges in Blinded Safety Monitoring 120
6.4 Related Regulatory Consideration 121
6.5 Frequentist, Bayesian, and Likelihood Approaches to Inference 122
6.6 Statistical Methods for Blinded Safety Monitoring 123
6.7 Some Practical Considerations in Implementing Blinded
    Safety Monitoring 127
6.8 Linking Blinded and Unblinded Safety Monitoring 127
6.9 Tools and Software for Quantitative Safety Monitoring 128
6.10 Illustrative Example of Quantitative Blinded Safety Monitoring 128
6.11 Conclusion 130
References...130

6.1 Introduction

In a broad sense, safety monitoring spans from individual trial level to program level, from unblinded to blinded, from expected to unexpected adverse events (AEs), and from patient profiles to aggregate safety monitoring and benefit-risk assessment. It also serves as a foundation for integrated safety summary preparation and benefit-risk analysis in the clinical overview and possible advisory committee meeting. A plethora of statistical methodological approaches for safety monitoring have been proposed for blinded and

DOI: 10.1201/9780429488801-8

unblinded analyses of safety data. This chapter will provide an overview of some statistical methods for assessing safety data in a blinded fashion.

6.2 Defining Blinded Safety Monitoring

Blinded safety monitoring is the process of looking at safety data without any knowledge of the treatment assignments. It can include masked treatment, for example, Treatment A versus Treatment B, without knowledge of what A and B are. Applications of blinded safety monitoring can be limited to one study or multiple studies. Drug development programs can continue for long periods of time, and as such, some safety information on the drug may be known from completed studies. It is important that this information is accounted for in some way. Similarly, historical control information from the same class or population should also be accounted for in some form in blinded safety monitoring activities. This is essential especially in ongoing clinical trial settings where there is some interest in continuous assessment of safety outcomes of special interest without knowledge of treatment assignments.

An immediate point to note is that blinded looks at safety data will not be as informative and as efficient as in an unblinded look at the data where the treatments are known. During a drug development program, many ongoing studies may hold important clues to the safety profile of a drug. Upon study completion, the data typically get unblinded for the assessment of both efficacy and safety data. Unblinding of study treatment prior to study completion is either part of a planned or unplanned set of interim analyses or the result of unblinding specific study subjects who have experienced serious adverse events (SAEs) that require knowledge of the treatment that they were assigned to help in risk mitigation of the event, as appropriate. A safety data monitoring committee (DMC) can make more informed clinical assessment about a drug program if they have some knowledge of the actual treatments received. However, this comes at the expense of no longer having a fully blinded study. Additionally, there is an ethical and moral responsibility to proactively assess drug safety on an ongoing basis in the interest of patients.

6.3 Challenges in Blinded Safety Monitoring

Assessing safety data in a blinded fashion comes with various challenges and issues, including ethical considerations, statistical considerations, and

TABLE 6.1

Advantages Versus Disadvantages of Blinded Safety Monitoring

Advantages	Disadvantages
• Identify potential safety issues ahead of scheduled DMC meetings.	• It may not be as informative and efficient as in an unblinded analysis.
• Identify safety issues that are or have potential to become a key concern.	• It will inevitably raise logistical questions regarding monitoring patient safety while at the same time maintaining the study blind.
• Drive decisions regarding an unblinded analysis or a decision to setup a DMC, or even stopping a trial or development altogether	
• Facilitate for an ethical and moral responsibility to proactively assess safety.	

design considerations. Ball and Piller (2011), Ball et al. (2011), and Ball (2011a, 2011b) have discussed these issues and challenges as they pertain to blinded safety monitoring. For example, they point out that proactive safety signal monitoring using blinded data in an ongoing clinical trial will inevitably raise logistical questions regarding monitoring of patient safety while at the same time maintaining the study blind. Another consideration that arises within the context of blinded safety monitoring is with regard to the available information on the safety concern as well as how to address blinded safety monitoring in the case where some studies are still ongoing. In particular, how does one harness information from the completed studies and those that are ongoing.

The above considerations raise the question: *how much information can blinded safety monitoring provide*? As it turns out, a good deal of safety information and learning can be gained from blinded data as suggested in the definition itself and some consequences of defining it this way. Table 6.1 below summarizes the advantages and disadvantages of blinded safety monitoring,

6.4 Related Regulatory Consideration

Regulatory guidances, such as CIOMS VI (Management of Safety Information from Clinical Trials, (2005)), ICH E2C (PBER: Periodic Benefit Risk Evaluation Report, (2012)), and the Food and Drug Administration (FDA) IND safety reporting guidance (2015), highlight the importance and provide recommendations on aggregate safety monitoring and blinded safety monitoring. The FDA Draft Guidance for IND Safety Reporting

Final Rule requires expedited reporting whenever aggregate analysis indicates that events occur more frequently in the investigational drug than in a concurrent or historic control group and requires use of all available data. Suspected adverse reactions should be evaluated in an ongoing basis. The guidance requires sponsors to periodically review accumulating safety data, integrated across multiple studies, both completed and ongoing, and provide a quantitative framework for measuring the evidence of an association for anticipated events or a clinically important increase for expected events and making a judgment about *reasonable possibility* for IND safety reporting. The FDA's preferred approach calls for a safety assessment committee (SAC) that regularly performs unblinded comparisons across treatment groups to detect numerical imbalances with appropriate steps taken to maintain overall study blinding. An alternative approach would be to only unblind treatment assignment if the overall rate for all treatment groups under blinded assessment for a specific event is substantially higher than a predicted rate. This requires sponsors to pre-specify predicted rates of anticipated events and expected events and guidelines for determining when an observed rate has exceeded the predicted rate. SAC activities present an opportunity to perform blinded assessments of safety data. Some drug development companies have implemented or are in the process implementing SAC type activities employing a variety of different statistical methods, along with logistical considerations and processes requiring cross-functional collaboration.

6.5 Frequentist, Bayesian, and Likelihood Approaches to Inference

Various statistical methods have been proposed for blinded safety monitoring. Broadly speaking, these methods can be subdivided into three broad categories including, frequentist methods, Bayesian methods, and likelihood methods. Frequentist methods revert to large sample theory to make inference. Bayesian methods make use of prior information and the likelihood leading to a posterior distribution from which inference is made. Likelihood-based methods provide a framework for presenting and evaluating likelihood ratios as measures of statistical evidence for one hypothesis over another. Of note, all three approaches use likelihood functions, but in different ways and for different immediate purposes (see for example, Royall (2004) and van der Tweel (2005). The likelihood principle stands out as the simplest and yet most far-reaching. It essentially states that all evidence, which is obtained from an experiment about an unknown quantity, say θ, is contained in the likelihood function of θ. From a question-based perspective, one may think of each one of these approaches as a camp, i.e., frequentist camp, Bayesian camp, and likelihood camp, with each camp addressing different questions as follows:

Camp	Question
Frequentist	• What would happen in repeated experiments? • Use likelihood functions to design experiments that are in some sense guaranteed to perform well in repeated applications in the long run, no matter what the truth may be.
Bayesian	• What should we believe? • Use likelihood functions to update probability distributions.
Likelihood	• What do the data say? • Use likelihood functions to characterize data as evidence.

6.6 Statistical Methods for Blinded Safety Monitoring

In practical applications, all approaches often utilize historical control data or other resources relative to the estimate obtained from the blinded data as part of the decision-making process. More specifically, this requires establishing a cut-off point for a safety metric of interest, say, θ^c representing a corresponding estimate from blinded data, say, θ^d and comparing these two on the basis of a rule or set of rules based on the statistical method. Generally, the idea is to make some inference about θ, for example, the AE rate θ^d on the basis of observed data. In general, θ can be a derived safety metric, for example, the crude event rate, exposure-adjusted adverse event rate (EAER), exposure-adjusted incidence rate (EAIR), risk difference (RD), relative risk (RR), odds ratio, and so on. Whether the statistical method used is frequentist, Bayesian, or likelihood-based, the ultimate objective is to make a decision on the basis of accrued blinded data. The statistical decision can be based on a point estimate or a set of estimates, confidence intervals (CIs), or credible intervals in the case of Bayesian methods, or some probabilistic assessment, or it can be a combination of a mixture of statistical measures and probabilistic assessments. The decision itself will require careful consideration from various stakeholders taking into account clinical discernment and benefit-risk with patient safety at the center. The decision making process will be essentially a cross-functional and collaborative activity. Table 6.2 below presents some simple example settings of how quantitative decisions could be made using the different statistical approaches.

TABLE 6.2

Example Settings for the Different Statistical Methods

Method	Setting
Frequentist	• Set criteria for decision-making, for example, estimate of lower bound θ^L of θ and a cut-off point θ^c which if exceeded by θ^L would lead to a decision being made, e.g., unblind study.
	• So we want to check if $\theta^L > \theta^c$ which would lead to a decision being made, e.g., unblind study.
	• Use large sample theory to estimate θ^L using blinded data and then compare this with θ^c.
Bayesian	• Set criteria for decision-making, for example, set cut-off point, θ^c which if exceeded by a particular percentage, say, 90% would lead to a decision being made, e.g., unblind study.
	• So, want to estimate a probability associated with θ, e.g., $P(\theta > \theta^c) > 90\%$.
	• Via Bayes theorem, use blinded data and appropriate prior to obtain posterior which is used to obtain $P(\theta > \theta_c)$.
Likelihood	• Set criteria for decision-making, for example, lower bound θ^L for θ and a cut-off point θ^c which if exceeded by θ^L would lead to a decision being made, e.g., unblind study.
	• So want to check if $\theta^L > \theta^c$ which would lead to a decision being made, e.g., unblind study.
	• Use likelihood methods, e.g., a $\frac{1}{8}$ support interval to estimate θ^L based on blinded data and compare this to θ^c.

Table 6.3 below presents a broad overview of some quantitative methods that can be used for blinded safety monitoring, including frequentist, Bayesian, and likelihood methods. Of note, some of the methods may not have been specifically designed or developed for blinded monitoring but can be adapted for this purpose and/or to help in the decision-making process.

TABLE 6.3

Some Quantitative Methods for Blinded Safety Monitoring

Method	Description
A	**Frequentist**
F1	Peace (1987), Peace and Chen (2011), and Chen and Peace (2011) (Direct comparison): Construct a two-sided $100(1-2\alpha)\%$ CI on the difference δ in the incidence rate relative to a comparator rate. A positive lower limit of the CI would be suggestive that the observed rates are different at the nominal significance level. Decisions can be based on the basis of magnitude of difference.
F2	Peace (1987), Peace and Chen (2011), and Chen and Peace (2011) (Indirect comparison): Construct a two-sided $100(1-2\alpha)\%$ for the incidence rate versus the rate for comparator. Compare lower limit and upper limit of CIs. Conclude that rates are different at the nominal significance level α if the lower limit of CI is greater than the upper limit of comparator.

TABLE 6.3

(Continued)

Method	Description
F3	Moye (2005): Assuming that the study plans to enroll N patients in trial. Suppose the accepted rate of AEs is anticipated to be θ. Given the observed rate at any juncture in the trial, we can estimate the expected number of patients with the AE at end of the trial. We can then check to see if the expected number will exceed a given threshold.
F4	Kramar and Bascoul-Mollevi (2009) and Bascoul-Mollevi *et al.* (2011): Focus on stopping rules after the occurrence of each SAE by comparing the number of patients included to the number of patients satisfying maximum SAE criteria.
F5	Yao *et al.* (2013) and Zhu *et al.* (2016): Based on current enrollment, assess if the rate will exceed a threshold once rest of the subjects are enrolled. This approach works on a single cohort or combined cohort. The safety boundary can be calculated in the combined arm.
F6	Duke *et al.* (2017): Two approaches: (i) Use of binomial probabilities to determine the probability of observing one or more events given the expected background rate of AE in disease population of interest and (ii) use of incidence rates to determine whether the observed incidence rate is expected given the background rate.
F7	Ye *et al.* (2020): Use of the Sequential probability ratio test (SPRT) and maximized SPRT (MaxSPRT), and Bayesian posterior probability threshold (BPPT) for evaluating specific safety signals that incorporate pre-specified background rates or reference risk ratios.
B	**Bayesian**
B1	Chuang-Stein (1993): Use of the beta-binomial model to summarize the medical event rates observed in different trials and use this to estimate parameters in a beta-binomial model. Also provides an estimate for the average event rate and uses a model to look for unusual event patterns in future trials.
B2	Huang (2005): Use of beta prior and binomial likelihood for safety assessment of a rare event. Requires a preset upper threshold for guidance, and available data to obtain posterior distribution are used to assess the probability of exceeding threshold by some pre-specified value.
B3	Ball (2011b): Early stopping rules that act as continuous safety screens for randomized clinical trials with blinded treatment information, under the Bayesian framework.
B4	Wen, *et al.* (2015): Blinded safety monitoring within a Bayesian framework. Method be applied to one or more AESIs. The idea is to evaluate probability that a clinical parameter of interest exceeds a pre-specified critical value, given observed blinded data. The critical value is selected based on historical data or medical judgment. If probability meets criteria of *big enough*, this would signal a potential safety concern, leading to other additional investigations.
B5	Sailer and Bailer (2015): Use of a beta-binomial model to examine if AE rate throughout ongoing trial is higher than expected from historical data. Posterior is used to describe an increasingly higher AE rate converging to a new AE rate. Alerts are issued if the posterior AE rate is greater than the predefined threshold with a certain trigger probability. Leverages weights such that prior is informative especially early on but trial data are not dominated.

(Continued)

TABLE 6.3

(Continued)

Method	Description
B6	Gould and Wang (2017): Bayesian approach to determine the likelihood of elevated risk for binomial or Poisson likelihoods that can be used regardless of the metric used to express the difference. The method is particularly appropriate when the AEs are not rare.
B7	Schnell and Ball (2016): Safety monitoring procedure for two-arm blinded clinical trials. The procedure incorporates a Bayesian hierarchical exposure-time model for using prior information and blinded event data to make inferences on the rate of AEs of special interest in the test treatment arm. The procedure is also appropriate for inferring the rate of AEs in multiarmed clinical trials with blinded data.
B8	Zhu *et al.* (2016) and Yao *et al.* (2013): Predictive approach in the Bayesian framework by combining experts' prior knowledge about both frequency and timing of the event occurrence and combining this with observed data. During each look, each event-free subject is counted with a probability that is derived using prior knowledge. Can be used where the incidence rate of an AE is continuously monitored.
B9	Gould (2016): Use of cumulative sum (CUSUM) control charts for monitoring AE occurrence within a Bayesian paradigm based on assumptions about the process generating the AE counts in a trial as expressed by informative prior distributions. The method uses control charts for monitoring AE occurrence based on statistical models and construction of useful prior distributions.
B10	Mukhopadhyay *et al.* (2018): Method for monitoring and detecting safety signals with data from blinded ongoing clinical trials, specifically for AESIs when historical data are available to provide background rates. The method is a two-step Bayesian evaluation of safety signals composed of a screening analysis followed by a sensitivity analysis. The modeling framework allows making inference on the RR in blinded ongoing clinical trials to detect any safety signal for AESIs.
B12	Goren *et al.* (2020): Two-stage Bayesian hierarchical models for safety signal detection following a pre-specified set of interim analyses that are applied to efficacy. At Stage 1, a hierarchical blinded model uses blinded safety data to detect a potential safety signal and at Stage 2, a hierarchical logistic model is applied to confirm the signal with unblinded safety data.
B13	Goren *et al.* (2020): Bayesian meta-analytical approach for detecting a potential elevated safety risk conferred by an investigational treatment, as compared to control, using blinded data from multiple trials
B14	Ye *et al.* (2020): Bayesian posterior probability threshold (BPPT) approach for evaluating specific safety signals that incorporate pre-specified background rates or reference risk ratios.
C	**Likelihood Methods**
L1	Blume (2002), van der Tweel (2005), and Herson (2015, 2016): Binomial setting - look at the pooled incidence rate for the AE for accumulation data. Obtain likelihood 1/8 (or 1/32) supporting a lower limit to see if this exceeds the threshold incidence rate.
L2	Blume (2002), van der Tweel (2005), and Herson (2015, 2016): Poisson setting - look at the pooled frequency rate of AE per 100 patient-year based on accumulation of AEs over time. Obtain likelihood 1/8 (or 1/32) supporting interval for a rate per 100 patient-years to see if the lower limit exceeds the threshold rate.

6.7 Some Practical Considerations in Implementing Blinded Safety Monitoring

Hendrickson (2018) discussed several questions that need to be considered in practical applications and implementation of blinded safety monitoring. They include: *Are the events only anticipated for a subset of subjects?*; *For known consequences of the disease, how specific is the event's association with the disease?*; *Is there overlap of the anticipated event with a potential risk of the study product?*; and *What is the capability to assess if the event is occurring at a higher rate than anticipated?*. All these questions and other considerations need to be thought through as appropriate in practical applications of blinded safety monitoring. Additionally, complexities can occur when dealing with multiple product indications as this dictates a need for multiple lists of anticipated AEs based on the different study populations and execution of monitoring activities (e.g., SAC activities) with varying lists of anticipated events for different cross-product studies. Another question that needs to be addressed is with regard to the granularity by which events should be reported. Individual SAEs are reported at the level of a MedDRA preferred term. Thus, a key question should be with regard to which anticipated events be listed instead at a medical concept level (e.g. specified preferred groupings, high level term, narrow or broad standardized MedDRA queries, or custom queries). Another challenge is with regard to the lack of reliable data regarding the background rates of the AEs of interest in the study population. It is unlikely that the full list of SAEs reported in clinical studies will be readily available. Additionally, rates of anticipated events can be influenced by various factors that differ among programs, for example, due to eligibility criteria. Reliance on electronic health records and claims databases to provide background reference rates may present issues for extrapolation to clinical trial databases. For example, the approach to recording diagnoses leading to hospitalization may be different for claims databases versus clinical trials. Last but not least, blinded safety monitoring is a labor-intensive activity requiring cross-functional resources and collaborations. Thus activities surrounding blinded safety monitoring should be well planned to ensure that various methodological and logistical considerations are addressed. This can be best facilitated with a charter.

6.8 Linking Blinded and Unblinded Safety Monitoring

In practical applications, there will be ongoing studies that are blinded along with those that have been unblinded. It is thus critical to have careful considerations on how to combine the accumulating safety data from both blinded and unblinded studies. Lin *et al.* (2019) proposed a two-stage framework for this purpose. In the first stage, there is periodic analysis of blinded safety data to detect and flag AEs that may have potential risk elevation. In the second stage, a planned unblinded analysis is conducted to quantify associations between the drug and AE(s) and to determine thresholds for referring AEs for medical reviews and possible safety reporting. They also discussed some practical considerations and related issues. Below is an example of how one sponsor has implemented blinded safety monitoring as part of their safety monitoring processes.

6.9 Tools and Software for Quantitative Safety Monitoring

The use of the quantitative methods discussed in this chapter can be facilitated through various tools and software resources. These include standalone programs developed for a specific method(s) and/or statistical packages developed for performing one or more of these methods. Standalone programs can be developed using general programs such as SAS, R, and Python and are sometimes made available by the authors of the methods. An example of a standalone program can be seen in the study by Duke *et al.* (2017) who presented a SAS program for implementing the method **F6** in Table 6.2. Similarly, Chen and Peace (2011) and Peace and Chen (2011) present R programs for methods **F1** and **F2** in Table 6.2. Similarly, Schnell (2018) also presented a series of R programs implementing the method **B7**. Blume (2002) presented three S-plus, bin.lik for binomial likelihoods, betabin.lik for profile beta-binomial likelihoods, and rr.lik for conditional likelihoods for the RR in implementations of the likelihood method for methods **L1** and **L2**. A more recent approach is to leverage packages that are available for performing these analyses along with visualizations. For example, the bdribs R package implements method **B10** (2018). Another package, the SAE package was developed by Bascoul-Mollevi *et al.* (2011) for implementing method **F4**. One can also leverage interactive functionality by putting packages as a backend to tools such as R Shiny, see for example, Mukhopadhyay and Waterhouse (2021) and Ye *et al.* (2020). Using these interfaces, one conduct analyses quantitative blinded safety monitoring based on the methods implemented without installing and writing any code (Figure 6.1).

6.10 Illustrative Example of Quantitative Blinded Safety Monitoring

For illustration purposes, we consider the example presented in Table 6.4 (adapted from Table 18.2, Page 371 from Peace and Chen (2011); also Table 11.1 Page 300 in Chen and Peace (2011)). The data provide reports of an AE

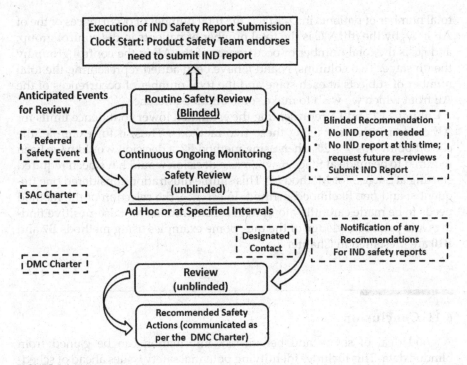

FIGURE 6.1

Example of an implementation of blinded assessment in safety monitoring (adapted from Hendrickson, 2018)

TABLE 6.4

Cumulative Entry and AE Occurrence across Treatments

Stage	N_iD_2	f_iD_2	N_iD_1	f_iD_1	N_iC	f_iC	N	f	$\theta^{F2\%}$	$\theta^{L1\%}$
1	15	1	15	0	15	0	45	1	<1	<1
2	30	2	31	0	29	0	90	2	<1	<1
3	50	4	48	1	49	0	147	5	1	1
4	70	4	68	3	72	2	210	9	2	2
5	95	6	94	4	95	3	284	13	2	3
6	120	9	119	5	120	4	359	18	3	3
7	140	11	138	5	141	6	419	22	3	3
8	150	12	150	6	150	7	450	25	4	4

from a parallel, randomized, double-blind clinical trial of two doses, D_1 and D_2 of a new drug compared to a control C. Entry of subjects in the trials is staggered. The data consist of the number of patients in the trial and the number having the AE in each group totaled sequentially at each stage. The columns at each stage are follows: N_iD_1 is the total number of patients in D_1 at the Stage i; f_iD_1 is the total number of occurrences of the AE in D_2; N_iD_2 is the total number of patients in D_2; f_iD_2 is the total number of occurrences of the of AE in D_2 by the ith; N_iC is the total number of patients in the control group; and f_iC is the total number of occurrences of the AE in the control group by the ith stage. Two columns, N and f, have been added representing the total number of subjects at each stage and the total number of occurrences of the AE from which we wish to make some blinded inference.

Columns **F2** and **L1** representing the estimated lower confidence limits for AE crude incidence θ using these two methods. Thus, if for example, the cut-off point was $\theta^c = 2\%$, then using method **F2**, a decision would have been required to be made at Stage 6, whereas a decision would have been required starting at Stage 5 for Method *L1*. This simple illustration considered one frequentist and one likelihood method. In essence, the selection of the method needs to be made carefully along with considerations for false-positive findings and prior knowledge of the rate. Some examples using methods **B7** and **B10** are presented in Chapter 7.

6.11 Conclusion

A good deal of safety and benefit-risk information can be gained from blinded data. This includes identifying potential safety issues ahead of scheduled DMC meetings, identifying safety issues that are, or have potential to become, a key concern, and addressing issues where there is an ethical and moral responsibility to proactively assess patient safety while maintaining the study blind. It is important to acknowledge and anticipate some of the challenges of blinded safety monitoring, such as logistical questions regarding monitoring patient safety while maintaining the study blind. It is recommended that when using blinded safety monitoring, the following should be considered carefully: criteria and decisions that should be made a priori preferably cross-functionally and there should be some careful thought as to the method to use among, frequentist, Bayesian, and likelihood approaches. Last, but not least, this chapter only considered methods that were included in the review at the time of writing of this chapter and the available tools. Presumably, there are other methods and tools that are not included in the chapter available or not yet developed at the time of writing of this chapter.

References

Ball, G. 2011a. "Continuous Safety Monitoring for Randomized Controlled Clinical Trials with Blinded Treatment Information – Part 3: Design Considerations." *Contemporary Clinical Trials* **32**: S8–S10.

Ball, G. 2011b. "Continuous Safety Monitoring for Randomized Controlled Clinical Trials with Blinded Treatment Information. Part 4: One Method." *Contemporary Clinical Trials* **32**: S11–S17.

Ball, G., and L. B. Piller. 2011. "Continuous Safety Monitoring for Randomized Controlled Clinical Trials with Blinded Treatment Information – Part 2: Statistical Considerations." *Contemporary Clinical Trials* **32**: S5–S7.

Ball, G., L. B. Piller, and M. H. Silverman. 2011. "Continuous Safety Monitoring for Randomized Controlled Clinical Trials with Blinded Treatment Information. Part 1: Ethical Considerations." *Contemporary Clinical Trials* **32**: S2–S42.

Bascoul-Mollevi, C., A. Laplanche, M. C. Le Deley, and A. Kramar. 2011. "SAE: An R Package for Early Stopping Rules in Clinical Trials." *Computer Methods and Programs in Biomedicine* 104 (2): 243–48.

Blume, J. D. 2002. "Likelihood Methods for Measuring Statistical Evidence." *Statistics in Medicine* **21**: 2563–99.

Chen, D-G., and K. E. Peace. 2011. *Analysis of Adverse Events in Clinical Trials, Chapter 11. In Clinical Trial Data Analysis Using R.* Boaca Raton: CRC Press, Taylor; Francis Group.

Chuang-Stein, C. 1993. "An Application of the Beta-Binomial Model to Combine and Monitor Medical Event Rates in Clinical Trials." *Drug Information Journal* **27**: 515–23.

Duke, S. P., C. Kleoudis, M. Polinkovsky, D. Bennett, D. Hill, and E. Lewis. 2017. "Quantitative Methods for Safety Monitoring of Rare Serious Adverse Events." *Pharmaceutical Medicine* **31**: 113–18.

Goren, E., L.-A. Lin, Blaustein R. O., and G. Ball. 2020. "Bayesian Meta-Analysis of Safety Outcomes Using Blinded Clinical Trial Data." *Therapeutic Innovation and Regulatory Science* **54**: 1557–65.

Gould, A. L. 2016. "Control Charts for Monitoring Accumulating Adverse Event Count Frequencies from Single and Multiple Blinded Trials." *Statistics in Medicine* **35**: 5561–78.

Gould, A. L., and W. B. Wang. 2017. "Monitoring Potential Adverse Event Rate Differences Using Data from Blinded Trials: The Canary in the Coal Mine." *Statistics in Medicine* **36**: 92–104.

Guidance, FDA, 2021, Sponsor Responsibilities - Safety Reporting Requirements and Safety Assessment for IND and Bioavailability/Bioequivalence Studies https://www.fda.gov/media/150356/download

Hendrickson, B. 2018. "Identifying Expected Adverse Reactions and Anticipated Events." https://healthpolicy.duke.edu/sites/default/files/2020-03/ind_slide_deck_2018_03_08_final.pdf

Herson, J. 2015. *Safety Monitoring, Chapter 11. In Statistical Methods for Evaluating Safety in Medial Product Development, by a Lawrence Gould.* New York: Wiley.

Herson, J. 2016. "Data and Safety Monitoring Committees in Clinical Trials." Chapman and Hall/CRC Biostatistics Series.

Huang, D. 2005. "A Bayesian Approach to the Design and Analysis of a Study for Handling a Rare Event." Presentation at 2005 BASS XII Meeting, https://www.bassconference.org/tutorials/BASS 2005 D_Huang.pdf

ICH Guideline. 2012. "ICH E2c (R2) Periodic Benefit-Risk Evaluation Report." https://www.ema.europa.eu/en/documents/regulatory-procedural-guideline/international-conference-harmonisation-technical-requirements-registration-pharmaceuticals-human-use_en-0.pdf van der Tweel, I. 2005. "Repeated Looks at Accumulating Data: To Correct or Not to Correct?" *European Journal of Epidemiology* **20**: 205–11.

Kramar, A., and C. Bascoul-Mollevi. 2009. "Early Stopping Rules in Clinical Trials Based on Sequential Monitoring of Serious Adverse Events." *Medical Decision Making* **29**: 343–50.

Moye, L. A. 2005. *Safety and Futility, Chapter 7*. In *Statistical Monitoring of Clinical Trials: Fundamentals for Investigators*. New York: Springer Verlag.

Mukhopadhyay, S. 2018. "Bdribs: Bayesian Detection of Potential Risk Using Inference on Blinded Safety Data." https://cran.r-project.org/web/packages/bdribs/index.html

Mukhopadhyay, S., and B. Waterhouse. 2021. "Bayesian Detection of Potential Safety Signal from Blinded Clinical Trial Data." http://www.bayesianscientific.org/wp-content/uploads/2021/06/BDRIBS-DIA-BSWG-KOL-2021-v2_BW4_SM1.pdf; http://bdribs.shinyapps.io/bdribs_ver2

Mukhopadhyay, S., B. Waterhouse, and A. Hartford. 2018. "Bayesian Detection of Potential Risk Using Inference on Blinded Safety Data." *Pharmacutical Statistics* **17**: 823–34.

Peace, K. 1987. "Design, Monitoring, and Analysis Issues Relative to Adverse Events." *Drug Information Journal* **21**: 21–28.

Peace, K. E., and D-G. Chen. 2011. *Monitoring Clinical Trials for Adverse Events, Chapter 18*. In *Clinical Trial Methodology*. Boaca Raton: CRC Press, Taylor; Francis Group.

Royall, R. 2004. *The Likelihood Paradigm for Statistical Evidence, Chapter 5*. In *Nature of Scientific Evidence: Statistical, Philosophical, and Empirical Considerations*, by Mark L. Taper and Subhash R. Lele. Chicago: University of Chicago Press.

Sailer, O., and T. Bailer. 2015. "Blinded Bayesian Adverse Event Monitoring." http://www.bayes-pharma.org/wp-content/uploads/2014/10/poster-sailer-basel-bayes2015.pdf

Schnell, P. M. 2018. "A Bayesian Exposure-Time Method for Clinical Trial Safety Monitoring with Blinded Data." https://github.com/schnellp/TIRS-2016-Blinded.

Schnell, P. M., and G. Ball. 2016. "A Bayesian Exposure-Time Method for Clinical Trial Safety Monitoring with Blinded Data." *Therapeutic Innovation & Regulatory Science* **50**: 833–38.

VI, CIOMS. 2005. "Management of Safety Information from Clinical Trials: Report of Cioms Working Group Vi." https://cioms.ch/wp-content/uploads/2017/01/Mgment_Safety_Info.pdf

Wen, S., G. Ball, and J. Dey. 2015. "Bayesian Monitoring of Safety Signals in Blinded Clinical Trial Data." *Annals of Public Health and Research* **2**: 1019.

Yao, B., L. Zhu, Q. Jiang, and H. A. Xia. 2013. "Safety Monitoring in Clinical Trials." *Pharmaceutics* **5**: 94–106.

Ye, J., S. Wen, J. Schoenfelder, and S. Islam. 2020. "Quantitative Safety Monitoring in Clinical Trials: Application of Multiple Statistical Methodologies for Infrequent Events." *Therapeutic Innovation & Regulatory Science* **54**: 1175–84.

Zhu, L., B. Yao, H. A. Xia, and Q. Jiang. 2016. "Statistical Monitoring of Safety in Clinical Trials." *Statistics in Biopharmaceutical Research* **8**: 88–105.

7

Bayesian Safety Methodology

Michael Fries

CSL Behring, King of Prussia, PA, USA

Matilde Kam

Center for Drug Evaluation and Research, US Food and Drug Administration, Silver Spring, MD, USA

Judy X. Li

Bristol-Myers Squibb, San Diego, CA, USA

Satrajit Roychoudhury

Pzifer Inc., New York, NY, USA

Mat Soukup

Center for Drug Evaluation and Research, US Federal Food and Drug Administration, Silver Spring, MD, USA

Brian Waterhouse

Merck & Co, Inc, Kenilworth, NJ, USA

CONTENTS

7.1 Introduction 134
    7.1.1 Bayesian Statistical Methods in Safety Analysis 134
    7.1.2 Bayesian Statistics 135
        7.1.2.1 Illustrative Example of Bayesian Analysis 138
        7.1.2.2 Hierarchical Models 140
    7.1.3 Chapter Flow 141
7.2 Bayesian Approaches to Safety Monitoring of Blinded Data 141
    7.2.1 Bayesian Framework 142
    7.2.2 Methods for the Event Rate
        (Number of Events/Exposure Time) 142
        7.2.2.1 Example 1: The Method of Schnell and Ball 144
    7.2.3 Methods for Proportion of Subjects with an Event 144
    7.2.4 Bayesian Detection of Risk Using Inference on Blinded
        Safety Data (BDRIBS) Method 145
        7.2.4.1 Example 2: BDRIBS Method 146
    7.2.5 Discussion 147

DOI: 10.1201/9780429488801-9

7.3　Bayesian Hierarchical Modeling Approaches to Safety
　　　Monitoring of Unblinded Data 148
　　　7.3.1　Literature Review 149
　　　7.3.2　Bayesian Hierarchical Model 150
　　　　　　　7.3.2.1　Three-level Bayesian Hierarchical Mixture Model 151
　　　　　　　7.3.2.2　Discussion 155
7.4　Bayesian Meta-Analytic-Predictive(MAP) Prior and Empirical
　　　Meta-Analytic-Predictive (eMAP) Prior Approach to Borrowing
　　　Historical Information 156
　　　7.4.1　MAP Prior 157
　　　7.4.2　Robust MAP (rMAP) Prior 158
　　　7.4.3　eMAP Prior 158
　　　7.4.4　Illustrative Example Applying MAP Priors 159
　　　　　　　7.4.4.1　Discussion 161
7.5　Chapter Discussion 162
References 162

7.1 Introduction

7.1.1 Bayesian Statistical Methods in Safety Analysis

There are many challenges in safety signal detection and safety assessment. For example, consider monitoring adverse events (AEs). At the start of the trial, we don't know how many different kinds of AEs we will see or even what they potentially are. When we collect the data, some AEs will be sparse and some will be plentiful. We may see an imbalance between treatment groups by simple random variation, but when is that the case and when is there really a signal? From a statistical viewpoint, this is in part a multiplicity problem. Any simple test to assess if a difference is random has to account for the fact that there are many different kinds of AEs being examined, and they are being examined multiple times throughout the trial. Furthermore, assessment of AEs during the trial takes place with the sponsor being blinded from treatment. Another challenge arises when we have prior experience with the disease in similar populations from our companies' trials from publications, and the incorporation of this information would be useful. For the working statistician, we provide methodologies to face these challenges.

In this chapter, we take advantage of the functionality found in Bayesian analysis to incorporate historical information, estimate many parameters in a multi-level model, and update information as it accrues to face these stated challenges that are applicable to safety statistics. Specifically, we will be looking at leveraging the advantages of Bayesian statistics in the framework of blinded safety monitoring, unblinded signal detection, and safety assessment of AEs.

There are other well-known benefits to a Bayesian approach that are especially useful from a safety perspective, including the ease of interpreting the resulting posterior distribution compared to either a frequentist testing approach (p-value) or confidence interval. However, these won't be discussed explicitly here.

7.1.2 Bayesian Statistics

Bayesian statistics represents a philosophy of statistical analysis which differs from the more traditional frequentist statistical philosophy in the fundamental notion of what probability is. A consequence of this is that the parameters of distributions are considered random variables. The underlying principles and implications of the philosophy have been discussed at length in the literature, but these philosophical discussions are not the principal focus here. Our focus is on the practical benefits of the application of Bayesian methods in the analysis and interpretation of safety data from clinical trials.

Many of the ideas presented in this chapter, such as incorporating historical information, estimating many parameters in a multi-level model, updating information as it accrues, and bringing historic information into consideration, are consequences of the Bayesian philosophy.

A key aspect of Bayesian statistics is that although there are true values for the distribution parameters, we are uncertain of what those values are. This uncertainty is captured by treating these variables as random variables with associated distributions themselves. Our knowledge, or belief in the value of the parameter, is encapsulated within its distribution. The observed quantities of interest are also random variables. The information from these observed values is captured within the likelihood function, referred to as "the likelihood." Then, an application of Bayes Theorem is used to incorporate the observed information (the likelihood) with the belief we had prior to an experiment (prior distribution of the parameter) to establish our belief we have post-experiment (posterior distribution of the parameter). This is encapsulated in an application of Bayes' theorem, as represented by the formula:

$$\text{Posterior} \propto \text{Likelihood} * \text{Prior} \tag{7.1}$$

Here, the "Posterior" refers to the distribution of parameters given the data, "Prior" refers to the distribution of the parameters before the data, and "Likelihood" is the information from the data. That is, the information is given as a likelihood function of parameter values conditional on the observed data. These distributions represent our belief in the value of the parameter. For example, a prior distribution for an AE rate describes our belief in what the AE rate is before observing the data, while the posterior describes our belief in what the AE rate is after observing the data. The proportionality is there to ensure that the posterior is indeed a distribution, with an area under the curve of 1. This prior could be derived from historical information (either a single trial, information from an ongoing trial as in Section 7.2, or combined information from several trials as in Section 7.4) or elicited from expert opinion as in Best, Dallow, and Montague (2020) or Oakley and O'Hagan (2010). Often, it is decided to use a "non-informative" prior. A non-informative prior is one that is of a weak nature, representing little information. When information is observed, the likelihood from those data will strongly influence the resulting posterior distribution. One way Bayesians use to gauge the strength of a prior is through an estimation of how many observations worth of data it represents. For example, a prior that is roughly one observation worth of data is not very informative and will easily be overwhelmed by a modest number of new observations. Meanwhile, a prior that represents 100 observations would be much more influential, requiring more information from the observed data to overwhelm it. There are various methods to understand how many observations a particular prior is "worth," (Neuenschwander et al. 2020), although a more detailed exposition is beyond the scope of this book.

To clarify our later notation, priors are sometimes denoted by their distribution and simply specifying their parameters, for example, Beta(α, β). At other times, it is useful to associate the prior with the parameter explicitly. For example, if we are discussing a beta prior on the parameter Θ, we may write Beta ($\Theta \mid \alpha, \beta$). The use we chose here varies depending upon the needs for the current exposition.

There is also the possibility of uncertainty in the parameters of the prior distribution. In that case, we can assign prior distributions to these parameters, which are named hyper-priors, which themselves have parameters, called hyper-parameters.

Incorporating prior information helps us assess the impact of the totality of evidence from multiple sources. This can reduce the information needed from the current trial to draw conclusions, resulting in reduced sample sizes or increased sensitivity. This reflects the "march of science," in which information from disparate sources can be quantitatively integrated to obtain knowledge. For example, a study that was "underpowered" for generating a p-value can still add information to future analysis and contribute to the overall understanding of a particular question.

While the actual prior beliefs should drive the shape of the prior, certain families of shapes are often selected as priors because they lead to mathematical simplifications that allow the calculation of posteriors directly, rather than through a simulation or numerical computation. Priors of such shape are called 'conjugate.' When conjugate priors are not available or do not sufficiently represent the prior belief, then a non-conjugate prior should be used and the posterior needs to be calculated. The class of algorithms generally used for this is called Markov Chain Monte Carlo (MCMC). Many software packages such as SAS, BUGS, JAGS, and STAN provide easy access to such algorithms.

Bayesian analysis naturally gives the ability to incorporate data as it is observed, because the likelihood function can be thought of as a series of products. To describe this more formally, define:

- The distribution of a parameter θ, conditional on data x, is $P(\theta \mid x)$, or unconditional $P(\theta)$
- the likelihood function of θ given data x is $l(\theta \mid x)$
- the data x can be broken into two distinct parts, $x = x_1 \cup x_2$

$$
\begin{aligned}
p(\theta \mid x) &\propto l(\theta \mid x) * p(\theta) \\
&= l(\theta \mid x_1) * l(\theta \mid x_2) * p(\theta) \\
&= l(\theta \mid x_1) * \big(l(\theta \mid x_2) * p(\theta)\big) \\
&\propto l(\theta \mid x_2) * p(\theta \mid x_1)
\end{aligned}
\tag{7.2}
$$

Therefore, one can first incorporate the information from the first sample of data $l(\theta \mid x_1)$ and then later incorporate the information from the second sample of data $l(\theta \mid x_2)$, and the resulting distribution is exactly the same as if you incorporated the information from all of the data all at once. This means that the assessment of partial information can easily be incorporated under the Bayesian paradigm and has no impact on the resulting distribution. This provides a platform not only for looking at partial information during a trial but also for incorporating historic information, or belief, into the quantitative assessments. We can ascertain information from accumulating data in a natural way.

However, there is a risk that incorrect decisions may still occur if they are made based on interim information. Bayesian analysis does provide natural protection against this kind of multiplicity, for example, through the use of a skeptical prior. See David J. Spiegelhalter, Abrams, and Myles (2004) for a more complete review.

Posteriors can also be used to predict future values. When doing so, one needs to consider both the uncertainty in the parameters and the variability of the distribution. The resulting distribution is a posterior predictive distribution. This is illustrated in Section 7.1.2.1, and we refer to David J. Spiegelhalter, Abrams, and Myles (2004) for more details.

It has been suggested that clinicians are naturally Bayesian (Gill, Sabin, and Schmid 2005). The authors suggest that reaching a diagnosis based on questions and test results is in fact Bayesian, whether or not the physician considers it as such. The physician may start with a belief level in a certain disease, and as information is added through discussions with the patient, physical examinations, and then test results, more and more confidence is generated. Or in the language above, the physician starts with a prior belief in a diagnosis, and each piece of information generates a likelihood, which results in a posterior. As information accrues by the addition of information, the posterior probability of some diagnoses decreases while others increase. This tells the physician what new test(s) to perform to provide the greatest information until sufficient probability is reached to determine a treatment strategy.

7.1.2.1 *Illustrative Example of Bayesian Analysis*

We present a short example to illustrate some of the terms presented so far. This example is fairly simple - we are trying to consider a binary event proportion. In this case, we will use conjugate priors. This is not just for illustration; conjugate priors are often used for binomial analysis.

We need to incorporate information from the data using a likelihood based on the binomial distribution.

$$l(p|k) = \binom{N}{k} p^k (1-p)^{N-k} \tag{7.3}$$

Here, we think of the counts k as the number of successes, N as the number of trials, and p as the probability of success. The conjugate distribution for the binomial is the beta distribution which has the probability density function with parameters $\alpha > 0$ and $\beta > 0$.

$$f_P(p:\alpha,\beta) = \frac{\Gamma(\alpha+\beta)}{\Gamma(\alpha)\Gamma(\beta)} p^{\alpha-1} (1-p)^{\beta-1} \tag{7.4}$$

where Γ is a gamma function. The beta distribution has a mean of $\alpha/(\beta + \alpha)$.

We start with a non-informative prior, reflecting a belief that we really don't know much about the rate at the start of the experiment. Two beta distributions are often used as non-informative priors, the Jeffreys prior (Beta(1/2, 1/2)) (Jeffreys 1946) and the Uniform distribution (Beta(1, 1)). These can be seen in Figure 7.1. They each have a mean of 1/2 and represent roughly 1 and 2 patients' worth of data, respectively. A convenient feature of the beta/binomial conjugacy is that we can relate the number of observations worth of data to the sum of the parameters of the beta distribution - so, for example, a Beta(x| 8, 12) can be thought of as representing 8 + 12 = 20 subjects worth of data.

Now let's suppose that we observe one patient worth of data, and they were a success. The likelihood based on this one observation is:

$$l(p) = \binom{1}{1} p^1 (1-p)^{1-1} = p \tag{7.5}$$

Using formula (7.1), and a uniform prior, this gives a Beta(2, 1) posterior.

As seen in Figure 7.1, our posterior represents a slight shift to the right in the distribution mass. In fact, the mean has shifted from 1/2 to 2/3.

Now suppose we observe the next 99 subjects, for a total of 100, and that we observed 72 successes and 38 failures. The posterior is now a Beta(74, 39), with a mean of 74/113, or about 0.655, as illustrated in Figure 7.1.

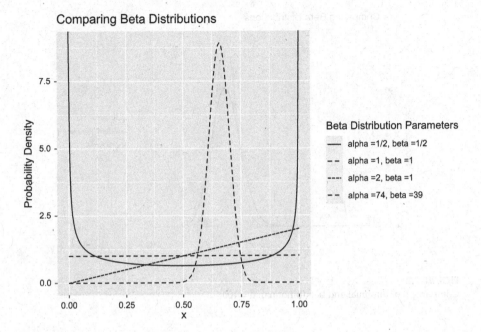

FIGURE 7.1
Beta distributions from examples

The higher peak and narrower distribution show a reduced variance, which we interpret as greater certainty in our belief. We could then answer questions of interest, such as "what is the probability that the rate is above 50%?" (0.9996), "what is the probability that the rate is greater than 75%?" (0.0129), etc.

As discussed in Section 7.1.2, posterior-predictive density can be used to predict future values. In this case, posterior-predictive density can be determined in a closed-form solution, and it follows a beta-binomial distribution. This distribution has greater uncertainty than a binomial distribution because it takes into account that the probability p is estimated through a beta distribution. The density requires three parameters: the α and β from the beta distribution and n, which represents the number of trials we will predict success for, and it provides the probability of observing x successes in those n trials. In Figure 7.2, we overlay the beta-binomial distribution of the probability of k successes in the next 100 patients arising from a binomial distribution with p set to the mean of our posterior on top of that binomial distribution. This shows the additional uncertainty resulting from the prediction.

While this example illustrated the use of a binomial distribution with a beta conjugate prior, other pairs are commonly used. In Section 7.2, a Poisson distribution is paired with its conjugate, a gamma distribution.

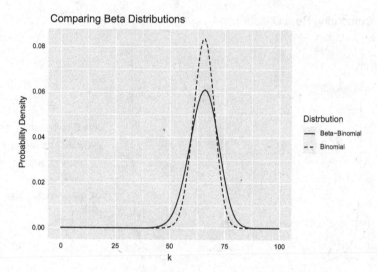

FIGURE 7.2
Comparison of binomial and beta-binomial, n = 100

7.1.2.2 Hierarchical Models

Estimating many parameters within the same statistical model is a challenge. However, when the safety data follow a certain hierarchical structure, we can take advantage of that structure within the model. This concept can be implemented in a Bayesian framework through a hierarchical model which "borrows" information across outcomes within groups so that signals may be easier to detect.

In these models, individual extreme values will be pulled back to the group mean - a property implied by the term "shrinkage estimator" for individual estimates. This provides better estimates than the ones garnered individually, and meanwhile, if there is a stronger tendency toward more events within a specific group, the borrowing effect will help to flag areas of possible risk.

An example of a hierarchical model discussed in Section 7.3 is based on the Medical Dictionary for Regulatory Activities (MedDRA) classification of AEs. AEs in the same body system are assumed to be more closely related to each other than those in other body systems (a strong assumption). Within a particular body system, we estimate rates in a way that relates them to each other. If one preferred term (PT) has a higher rate, it will get pulled down by the other terms in the body system, while it also pulls up the rates for other terms within the same body system.

A key assumption that is often made within a level of hierarchical models is exchangeability. Exchangeability is the assumption that if we had a sequence of random variables, and we permuted the order of a finite subset of them, they have the same joint distribution as before the permutation. Although this may not seem very intuitive, this concept can be thought of as imposing a level of similarity on observations, allowing us to predict future observations based on current ones. Exchangeability has theoretical implications related to foundational notions of Bayesian Statistics which are beyond the scope of this book. A simpler and related intuitive notion is that if we consider ten flips of a coin to understand how likely we are to see heads, the order of the flips doesn't matter; what matters is merely how many heads we observed.

7.1.3 Chapter Flow

In Section 7.2, the ability to include what we have learned before an experiment about a parameter (through a prior) and the ability to assess accumulating information throughout the trial are leveraged for the monitoring of blinded safety data during a trial.

In Section 7.3, the ability to estimate many parameters in a multi-level model is leveraged to simultaneously assess all AE data for safety signal detection. This is borrowing information within groups based on the hierarchy, which ultimately results in shrinkage toward the overall mean in the setting where unblinded safety data are available.

In Section 7.4, the ability to adjust how much historic information is incorporated into the model through hyper priors is leveraged to adjust models when the historic data are not aligned with the current observed data, allowing models to self-adjust to the safety environment changes over time.

Section 7.5 provides an overall summary discussion for the whole chapter.

7.2 Bayesian Approaches to Safety Monitoring of Blinded Data

Ongoing safety monitoring of clinical trials is an essential aspect of sponsors' responsibility to patient safety. While unblinded data monitoring committees (DMCs) are sometimes used for large clinical trials or programs, blinded monitoring by sponsors is an important component of their safety monitoring and pharmacovigilance activities. Many Bayesian methodologies for safety monitoring have been suggested in the literature, a sampling of which in the blinded monitoring setting will be overviewed in this section. Additional methods of blinded safety monitoring including frequentist- and likelihood-based approaches, as well as a further discussion on the utility of blinded safety monitoring, are presented in Chapter 6.

7.2.1 Bayesian Framework

The basic tenet of Bayesian theory readily lends itself to the problem of blinded safety monitoring. We start with an expectation of our drug's safety (our prior) and update our understanding of the drug's safety after observing the blinded data. We can use the resulting posterior distribution to assist in decision-making.

Each of the methodologies that follow can be viewed using the same basic Bayesian framework:

1. Historical or epidemiological data are used to obtain prior information for the statistic of interest.
2. A critical value for the statistic of interest is determined.
3. Blinded data from the ongoing clinical trial(s) are used to update the prior distribution and obtain the posterior.
4. The posterior probability that the statistic of interest is greater than the critical value is determined.
5. If this probability is greater than some threshold value, the next steps are taken. The next steps could range from getting a group of experts together to discuss the cases, to communicating a concern to the DMC, and to unblinding specific safety data in the clinical trial(s).

This framework was described by Wen, Ball, and Dey (2015) and formulated as follows for a Bayesian posterior probability:

$$\Pr(\theta > \theta_c | \text{blinded safety data}) > \text{P cut-off}$$

where θ is a particular clinical parameter or metric of interest (such as pooled proportion, risk difference, or odds ratio), θ_c represents the critical value for comparison, and P cut-off is a probability threshold (such as 90% or 99%) representing the desired confidence needed to identify a potential safety signal. (Wen, Ball, and Dey 2015).

7.2.2 Methods for the Event Rate (Number of Events/Exposure Time)

When the statistic of interest is the event rate of a particular AE or group of AEs where it is assumed that the events follow a Poisson process, we can leverage the gamma conjugate prior with the Poisson likelihood function. For illustration, we can look at a simplistic approach before examining the more complicated versions found in the literature. Suppose that the historical event rate of the control, λ_0, is represented by a gamma distribution with shape parameter, E_0, representing the number of events, and rate parameter, η_0, representing patient-years exposure. With proper weighting, this can serve as an appropriate prior distribution for the event rate, λ_T, for the event of interest in an ongoing clinical trial (step 1). The critical value for comparison could be the historical event rate, E_0/η_0 (step 2). If e events are observed after E patient-years in the ongoing clinical trial, the posterior distribution is $Gamma(E_0 + e, \eta_0 + E)$ (step 3). We can then use this posterior distribution to compute $\Pr(\lambda_T > E_0/\eta_0 \,|\, \text{blinded safety data})$ (step 4) and compare it to a pre-determined probability threshold to decide the next steps (step 5).

Whereas in the simple example above we use the historical information as a prior for the event rate in the clinical trial (and implicitly assume that this is an appropriate prior for both the control and active treatments), methods have been proposed which separate the prior information between the control and active treatments. One such method, proposed by Schnell and Ball (2016), is described below.

AEs are assumed to follow a Poisson process. This means that the events are independent of each other and that while the amount of time between events is variable, the average event rate is constant. Another assumption of a Poisson process is that events will not occur at the same time. For this reason, this process is often associated with events that are more rare; but for practical purposes, the definition of rare depends on the size of the clinical trial. Conjugate gamma priors are assumed for the active treatment and the control group. The prior for the control group is obtained from historical information. The prior for the treatment under study may be informed by historical information (e.g., phase 2 data) or may be elicited from medical experts using, for example, the method of quantiles. The full model is

$$Y_i \,|\, a_i, \lambda_0, \lambda_1 \sim \text{Poisson}(\lambda_{ai}), \, i = 1, \ldots, N$$
$$a_i \sim \text{Bernoulli}(r), \, i = 1, \ldots, N$$
$$\lambda_0 \sim \text{Gamma}(\alpha_0, \beta_0)$$
$$\lambda_1 \sim \text{Gamma}(\alpha_1, \beta_1)$$

where $\alpha_0, \beta_0, \alpha_1$, and β_1 are elicited hyper-parameters.

If $\vec{t}$ is the vector of exposure times for the patients that have been observed at an interim look and $\vec{y}$ is the indicator vector for whether a patient experienced an event, then the joint posterior distribution, $\pi(\lambda_0, \lambda_1 \,|\, \vec{t}, \vec{y})$, can be calculated as described by Schnell and Ball. The authors suggest that simulations should be used to determine an appropriate probability threshold for $\pi(\lambda_1)$, the posterior probability of an event occurring on the treatment under study, so that the procedure has desirable operating characteristics. The authors provide R functions and programs in order to facilitate the use of this method. An example of this method and the use of the R functions and programs is provided below with further details at https://community. amstat.org/biop/workinggroups/safety/safety-home.

7.2.2.1 Example 1: The Method of Schnell and Ball

Schnell and Ball described a scenario where a hypothetical trial in Alzheimer patients is under investigation. In this trial, the randomization ratio is 2:1 (active: placebo), and 1500 subjects are to be randomized. Three hundred and thirty-nine events of interest were observed on placebo in a meta-analysis of previous trials in this disease area over 26064 months of exposure (or 0.013 events per patient month of exposure). In the phase 2 trial of the active drug, four events were observed over 167 months of patient exposure. After discussion with clinical colleagues, it is determined that an appropriate prior for the active treatment takes into account a down-weighting of the placebo rate and updates that information with the phase 2 data. This resulted in a prior belief in the rate of $\lambda_0 \sim$ Gamma (339/100 + 4, 26064/100 + 167) centered at 0.0173 events per patient month of exposure. In addition, an increase of 20% above the historical placebo rate of 0.013, or 0.0156, is determined to be the critical value for which action should be taken. Simulations (which can be performed by running the find.probability.threshold function and are demonstrated in file 2-simulation-study.R from the zipped folder that the authors provide) show that a probability threshold of 0.62 provides desirable operating characteristics, alerting the team of an issue at most half the time when the event rate is exactly equal to the critical value.

Suppose that we investigate the event of interest after the trial has gone on for four months. Since we require both the number of events and the exposure time for each subject, we simulate the data for this example so that the reader can reproduce it. For this simulation, we assume that the placebo rate matches the historical rate and that the active rate is 0.018. Code for this simulation and the remaining computations is provided at https://community.amstat.org/biop/workinggroups/safety/safety-home.

From the simulation, the number of events observed in the first four months is 49 with 2988.5 total months of follow-up.

As shown in https://community.amstat.org/biop/workinggroups/safety/safety-home, the function gibbs.sample computes the posterior distribution based on the prior parameters and the updated data. The number of simulations and the amount of burn-in are the first two parameters of the function. Using the posterior distribution, we determine that the posterior probability that the active treatment exceeds the critical value is 0.709. This would result in an alert to the team as the probability threshold of 0.62 is exceeded. Note that the posterior mean event rate for the placebo arm is 0.0131, and the posterior mean event rate for the active treatment arm is 0.0171.

7.2.3 Methods for Proportion of Subjects with an Event

When the statistic of interest is the proportion of subjects that experience a particular AE or a group of AEs, we have a Bernoulli process and can leverage the beta conjugate prior with the binomial likelihood function. In particular, the historical event rate of the control π_0 is represented by a beta distribution, Beta(x_0, $N_0 x_0$), where x_0 is the number of historical subjects with an event and N_0 is the total number of historical control subjects. As a simple illustration, this could serve as the prior distribution for the probability, π, that a subject in a clinical trial (active + control arms) experiences the event of interest (step 1). The critical value for comparison could be x_0/N_0 or some scaled factor of x_0/N_0, such as 1.5 or 2 times (step 2). If m subjects out of M experience this event in the ongoing clinical trial, the posterior distribution is Beta($x_0 + m$, $N_0 x_0 + (M-m)$) (step 3). We can then use this posterior distribution to compute $\Pr(\pi > x_0/N_0$ blinded safety data) (step 4) and compare it to a pre-determined probability threshold to decide the next steps (step 5).

More complex versions of this approach have also been proposed. Yao et al. (2013) described a scenario in which the beta distribution could be used to investigate whether the proportion of subjects in a trial that experienced a specific AE was sufficiently greater than expected to warrant alerting a DMC.

In this scenario, the statistic of interest is the event rate as represented by the proportion of subjects that experience an event. The standard control therapy event rate, π_S, is represented by a beta distribution, Beta(a_S, b_S), where a and b are determined based on historical data or expert opinion. A prior distribution on the proportion of total (control and active combined) subjects that experience an event, π, is given by Beta(a, b), where a and b are chosen to be relatively small numbers in the informative prior case or 0 or 0.5 (Jeffreys' prior) in the non-informative prior case. Similar to the simple scenario, the posterior distribution of π after observing i events in I subjects is Beta($a + i$, $b + I - i$). During the ongoing trial, the parameters characterizing the distribution of π get updated, but π_S remains the same. A minimally clinically significant threshold, δ, is identified, and the $\Pr(\pi - \pi_S > \delta$ blinded safety data) is determined through MCMC sampling. If this probability is greater than p_U, a large pre-specified value (e.g., 0.9), then we conclude that greater risk is seen in the ongoing trial. If this probability is less than p_L, a small value (e.g., 0.05), we conclude that it is unlikely that the experimental treatment is causing an increased risk of at least δ over π_S. Otherwise, continue the trial. Note that in the manuscript the authors wrote these rules in terms of a monitoring guideline based on the number of subjects with an event. For consistency of presentation, the rules have been rewritten based on probability thresholds.

7.2.4 Bayesian Detection of Risk Using Inference on Blinded Safety Data (BDRIBS) Method

Beyond beta-binomial and the gamma-Poisson conjugate relationships, more sophisticated models are available. We now examine one such example, BDRIBS, the method by Mukhopadhyay, Waterhouse, and Hartford (2018). The method models the relative risk of an investigational drug versus control and considers a single event category at a time (e.g., MACE). It is assumed that events follow a Poisson process (see Methods for the Event Rate (Number of Events/Exposure Time)). Furthermore, a historical background event rate should be available for the control arm. This background event rate can be a fixed number or it is assumed to follow a gamma distribution with the prior distribution given by the historical number of events and patient-years of exposure. Also, the prior for the probability (p) that an observed event occurs on the investigational drug is assumed to follow a beta distribution. In most cases, this prior will be non-informative. These two priors induce a prior on the relative risk. With this induced prior and taking into account the randomization ratio, the total number of observed events and patient exposure from the ongoing blinded trial is used to compute the posterior probability of the relative risk. With the posterior distribution in hand, the probability that the relative risk is greater than 1 or some other clinically meaningful value can be used to determine if an AE needs further discussion with the clinical team. The authors have provided an R package to facilitate the use of this method, and an example of this method is provided below with code at https://community.amstat.org/biop/workinggroups/safety/safety-home. The parameters for the R package are specified in the examples. In addition, an R-Shiny application is available at http://bdribs.shinyapps.io/bdribs_ver2 that can facilitate the use of this methodology. The authors have also provided some examples of applying the method under various scenarios Waterhouse B, Hartford A, Mukhopadhyay S, Ferguson R, Hendrickson BA, 2021.

7.2.4.1 Example 2: BDRIBS Method

To illustrate this method, we will use the same hypothetical clinical trial in Alzheimer patients that was outlined by Schnell and Ball and described in Example 1. Recall that in this trial, the randomization ratio is 2:1 (active: placebo), and 1500 subjects are to be randomized. 339 events of interest were observed on placebo in a meta-analysis of previous trials in this disease area over 26064 months or 2172 years of patient exposure. This yields an event rate of 15.6 E/100 PYs for placebo. The prior information from the phase 2 trial was four events over 167 patient months or 28.7 E/100 PYs. We will use the same simulated clinical trial data as in the previous example with 49 events and 2988 patient-months or 249 patient-years of exposure. A priori the team decides that further discussion of a possible signal is warranted if the posterior probability that the relative risk is greater than one exceeds 90%.

7.2.4.1.1 BDRIBS Assuming the Historical Event Rate Is Fixed

The BDRIBS package comes with a function to calculate the posterior distribution of the relative risk given the number of observed events (y), the observed patient-years of exposure (pyr), the randomization ratio (k), the historical background rate (bg.rate), and the priors for p (p.params). In our example, y = 49, pyr = 249, k = 2, bg.rate = 15.6/100, and p.params = (0.5, 0.5). See code at https://community.amstat.org/biop/workinggroups/safety/safety-home.

Using the BDRIBS function, the probability that the relative risk is greater than 1 is 0.92255. Rather than simply using the placebo historical control and the observed event rate, we could also incorporate the phase 2 trial results via the prior distribution of p, the probability that an observed event occurs in the treatment arm. Recall that the event rate for the study treatment was 28.7 E/100 PYs. With the historical placebo event rate, we can compute an estimate of p as 28.7/(28.7 + 15.6) = 0.65. If we decide that the phase 2 data are equivalent to ten patients worth of information, we can assume p Beta(6.5, 3.5) and compute the probability that the relative risk is greater than one as 0.91195. In either case, the agreed to threshold of 90% is exceeded.

7.2.4.1.2 BDRIBS Assuming the Historical Event Rate Follows a Gamma Distribution

In order to incorporate the variability of the historical event rate in the modeling of the relative risk, the BDRIBS function can take on the following two parameters which replace the fixed background rate bg.rate: the background number of events, bg.events (=339 events), and the background patient-years of exposure, bg.pyr (=2172 patient years). See code at https://community.amstat.org/biop/workinggroups/safety/safety-home.

Here, the posterior probability that the relative risk is greater than one is 92.1% which exceeds the previously agreed threshold of 90%, and this AE warrants further discussion.

Using a gamma distribution instead of the fixed rate assumption may seem more attractive; however, there are a few reasons to start with the fixed-rate assumption. The first is that the actual number of events and exposure may not be available for the historical event rate depending on the source of this information. The second is that considering the variability of the historical event rate has the side effect of increasing the variability of the posterior distribution of the relative risk. This impacts the probability calculation for whether the critical rate is exceeded so that it is often smaller under the gamma assumption than it is under the fixed event rate assumption. This could be seen as less conservative.

7.2.5 Discussion

The examples discussed are not meant to be an exhaustive look at the Bayesian methods available for blinded event monitoring; this is an area of active research, and new methods (in this and other areas) are always being developed. We have provided a basic framework to aid the reader and examples to allow the reader to more easily adopt and apply Bayesian methods for blinded signal detection including more complex methods. In general, the authors recommend using the methods dealing with proportions if there is strong epidemiological evidence to allow the correct assessment of the resulting proportions. In most cases, the event rate methods are easier to implement and interpret because event rates per 100 patient-years are commonly available in the literature. The more sophisticated methods discussed benefit from better precision due to incorporating study design characteristics in their models, but often at the cost of more required assumptions and greater computational and operational burden. However, software is often available for these methods to mitigate this burden, as shown in the examples at https://community.amstat.org/biop/workinggroups/safety/safety-home.

7.3 Bayesian Hierarchical Modeling Approaches to Safety Monitoring of Unblinded Data

The assessment of safety for a specific compound often consists of evaluating expected safety outcomes (i.e., AEs of special interest) as well as unexpected safety outcomes (Crowe et al. 2014). The evaluation of unexpected safety outcomes typically occurs in later stages of compound development as exposures to the compound increase. This evaluation usually takes place after unblinding of the data and requires the identification of potential safety outcomes (i.e., those with differential effects) and determining if the observed differences are real signals or noise.

One data source that is used to identify safety signals is AE data collected through routine AE reporting which typically relies on using a medical dictionary to convert verbatim adverse experience(s) of the subject into dictionary terms in the unblinded setting. The most common dictionary used in clinical trial research intended for regulatory submissions is MedDRA which is a dictionary with a hierarchical structure where terms are mapped from narrow/granular terms (e.g., PT) into broader groupings of terms (e.g., system organ class [SOC]). Typically, AE signal detection is performed at the MedDRA dictionary's PT level, consisting of nearly 24,000 unique PTs.

The exercise of identifying safety signals using AE data is not solely a statistical exercise. However, consideration of statistical methods can help interpret whether observed differential effects are real signals or noise. This is because the number of unexpected safety outcomes identified using conventional hypothesis testing approaches applied to AE data can be quite large, making it difficult to determine if they are signals or noise when using simple statistical hypothesis tests that do not account for the multiple hypotheses being tested. Several guidance documents discuss multiplicity (FDA 2017; EMA 2016) in the context of safety evaluation but provide no clear recommendation on how to overcome this issue. Unadjusted analyses may result in many false-positive results while using simple adjustments (e.g., Bonferroni procedures) is generally too conservative and counterproductive for considerations in safety signal detection as they would not be sensitive to identifying the true safety signals.

7.3.1 Literature Review

Several frequentist and Bayesian methods have been proposed to address this issue. There are four considerations when a type of AE should be flagged, namely, (1) actual significance levels; (2) the total number of types of AEs being considered; (3) rates for those AEs not considered for flagging, including their similarity with those that are being considered; and (4) biological relationships among the various AEs (Berry and Berry 2004). The first two are standard considerations in the frequentist approach to multiple comparisons while the second two are only relevant in the Bayesian approach.

False discovery rate (FDR) procedures are frequentist methods for screening frequent AEs designed to control the expected proportion of incorrectly rejected null hypotheses ("false positives"). The double FDR method is a two-step adjustment procedure, first on the higher group level (e.g., SOC) and then on the analysis level (e.g., PT) (Benjamini and Hochberg 1995; Devan V. Mehrotra and Heyse 2004; Devan V. Mehrotra and Adewale 2012).

The Bayesian approach for safety analysis incorporates prior knowledge about the treatment's safety profile and updates knowledge based on accumulating data. The potential advantages of the Bayesian approach relative to classical frequentist statistical methods include the flexibility of incorporating the current knowledge of the safety profile originating from multiple sources into the decision-making process (e.g., known class effects). However, it can be difficult to specify the prior distributions reliably, and empirical application of the methods to a variety of real clinical trial settings has not been evaluated.

Some of the Bayesian methods proposed for clinical trial signal detection include Bayesian hierarchical models (BHMs), multivariate Bayesian logistic regression (MBLR), and the Bayesian unified screening approaches. Berry and Berry (2004) proposed a three-level hierarchical mixed model to account for multiplicities in AE assessment. In their model, the conclusion that one type of AE is affected by treatment depends on data from other AE types, especially from within the same body system (i.e., MedDRA SOC). More traditional multiple comparison methods base conclusions only on the number of AE types under consideration. Xia, Ma, and Carlin (2011) expanded Berry and Berry's method into a hierarchical Poisson mixture model which accounts for the length of the observation of subjects and improves the characteristics of the analysis for rare events. They provide guidance on choosing a signal detection threshold to achieve a fair balance between false-positive error rates and false-negative error rates via a simulation study.

DuMouchel (2012) proposed a MBLR method to analyze safety data when there are rare events and sparse data from a pool of clinical trials. It is designed to compromise between performing separate analyses of each event and a single analysis of a pooled event. As with the Berry and Berry method, MBLR assumes that the events are classified into similar medical groupings in order to use a shrinkage model to allow borrowing strength across similar events. It is exploratory and examines the relationship of the AE frequencies to multiple covariates and treatment by covariate interactions.

Gould (2008, Gould (2013)) proposed an alternative Bayesian screening approach to detect potential safety issues when event counts arise from binomial and Poisson distributions. The method assumes that the AE incidences are realizations from a mixture of distributions and seeks to identify the element of the mixture corresponding to each AE. One of these distributions applies to events where there is no true difference between treatments, and one applies to events where there is a treatment effect. The mixture components are specified a priori, and the calculations then determine the posterior probability that the incidences for each AE are generated from one or the other of the mixture components. It directly incorporates clinical judgment in the determination of the criteria for treatment association.

Gould (2018) later proposed a unified Bayesian screening approach which is suitable for clinical trials and large observational databases that do not test hypotheses, is self-adjusting for multiplicity using 2-component mixture distribution for metrics for between-group differences corresponding to the individual events, provides a direct assessment of the likelihood of no material drug-event association, and quantifies the strength of the observed association. The method provides both the probability given the data that the relationship between a test agent and an AE is "medically unimportant" as opposed to "medically important" based on explicit criteria and an assessment of the magnitude of potential risk elevations using an appropriate metric such as the arithmetic difference, ratio, or odds ratio. The need to define "medically important" and "medically unimportant" brings medical and regulatory considerations explicitly into safety evaluation.

Bayesian methods provide an efficient alternative approach to safety assessment in medical product development. Price et al. (2014) have highlighted many of the challenges associated with safety trials and provided recommendations for key areas in which Bayesian methods can be leveraged to enhance the design, analysis, and interpretation of safety trials. In the next section, we will focus on the use of the BHM in safety monitoring of unblinded data and as well as the properties of the BHM.

7.3.2 Bayesian Hierarchical Model

BHMs are widely used when data arise from different strata. A hierarchical model is written in terms of sub-models and is a useful tool when information is available on several different levels. Moreover, the idea to exploit the similarity and borrow information across strata using hierarchical models has a long tradition (Efron and Morris 1975; Carlin and Louis 1996), originating from Stein's ground-breaking work (Stein 1956; James and Stein 1961). The methods are known to improve estimation accuracy over estimates obtained from stratification or complete pooling. The key assumption for hierarchical models is the exchangeability of parameters which is expressed via a random-effect model of parameters. It is particularly useful for safety analysis as it exploits the correlation between AEs with common biological features by grouping them together.

For safety analysis, AEs are routinely coded in the MedDRA dictionary terms. MedDRA has a hierarchical structure where an AE is coded in a lower level term, the PT, the high-level term, high-level group term, and SOC. AEs are also often grouped using Standardized MedDRA Queries (SMQs), which are used to support signal detection and monitoring. Some SMQs are a simple set of PTs while other SMQs are hierarchical containing subordinate SMQs. The BHM is a natural choice to model the biological relationships among various AEs using the intrinsic medical coding structure. For simplicity, we focus on SOCs in this chapter. However, similar techniques can be applied to SMQs. Some notable work in this area includes Berry and Berry (2004), Devan V. Mehrotra and Heyse (2004), Xia, Ma, and Carlin (2011), and Rosenkranz (2010). For multiplicity adjustment, the BHM assumes exchangeablity for the AEs that share a biological feature or a body system, and the body systems themselves are assumed exchangeable at the next level of hierarchy. Further extension of the BHM framework in safety meta-analysis is discussed by Odani, Fukimbara, and Sato (2017), and Ohlssen et al. (2014). For the rest of this section, we focus on the three-level models proposed by Berry and Berry (2004) and Xia, Ma, and Carlin (2011). Chen et al. (2013) have extended the BHM framework for group sequential when safety data are analyzed repeatedly during a clinical trial.

7.3.2.1 Three-level Bayesian Hierarchical Mixture Model

Two types of analysis for AEs are commonly used in a clinical trial. The first analysis calculates the *subject incidence rate*. This crude estimate for an AE is calculated as the ratio of the number of patients that experienced the AE to the number of patients receiving the assigned study treatment. While such analysis is used widely, it has certain limitations. This measure has meaningful interpretation when all subjects have similar follow-up. However, when patients are exposed to study treatment and followed up differentially, the interpretation of such analysis is questionable. In this case, a *subject-year-adjusted subject incidence rate* is useful. In this analysis, the denominator is replaced by *total exposure*. Total exposure time for a treatment arm is defined as the sum of individual drug exposure durations and adjusted to

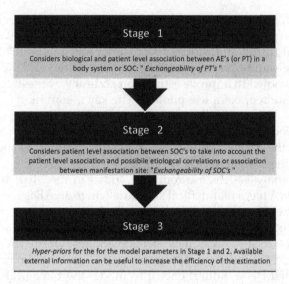

FIGURE 7.3
Three-level BHM model exploring the associations introduced by MedDRA

the subject-year (see Xia, Ma, and Carlin (2011) for more details). Neither analysis requires patient-level data; summary level data suffice. Other, more sophisticated analyses include adjustments for covariates, time to event, and longitudinal analysis. However, these analyses require patient-level information.

Models proposed by Berry and Berry (2004) and Xia, Ma, and Carlin (2011) handle the subject incidence rate and exposure-adjusted subject incidence rate. The first type of analysis is handled using a binomial likelihood, and the second type of analysis is handled using a Poisson likelihood. Note that one patient may experience multiple AEs from the same SOC or they may experience AEs across different SOCs. Therefore, these counts are not mutually independent. For both models, the AEs that share a body system or SOC are assumed similar or *exchangeable*. Moreover, the SOCs themselves are assumed to be exchangeable at the next level of hierarchy. This takes care of the biological association and patient-level correlations. Figure 7.3 shows different labels of the three-level BHM.

The parameter models for stage 1 and stage 2 of the binomial and Poisson models are based on the log odds and log hazard rates, respectively. In addition, Berry and Berry (2004) allow a mixture model which allows a point mass on equality of the treatment and the control rates for an AE. Other variations are also proposed by Xia, Ma, and Carlin (2011). In practice, one can fit multiple models to a data set and choose the best model with the lowest deviance information criterion (DIC) (David J. Spiegelhalter et al. 2002). The DIC is similar to Akaike's information criterion (AIC) in combining a measure of goodness-of-fit and with a measure of complexity. It is used extensively for practical model comparison in many disciplines.

The posterior distributions of odds ratio and relative risk (difference in hazard rates) are the corresponding summary measures to describe the treatment effect estimand. Extensive simulation studies are conducted by Berry and Berry (2004) and Xia, Ma, and Carlin (2011) to show the favorable operating characteristics (probability of false detection) of each proposed model under different scenarios.

Analytical forms of the posterior distributions of model parameters are not available except in some simple cases. MCMC sampling is required for estimating the trial-specific parameters. This can be achieved using modern computation tools, for example, R package *c212* (Carragher 2019), WinBUGS, JAGS, STAN, etc.

7.3.2.1.1 *Choice of Prior*

The BHM described requires prior specification for several parameters. This includes overall SOC level parameter effect (stage 2), probability of point mass (parameter in stage 1: the probability that the treatment and control have the same AE rates (i.e., no risk increase)), and between PT and SOC heterogeneity parameters. The heterogeneity parameters require special attention when using this model in practice. If the numbers of AEs and SOCs are small in a data set, it is not feasible to estimate between PT heterogeneity. Therefore, prior judgments are required. We suggest a weakly informative priors covering a wide range of plausible values for between PT and SOC heterogeneities. Both Berry and Berry (2004) and Xia, Ma, and Carlin (2011) have used non-informative gamma priors for these model parameters. Alternatively, half-normal, half-t, half-Cauchy, and log-normal priors can also be used (Neuenschwander et al. 2016a, 2016b). For example, in the binary case a half-normal prior with scale parameter 0.5 implies a prior median of 0.34 and 95% prior interval (0.016, 1.12). This covers small to large between PT heterogeneity. A larger value of the scale parameter for a half-normal distribution allows larger between PT or SOC heterogeneity and higher discounting.

7.3.2.1.2 *Example*

We provide one hypothetical example from the R package *c212* (Carragher 2019) to illustrate the utility of three-level BHM in practical applications. The data set has eight SOCs and 45 AEs or PTs. For simplicity, we use the Berry and Berry (2004) model only for this application. We fit the model using the *c212.BB* function. However, the package allows the implementation of other models as well. Finally, we have used the priors setting from Berry and Berry (2004).

Figure 7.4 shows the simple summary and posterior distribution of the odds ratio for the hypothetical data. The side by side plot is helpful for non-statisticians to understand the results from the BHM. The model flags AEs in

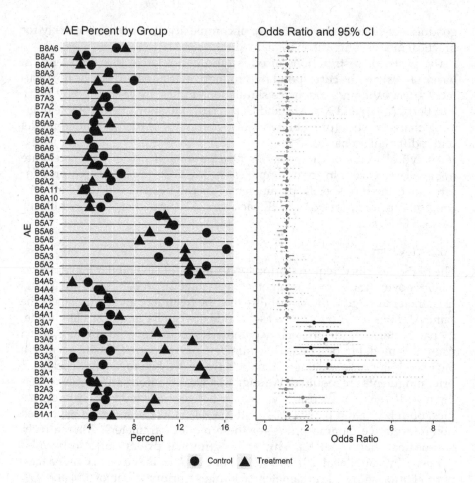

FIGURE 7.4

Posterior summary of the odds ratio using three-level BHM. Solid lines and circles (colored in black and dark gray) represent the posterior mean and 95% credible interval of odds ratio. Posterior mean odds ratio > 2 is marked black to indicate potential safety signals. The light gray circle and dotted line represent naive frequentist estimates using normal approximation. Each Bi represents a SOC, and each Aj represents an AE within that SOC.

SOC 3 as the posterior estimate of the odds ratio is more than 2. The borrowing helps in this case where the occurrence rate of the AE is low overall. The BHM gains precision and yields narrower CIs than the frequentist approach (gray dotted line in Figure 7.5), as the data's biological structure has been taken into consideration. This helps further decision-making.

We also look into the posterior probabilities of the point mass (see Figure 7.5). The posterior probability that the event rate on treatment is greater than on control is low to moderate (< 50%) for most of the SOCs. The only

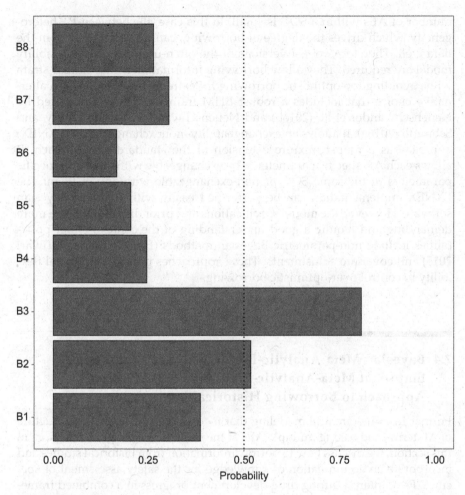

FIGURE 7.5
Risk probabilities. (P(OR>1 | data)). B1–B8 represent SOC 1 through 8.

exceptions are SOC 2 and 3. The only SOC with a high probability of being associated with treatment is SOC 3, with probability 0.808.

7.3.2.2 Discussion

Although the properties of the BHM are attractive, there are caveats. First, borrowing and precision gains are only advantageous if the exchangeability assumption is reasonable. This may not be the case when there are outlying observations or more than one cluster within a SOC. Second, unknown between-PT heterogeneity within a SOC poses a major problem when the number of AEs within a SOC is small. In this case, the between-PT hetero-geneity, which drives the degree of borrowing, cannot be inferred from the data well. Therefore, robust versions of the often-used full exchangeability model are required. They allow borrowing information across similar strata while avoiding too optimistic borrowing for extreme observations. An alter-native choice that includes a robust BHM framework was introduced by Neuenschwander et al. (2016b) and Neuenschwander, Roychoudhury, and Schmidli (2016a). It allows an exchangeability–nonexchangeability (EXNEX) approach as a robust mixture extension of the standard EX approach. It allows each AE-specific parameter to be exchangeable with other AE-specific parameters in the same SOC or non-exchangeable with any of them. The EXNEX implementation can be performed easily with standard Bayesian software. However, the model specifications and prior distributions are more demanding and require a good understanding of the context. Other possi-bilities include non-parametric Bayesian methods (Hu, Huang, and Tiwari 2015) and covariate adjustments. These approaches provide additional flex-ibility to control over-optimistic borrowing.

7.4 Bayesian Meta-Analytic-Predictive(MAP) Prior and Empirical Meta-Analytic-Predictive (eMAP) Prior Approach to Borrowing Historical Information

Further from hierarchical modeling that focuses on the hierarchical structure of AE terms and testing multiple AEs at the same time for signal detection, in this section, we discuss how to borrow information from historical studies and incorporate an accumulation of knowledge for the safety assessment of spe-cific AEs of interest during drug development progress in a combined frame-work of Bayesian and meta-analysis. Such an approach is particularly relevant in situations where the interpretation of the current study data is limited by the small sample size or low event rate of the current study. We first introduce the MAP prior approach, where the basic concept is to obtain a predictive poste-rior distribution from historical data, and then use it as the prior for the concur-rent study data. When previous study information is used to form a prior, an implicit assumption is the historical and current information's exchangeability. However, this might not always be the reality, leading to the conflict between prior and the data. Two strategies are discussed to address the challenges men-tioned above: a robust MAP approach that combined MAP with a vague con-jugate prior and an empirical MAP approach, where an empirical parameter is introduced to define the prior data conflict and control the borrowing from historical information empirically. Finally, we illustrated how the MAP prior approach could be applied to analyze specific AE rates in Section 7.4.4.

7.4.1 MAP Prior

Bayesian approaches and meta-analyses both have the advantages of utilizing historical information and combining relevant information. To combine the two, Neuenschwander et al. (2010) and Schmidli et al. (2014) proposed a MAP prior approach, in which the historical information is combined with the concurrent study using random-effect meta-analysis to obtain the prior information for the concurrent study. The basic concept of the MAP prior approach is to utilize the historical information to obtain the predictive posterior distribution of the parameter of interest and subsequently use that as part of the informative prior to analyze the new study data. Weaver, Ohlssen, and Li (2016) proposed to use the MAP prior approach to assess AEs and demonstrate it using safety data regarding ADASUVE, a drug/devise product used to treat acute agitation in schizophrenic and bipolar patients as an illustrative example.

Consider the situation of assessing the proportion of specific AEs for a particular treatment group. We have H available historical study data sets Y_H and y_i is the AE count for the ith historical study data with event rate θ_i and sample size n_i. A BHM could model the structure of Y_H as follows:

$$
\begin{aligned}
& Y_i \sim \text{Binomial}\left(n_i, \theta_i\right), \text{for } i = 1, 2 \ldots, H \\
& \Psi_i \sim \text{N}\left(\mu, \sigma^2\right) \text{ for } i = 1, 2 \ldots, H \\
& \mu \sim \text{N}\left(0, 1000\right) \\
& \sigma \sim \text{N}\left(0, 1\right) \quad \text{and} \quad \sigma > 0 \text{ A half Normal distribution}
\end{aligned}
\tag{7.6}
$$

where $\Psi_i = \text{logit}(\theta_i)$ for $i = 1, 2, \ldots, H$. The MAP prior $p(\theta_* \mid Y_H)$ can be derived from the above hierarchical model based upon historical data. Different methods could be applied to describe the MAP prior, such as kernel-density estimation or Rao-Blackwellized density estimation (Gelfand and Smith 1990). Schmidli et al. (2014) proposed to approximate the MAP prior by a mixture of k conjugate distributions as below, where k is a number that will not exceed the number of historical studies and can be selected based on AIC or Bayesian Information Criterion (BIC). w_i are positive mixing weights that sum up to one.

$$
\hat{p}_H\left(\theta\right) = \sum_{k=1}^{K} w_k \text{Beta}\left(\theta_* \mid a_k, b_k\right)
\tag{7.7}
$$

Subsequently, the approximated MAP prior is used as the predictive distribution for the new study's AE rate and combined with the new data via Bayes theorem.

FIGURE 7.6
MAP Prior process

The flow chart in Figure 7.6 demonstrates how MAP prior is constructed to borrow information from the historical data.

7.4.2 Robust MAP (rMAP) Prior

Regardless of the careful selection of the historical trials, historical information might not be relevant for the current study due to different study designs, conduct, or patients' population (ICH 2000). The resulting incongruity between the data and the prior is known as prior-data conflict. Considering prior-data conflict, a robust version of the MAP prior incorporates a vague conjugate prior $p_\nu(\theta_*)$ as a robustness feature that allows for a discounting of the informative prior $\hat{p}_H(\theta_*| Y_H)$ (Schmidli et al. 2014). The robust MAP prior incorporates a weight w_0 as the probability that the new study differs systematically from the historical studies, such as:

$$\hat{p}_{HR}(\theta_*) = (1-w_0)\hat{p}_H(\theta_*| Y_H) + w_0 p_v(\theta_*).$$

For binary endpoints, a good choice of non-informative prior could be a Beta (1,1) or a Beta (1/3,1/3) (Weaver, Ohlssen, and Li 2016).

7.4.3 eMAP Prior

In consideration of a potential issue of prior-data conflict, Li, Chen, and Scott (2016) proposed an eMAP prior approach as a further extension of the MAP prior and the robust MAP prior approaches. The proposed eMAP approach was then applied to safety data of hemophilia A studies as an illustrative example. The eMAP prior approach addresses the prior-data conflict by incorporating an empirical parameter to the MAP prior setting. The advantage of the eMAP prior approach is to avoid arbitrary determination of the choice of w_0 from the rMAP approach, which could be subjectively determined based on the degree of confidence that the clinical study team has in the relevance of the historical data. The eMAP prior approach introduces an empirical parameter τ to determine how quickly the informative prior is discounted with increasing prior-data conflict:

$$\hat{p}_{HE}(\theta_*) = \sum_{k=1}^{K} w_k \text{Beta}\left(\theta_* \Big| \frac{a_k}{\tau}, \frac{b_k}{\tau}\right)$$

The empirical parameter τ is determined empirically by maximizing the marginal likelihood according to the new study's outcome. The value of τ controls the variance of the eMAP prior and adds robustness to the synthesis of the historical data with a new study when they conflict. When $\tau = 1$, the eMAP prior and the MAP prior are exactly the same priors. When $\tau > 1$, the eMAP prior is less informative than the MAP prior but more robust in the presence of increasing prior-data conflict. When $0 < \tau < 1$, the eMAP prior is more informative than the MAP prior, indicating less prior-data conflict (Li, Chen, and Scott 2016).

7.4.4 Illustrative Example Applying MAP Priors

This section illustrates how MAP priors can be derived from historical data by applying the MAP prior approach to hemophilia A clinical safety data. The analysis was conducted using R Bayesian evidence synthesis tools –R package *RBesT* (Weber et al. 2019).

Hemophilia A is a blood clotting disorder caused by a Factor VIII (FVIII) gene mutation. The standard of care for patients with moderate to severe hemophilia A is FVIII replacement therapy, and several recombinant and plasma-derived FVIII products have been approved in the United States. One of the major safety concerns about the replacement therapy is developing inhibitory antibodies (Wight and Paisley 2003). Table 7.1 summarizes the public data of four historical studies for patients treated with various FVIII preparations who developed inhibitors (Gouw et al. 2013).

In this illustration, we chose N (0, 2) and half normal distribution with a standard deviation of 1 for the hyper-parameters of the BHM. The random-effect meta-analysis was applied using the MCMC algorithm (Stan Development Team 2018), and Figure 7.7 shows the per-trial point estimates

TABLE 7.1

Summarized Results of Historical Studies using Recombinant FVIII Products

Study name	Sample size	Inhibitor positive	Percent
RODIN	486	148	30.5
FrenchCoag (RODIN excluded)	303	114	37.6
UKHCDO (RODIN excluded)	319	85	26.6
CANAL	181	53	29.3
Total	1289	400	31.0

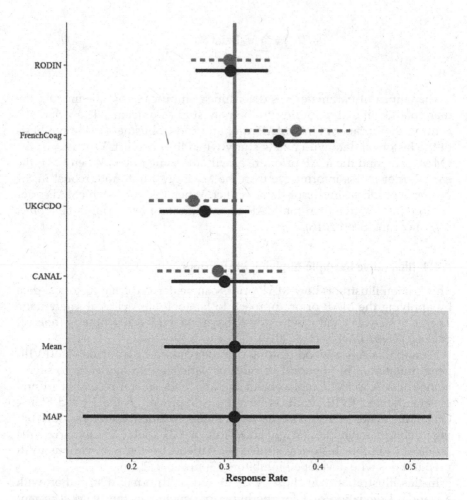

FIGURE 7.7
Forest plot augmented with meta-analytic shrinkage estimates per trial.

(light dot) as well as the model-derived median (dark dot) of the meta-analytic model (R Package *gMAp*). The selected beta mixture model based on the posterior predictive samples for the MAP prior is (0.496) Beta (12.286, 27.247) +(0.341) Beta (85.896, 189.72) +(0.163) Beta (2.4953, 4.9288).

Figure 7.8 overlays the MCMC histogram of the MAP prior and the respective parametric mixture approximation, which shows a good fit of the two. Now that we have obtained the MAP prior based on the historical data, it can be used as the prior and combined with the new trial data to obtain the posterior probability of the inhibitor rate, to provide the safety assessment needed.

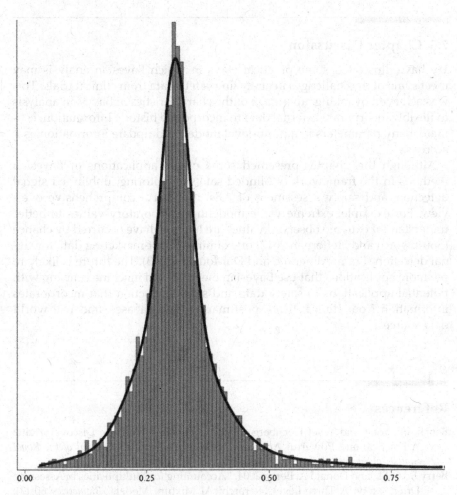

FIGURE 7.8
Mixture model approximation of the MCMC-represented MAP prior.

7.4.4.1 Discussion

This section illustrated ways to create priors based on combining information through a Bayesian meta-analytic approach and how to adjust these methods to mitigate the risk of prior-data conflict. These methods facilitate the creation of informative priors to be used in the analysis to reduce the amount of subject data required to draw conclusions within an analysis. These could be applied to any relevant analysis, such as AEs, laboratory data, or even efficacy assessments. Such an approach is particularly relevant in situations where the interpretation of the current study data is limited by the small sample size or low event rate of the current study.

7.5 Chapter Discussion

We have illustrated some practical ways in which Bayesian analysis may meet some of the challenges in analyzing safety data from clinical trials. This was achieved by taking advantage of the characteristics of Bayesian analysis to flexibly incorporate external data to incorporate historic information, estimate many parameters in a multi-level model, and update information as it accrues.

Although this chapter presented some of the applications of Bayesian methods in the framework of blinded safety monitoring, unblinded signal detection, and safety assessment of AEs, it is not a comprehensive overview. For example, extreme value modeling of laboratory values to better understand if extreme observed values are likely to have occurred by chance (Southworth and Heffernan 2012) or examining post-marketing data for signal detection (Fram, Almenoff, and DuMouchel 2003). The future is likely to see more applications that use Bayesian methods and machine learning with potential applications to safety data and signal detection that incorporates information from clinical trials, post-marketing databases, and real world data sources.

References

Benjamini, Yoav, and Yosef Hochberg. 1995. "Controlling the False Discovery Rate: A Practical and Powerful Approach to Multiple Testing." *Journal of the Royal Statistical Society: Series B (Methodological)* 57 (1): 289–300.

Berry, Scott M., and Donald A. Berry. 2004. "Accounting for Multiplicities in Assessing Drug Safety: A Three-Level HierarchicAl Mixture Model." *Biometrics* 60 (2): 418–26.

Best, Nicky, Nigel Dallow, and Timothy Montague. 2020. "Prior Elicitation." In *Bayesian Methods in Pharmaceutical Research*, 87–109. https://www.taylorfrancis.com/chapters/edit/10.1201/9781315180212-5/prior-elicitation-nicky-best-nigel-dallow-timothy-montague

Carlin, B. P., and T. A. Louis. 1996. *Bayes and Empirical Bayes Methods for Data Analysis*. London: Chapman & Hall.

Carragher, Raymond. 2019. *C212: Methods for Detecting Safety Signals in Clinical Trials Using Body-Systems (System Organ Classes)*. https://CRAN.R-project.org/package=c212.

Chen, Wenfeng, Naiqing Zhao, Guoyou Qin, and Jie Chen. 2013. "A Bayesian Group Sequential Approach to Safety Signal Detection." *Journal of Biopharmaceutical Statistics* 23 (1): 213–30.

Crowe, Brenda, H. Amy Xia, Mary Nilsson, Seta Shahin, Wei Wang, and Qi Jiang. 2014. "Program Safety Analysis Plan." *Quantitative Evaluation of Safety in Drug Development: Design, Analysis and Reporting* 67: 55.

DuMouchel, William. 2012. "Multivariate Bayesian Logistic Regression for Analysis of Clinical Study Safety Issues." *Statistical Science* 27 (3): 319–39. https://doi.org/10.1214/11-STS381.

Efron, Bradley, and Carl Morris. 1975. "Data Analysis Using Stein's Estimator and Its Generalizations." *Journal of the American Statistical Association* 70 (350): 311–19.

EMA. 2016. "Guideline on Multiplicity Issues in Clinical Trials." *Guidance, European Medicines Agency* 44: 1–15.

FDA. 2017. "Multiple Endpoints in Clinical Trials Guidance for Industry." *Guidance (Draft), Center for Biologics Evaluation and Research (CBER)*.

Fram, David M., June S. Almenoff, and William DuMouchel. 2003. "Empirical Bayesian Data Mining for Discovering Patterns in Post-Marketing Drug Safety." In *Proceedings of the Ninth ACM SIGKDD International Conference on Knowledge Discovery and Data Mining*, 359–68.

Gelfand, Alan E, and Adrian F.M. Smith. 1990. "Sampling-Based Approaches to Calculating Marginal Densities." *Journal of the American Statistical Association* 85 (410): 398–409.

Gill, Christopher J, Lora Sabin, and Christopher H. Schmid. 2005. "Why Clinicians Are Natural Bayesians." *BMJ* 330 (7499): 1080–83.

Gould, A. Lawrence. 2008. "Detecting Potential Safety Issues in Clinical Trials by Bayesian Screening." *Biometrical Journal: Journal of Mathematical Methods in Biosciences* 50 (5): 837–51.

Gould, A. Lawrence. 2013. "Detecting Potential Safety Issues in Large Clinical or Observational Trials by Bayesian Screening When Event Counts Arise from Poisson Distributions." *Journal of Biopharmaceutical Statistics* 23 (4): 829–47.

Gould, A Lawrence. 2018. "Unified Screening for Potential Elevated Adverse Event Risk and Other Associations." *Statistics in Medicine* 37 (18): 2667–89.

Gouw, Samantha C, Johanna G. Van Der Bom, Rolf Ljung, Carmen Escuriola, Ana R. Cid, Ségolène Claeyssens-Donadel, Christel Van Geet, et al. 2013. "Factor VIII Products and Inhibitor Development in Severe Hemophilia a." *New England Journal of Medicine* 368 (3): 231–39.

Hu, Na, Lan Huang, and Ram C. Tiwari. 2015. "Signal Detection in FDA AERS Database Using Dirichlet Process." *Statistics in Medicine* 34 (19): 2725–42.

ICH. 2000. "E10, Choice of Control Group and Related Issues in Clinical Trials."

James, W., and Charles Stein. 1961. "Estimation with Quadratic Loss." In *Proceedings of the Fourth Berkeley Symposium on Mathematical Statistics and Probability, Volume 1: Contributions to the Theory of Statistics*, 361–79. Berkeley, CA: University of California Press.

Jeffreys, Harold. 1946. "An Invariant Form for the Prior Probability in Estimation Problems." *Proceedings of the Royal Society of London. Series A. Mathematical and Physical Sciences* 186 (1007): 453–61.

Li, Judy X., Wei-Chen Chen, and John A. Scott. 2016. "Addressing Prior-Data Conflict with Empirical Meta-Analytic-Predictive Priors in Clinical Studies with Historical Information." *Journal of Biopharmaceutical Statistics* 26 (6): 1056–66.

Mehrotra, Devan V., and Adeniyi J. Adewale. 2012. "Flagging Clinical Adverse Experiences: Reducing False Discoveries Without Materially Compromising Power for Detecting True Signals." *Statistics in Medicine* 31 (18): 1918–30.

Mehrotra, Devan V., and Joseph F. Heyse. 2004. "Use of the False Discovery Rate for Evaluating Clinical Safety Data." *Statistical Methods in Medical Research* 13 (3): 227–38.

Mukhopadhyay, Saurabh, Brian Waterhouse, and Alan Hartford. 2018. "Bayesian Detection of Potential Risk Using Inference on Blinded Safety Data." *Pharmaceutical Statistics* 17 (6): 823–34.

Neuenschwander, Beat, Gorana Capkun-Niggli, Michael Branson, and David J. Spiegelhalter. 2010. "Summarizing Historical Information on Controls in Clinical Trials." *Clinical Trials* 7 (1): 5–18.

Neuenschwander, Beat, Satrajit Roychoudhury, and Heinz Schmidli. 2016a. "On the Use of Co-Data in Clinical Trials." *Statistics in Biopharmaceutical Research* 8 (3): 345–54.

Neuenschwander, Beat, Simon Wandel, Satrajit Roychoudhury, and Stuart Bailey. 2016b. "Robust Exchangeability Designs for Early Phase Clinical Trials with Multiple Strata." *Pharmaceutical Statistics* 15 (2): 123–34.

Neuenschwander, Beat, Sebastian Weber, Heinz Schmidli, and Anthony O'Hagan. 2020. "Predictively Consistent Prior Effective Sample Sizes." *Biometrics* 76 (2): 578–87.

Oakley, Jeremy E, and Anthony O'Hagan. 2010. "SHELF: The Sheffield Elicitation Framework (Version 2.0)." *School of Mathematics and Statistics*, University of Sheffield, UK. http://Tonyohagan.Co.Uk/Shelf.

Odani, Motoi, Satoru Fukimbara, and Tosiya Sato. 2017. "A Bayesian Meta-Analytic Approach for Safety Signal Detection in Randomized Clinical Trials." *Clinical Trials* 14 (2): 192–200.

Ohlssen, David, Karen L. Price, H. Amy Xia, Hwanhee Hong, Jouni Kerman, Haoda Fu, George Quartey, Cory R. Heilmann, Haijun Ma, and Bradley P. Carlin. 2014. "Guidance on the Imple- mentation and Reporting of a Drug Safety Bayesian Network Meta-Analysis." *Pharmaceutical Statistics* 13 (1): 55–70.

Price, Karen L., H. Amy Xia, Mani Lakshminarayanan, David Madigan, David Manner, John Scott, James D. Stamey, and Laura Thompson. 2014. "Bayesian Methods for Design and Analysis of Safety Trials." *Pharmaceutical Statistics* 13 (1): 13–24.

Rosenkranz, Gerd K. 2010. "An Approach to Integrated Safety Analyses from Clinical Studies." *Drug Information Journal* 44 (6): 649–57.

Schmidli, Heinz, Sandro Gsteiger, Satrajit Roychoudhury, Anthony O'Hagan, David Spiegelhalter, and Beat Neuenschwander. 2014. "Robust Meta-Analytic-Predictive Priors in Clinical Trials with Historical Control Information." *Biometrics* 70 (4): 1023–32.

Schnell, Patrick M., and Greg Ball. 2016. "A Bayesian Exposure-Time Method for Clinical Trial Safety Monitoring with Blinded Data." *Therapeutic Innovation & Regulatory Science* 50 (6): 833–38.

Southworth, Harry, and Janet E. Heffernan. 2012. "Extreme Value Modelling of Laboratory Safety Data from Clinical Studies." *Pharmaceutical Statistics* 11 (5): 361–66.

Spiegelhalter, David J., Nicola G. Best, Bradley P. Carlin, and Angelika Van Der Linde. 2002. "Bayesian Measures of Model Complexity and Fit." *Journal of the Royal Statistical Society: Series B (Statistical Methodology)* 64 (4): 583–639.

Spiegelhalter, David J., Keith R. Abrams, and Jonathan P. Myles. 2004. *Bayesian Approaches to Clinical Trials and Health-Care Evaluation*. Vol. 13. John Wiley & Sons.

Stan Development Team. 2018. "RStan: The r Interface to Stan." *R Package Version 2.17. 3.*

Stein, Charles. 1956. "Inadmissibility of the Usual Estimator for the Mean of a Multivariate Normal Distribution." In *Proceedings of the Third Berkeley Symposium on Mathematical Statistics and Probability, Volume 1: Contributions to the Theory of Statistics*, 197–206. Berkeley, CA: University of California Press.

Waterhouse, Brian, Alan Hartford, Saurabh Mukhopadhyay, R Ferguson, and Barbara Hendrickson. Using the Bayesian detection of potential risk using inference on blinded safety data (BDRIBS) method to support the decision to refer an event for unblinded evaluation. *Pharmaceutical Statistics* 2021;1-14. doi:10.1002/pst.2175.

Weaver, Jerry, David Ohlssen, and Judy X. Li. 2016. "Strategies on Using Prior Information When Assessing Adverse Events." *Statistics in Biopharmaceutical Research* 8 (1): 106–15.

Weber, Sebastian, Yue Li, John Seaman, Tomoyuki Kakizume, and Heinz Schmidli. 2019. "Applying Meta-Analytic Predictive Priors with the r Bayesian Evidence Synthesis Tools." *arXiv Preprint arXiv:1907.00603*.

Wen, Shihua, Greg Ball, and Jyotirmoy Dey. 2015. "Bayesian Monitoring of Safety Signals in Blinded Clinical Trial Data." *Annals of Public Health and Research* 2 (2): 1019–22.

Wight, J., and S. Paisley. 2003. "The Epidemiology of Inhibitors in Haemophilia a: A Systematic Review." *Haemophilia* 9 (4): 418–35.

Xia, H. Amy, Haijun Ma, and Bradley P. Carlin. 2011. "Bayesian Hierarchical Modeling for Detecting Safety Signals in Clinical Trials." *Journal of Biopharmaceutical Statistics* 21 (5): 1006–29.

Yao, Bin, Li Zhu, Qi Jiang, and H Xia. 2013. "Safety Monitoring in Clinical Trials." *Pharmaceutics* 53 (1): 94–106.

8

Likelihood-Based Methods for Safety Monitoring

Ed Whalen
Pzifer Inc., New York, NY, USA

Krishan Singh
(retired) GlaxoSmithKline, Brentford, UK

Li-An Lin
Merck & Co, Inc, Kenilworth, NJ, USA

Melvin Munsaka
AbbVie, Inc., North Chicago, IL, USA

William Wang
Merck & Co, Inc, Kenilworth, NJ, USA

CONTENTS

8.1 Introduction: Importance of Safety Monitoring during
  Development and Post-Approval Pharmacovigilance 168
  8.1.1 Experiences That Led to Further Enhancement
    of Monitoring 169
  8.1.2 The State of Monitoring 170
  8.1.3 Looking Forward 170
8.2 General Framework for Statistical Methods in Dynamic or
  Continuous Safety Monitoring 171
  8.2.1 A Framework in the Context of Randomized Studies 171
  8.2.2 Excessive and Untenable Levels of Risk and Levels
    of Evidence 172
  8.2.3 Probability of Misleading Evidence: P-Values in
    Repeated Safety Testing 173
  8.2.4 A Universal Bound 174
8.3 Overview of Likelihood-Based Statistical Methods 175
  8.3.1 The Likelihood as a Basis for Comparison of Multiple
    Sets of Conditions 175
  8.3.2 Wald Sequential Probability Ratio Test (SPRT) 176

DOI: 10.1201/9780429488801-10

8.3.3   Maximized Sequential Probability Ratio Test (MaxSPRT)....... 176
8.3.4   Conditional Maximized Sequential Probability Ratio Test
(CMaxSPRT)... 177
8.3.5   Sequential Likelihood Ratio Test (SeqLRT)............................... 178
8.3.6   Longitudinal Likelihood Ratio Test (LongLRT) 178
8.3.7   Sequential Generalized Likelihood Ratio Test (SGLRT)........... 179
8.3.8   Summary of Likelihood Ratio Methods 179
8.4   Examples of Several Likelihood-Based Methods in Safety
Monitoring 179
8.4.1   Monitoring for Fever and Neurological Effects of a Vaccine..... 181
8.4.2   Rotavirus Vaccine Monitoring for a Serious Intestinal Event..... 181
8.4.3   Example of Possible Applications of LongLRT 182
8.5   Comments on Other Methods and Software for LRT Methods.......... 182
8.5.1   Non-LR Methods for Pre- and Post-Marketing Surveillance...... 182
8.5.2   A Brief Note on Software for LRT Methods............................... 183
8.6   Summary and Discussion... 184
References... 185

8.1 Introduction: Importance of Safety Monitoring during Development and Post-Approval Pharmacovigilance

Worldwide regulators have emphasized safety monitoring as key to approving and maintaining a marketed drug. Before the drug reaches the market, it has to clear well-defined safety requirements and, while on the market, maintain or enhance the accumulated safety profile, as outlined in industry harmonizing guidances or approving agency documents – for example, the statistical guidance in ICH E9 section 7 or for post-approval safety surveillance, https://www.fda.gov/media/81015/download. Monitoring of safety occurs throughout the development program both at the study level and the program level as studies accumulate. Different organizations may have different levels of review during the pre-submission phases. Regardless, emerging safety issues can best come to light through ongoing monitoring. For approval, a summary of clinical safety (SCS) summarizes, analyzes, and describes in detail the accumulated knowledge of the drug's safety profile and issues based on all data available at the time of submission. During review, more questions typically arise from regulators to which sponsors provide further analyses or summaries. The sponsor's interpretation of and contextualizing of the study results develop through the collaboration of clinicians (both safety and project lead), pharmacologists, and statisticians discussing the safety results and other important aspects in both the SCS and the review process.

Most readers have some familiarity with monitoring methods based on either frequentist (think estimated rates and p-values) or Bayesian (think 'updated' rates) approaches. Those methods have been discussed in detail in Chapters 6 and 7. In this chapter, we focus on a third way of thinking about the evolving safety data in monitoring for drug side effects, especially in the post-approval safety surveillance setting. Frequentists and Bayesian methods differ from each other essentially in whether they view as a given, the data that we observe (Bayesian), or the parameters that we wish to estimate with those data (frequentist). Both can use, at different stages, the likelihood function or ratios of likelihoods. Taken in their own right, the likelihood-based methods focus on the likelihood ratio (LR) as a measure of alternative explanations for the observed data. The likelihood principle of section 8.2 provides a framework to measure the strength of statistical evidence for one hypothesis over another. The remainder of this chapter focuses on the likelihood paradigm and methods based on it, except for a bit more detail on the differences between it and the frequentist and Bayesian methods in section 8.5.1.

8.1.1 Experiences That Led to Further Enhancement of Monitoring

Several major examples have arisen over the decades of products that had safety signals detected pre- or post-approval. Follow-up on the signals established that the products did lead to serious adverse drug reactions that sponsors could not feasibly mitigate. As a result, the products subsequently had market authorization withdrawn, such as the withdrawals of mibefradil in 1998 and cerivastatin in 2001.

Mibefradil gained market approval in 1997 as a treatment for hypertension and angina. The initial product label listed three drug–drug interactions. During post-approval monitoring, more were discovered with some mediated through its inhibition of cytochrome P450 3A4 (Po & Zhang 1998). As its use grew in the real-world setting; so did reports of serious adverse events (AEs) associated with the drug–drug interactions. As a result, the sponsor removed mibefradil from the market in June of 1998. Although some drug–drug interaction risks were known at approval, monitoring of such drug–drug interactions did not reach levels to suggest they were of grave concern until after marketing, leaving open the question of what level of pre-approval analyses, simulation, modeling, and patient-level monitoring could have described the potential effects on an exposed post-approval population.

In another case of an approved drug, enhanced methods for repeated surveillance allowed for repeated looks at accumulating safety data and detection of post-approval safety concerns (in this case the multi-item gamma Poisson shrinker – not covered in this chapter – see Szarfman et al. 2002). Cerivastatin obtained approval for hypercholesterolemia in 1996 and later expanded its label to include higher doses. Cases of rhabdomyolysis occurred at a much higher level than for other drugs in the 'statin' class in the FDA's pharmacovigilance data of spontaneous reports. The FDA's systematic monitoring of spontaneous events versus similar-type drugs led to the detection of a tenfold higher rate for cerivastatin versus other statins. The sponsor subsequently withdrew the drug based on these findings.

8.1.2 The State of Monitoring

Once on the market, sponsors and regulators continue to monitor the drug for any safety issues. They do this through a wide set of data sources which include further clinical study data and observational data sources of various types. Uncertainty around specific events may lead to the conduct of a post-approval safety study or further investigation by regulators using systems (e.g. SENTINEL) not available a couple of decades ago and most commonly based on observational, real-world data sources.

The SENTINEL system provides one notable source of data and safety monitoring on which protocols can run to further investigate questions raised based on an observational review or that arise post-approval. Spontaneous reporting systems, such as FAERS, also have evolved and continue to provide information through constant monitoring and analysis for 'signals.' See Chapter 1 for more on signals in general. Finally, sponsors have the option to utilize many other sources such as registries, working with collaborative groups (e.g. Strober et al. 2018), or fully designed studies working within a healthcare system (e.g. Vestbo et al. 2016).

8.1.3 Looking Forward

The advances made in the last two decades have come from regulators, academia, and industry both alone or in collaboration as seen with groups such as the Sentinel initiative or the work of the Reagan-Udall Foundation to name two. Note that Reagan-Udall is a non-profit organization that includes a study-oriented group called the Innovation in Medical Evidence Development and Surveillance. Many other such organizations exist and will indeed push the capabilities of safety monitoring far beyond that of the current state as of this writing.

Regardless of what the future brings, some basic methods form the foundation of current and future work. During drug development and post-marketing safety monitoring, we have a dynamic and continuous collection of safety data. These data can come from many different sources, randomized controlled trials, observational studies, a multitude of health care record sources, and spontaneous reporting systems. We need statistical methodologies that have the robustness to handle these growing and varied data sources. LR approaches have a foundation in theory and practice that makes them a good option for that purpose. We now turn our attention to one group of methods known as likelihood-based methods and procedures.

Subsequent sections will cover the general framework, specific likelihood-based methods, some examples of their application, and a discussion of their relation to other methods.

8.2 General Framework for Statistical Methods in Dynamic or Continuous Safety Monitoring

The likelihood-based methods can apply to either experimental or observational data. Some methods, described later, such as the longitudinal LR test, have been applied to spontaneous report data as well as clinical trial data. All these methods allow for the continual monitoring of data as it accumulates, which forms a defining feature of them. Thus, they provide an opportunity for a more dynamic approach to safety monitoring. We will return to this theme throughout the chapter with a focus mostly on experimental data.

8.2.1 A Framework in the Context of Randomized Studies

Randomized controlled trials are typically designed to produce reliable results on the efficacy of interventions (Singh & Loke 2012). In designing these trials, efficacy endpoints are identified in advance, sample sizes are estimated to permit an adequate assessment of the primary efficacy endpoint, and planned analyses for interim looks at data are established in advance to control multiplicity and preserve the type I error. Therefore, the frequentist-defined hypothesis test methods are adequate for proof of the efficacy of intervention products. However, the same approach is not appropriate for the evaluation of the safety of intervention products. For a clinical trial, there could be many different safety outcomes to characterize. Safety includes outcomes such as AEs and laboratory test results that are not known prior to study conduct (though many can be anticipated), and, some of the important events may not have yet occurred because of their lower rate of occurrence. Thus, they cannot be predefined in the design of an earlier stage clinical trial. Inductive reasoning (in the sense of an evolving understanding of the truth) is more suitable to explore the potential safety concerns and evaluate the statistical evidence presented in the data. Confirming the safety signal as product risk (hypothesis) usually requires multiple sources of safety evidence, including the biological mechanism of the treatment, reproducibility of the safety signal across studies, and causality assessment. Providing the p-value with deductive reasoning (in the sense of a conclusive truth) for each safety endpoint could be misinterpreted because of excessive testing for any statistically significant differences between treatment groups across many events. A uniform and non-discretionary application of hypothesis testing, for post-hoc and unplanned analyses, drives straight against the Neyman–Pearson theory (Wasserstein & Lazar 2016).

8.2.2 Excessive and Untenable Levels of Risk and Levels of Evidence

The purpose of safety evaluation in a clinical trial is to identify and confirm the hazard or excessive risk of drugs. A given clinical trial may not be designed for safety endpoints, in particular, considering the rarity of many safety endpoints. The trial's data are usually underpowered to detect important treatment differences of many safety endpoints even when combined across several trials. By frequentist methods, failure to reject the null hypothesis of no excessive risk cannot prove that a drug is safe. To prove that a drug is safe, an estimate of the magnitude of the safety risk is more appropriate than concluding no risk based on failure to reject the null hypothesis – a somewhat common error in reasoning. However, it is misleading to interpret the p-value as a measure of the size of an effect or the importance of a result (Wasserstein & Lazar 2016). Many researchers think that a p-value of 0.05 means that the null hypothesis has a probability of only 5%. This is an understandable but categorically wrong interpretation because the p-value is calculated on the assumption that the null hypothesis is true. It cannot, therefore, be a direct measure of the probability that the null hypothesis is false (Goodman 1999).

Unlike the standard frequentist method, the likelihood principle directly uses LRs to measure the strength of statistical evidence. In essence, the principle assumes that the evidence in the data in favor of a hypothesis can be measured by a formula that relates the hypothesized model to the observations. Specifically, the formula expresses how likely it is that the observations have arisen from the hypothesized model. Under this paradigm, the assessment of accumulating evidence is not constrained by the study design. For example, the LR, unlike a p-value, is unaffected by the number of examinations of the data. This is an appealing feature for dynamic safety monitoring which requires continuous examination of the clinical trial data as the trial accumulates data. Furthermore, p-values are not able to distinguish whether the non-significant result is because of limited sample size or no excessive risk. P-values are particularly uninformative when no significant differences are found. LRs, by definition, evaluate the statistical evidence of one hypothesis over another. If the LR is greater than one, the evidence favors the alternative hypothesis over the null hypothesis of no difference (no increased safety risk). If the LR is less than one, the evidence favors the null hypothesis over the alternative of no risk. Therefore, three conclusions can be drawn from the likelihood approach:

1) there is sufficient evidence for the alternative hypothesis that, for example, an intervention has excessive risk;

2) there is sufficient evidence for the null hypothesis that the intervention has no effect over the alternative; and

3) the data are insensitive in distinguishing the hypotheses.

By these three conclusions, we can screen out the AEs that are clearly not affected by the treatment and focus more on the AEs showing stronger associations with the treatment intervention and the inconclusive AEs that need more data to quantify.

In the sense of the preceding three conclusions based on the LR, the LR works more like a summary statistic capturing the magnitude of the effect of each treatment rather than a straight yes/no logic of a hypothesis test. However, the LR has its own probability distribution which can produce a p-value if needed – either analytically or through techniques such as bootstrap. Thus, the same statistical procedures can help to confirm any suggested risks or, in the case of continuous testing, provide a framework to guard against excessive testing. The sequential testing procedures of section 8.3 allow for such testing and are described further in that section.

Bayesian methods also allow for a re-examination of accumulating data for evidence in favor or against the risk of a drug for any given safety event. These procedures are covered in Chapter 7. Here, we focus on a clearer exposition of the likelihood approach.

8.2.3 Probability of Misleading Evidence: P-Values in Repeated Safety Testing

Whether using a likelihood method or not, the ultimate safety goals center on establishing for each potential safety concern whether it is likely causally related to drug intake (rule in), is not causally related to drug intake (rule out), or requires further study. The three interpretations of LR results align with the safety goal. The causality portion will depend on the particulars of the study designs that generate the body of data. In the case of clinical trials, the ability to establish a level of causality usually features in their design. For data generated from observational sources, the causality issue requires much more consideration including the etiology of the disease and other clinically related perspectives such as with Hills Criteria discussed in Chapter 2. Here, we focus primarily on statistical evidence generated from interventional clinical trial data sources.

As the strength of the evidence increases (association in observational setting, causal relatedness where randomization is used), through more data accumulation, the data and their analyses can convert a skeptic into a believer or a tentative suggestion into an accepted truth. To interpret and communicate the strength of evidence, it is useful to divide the continuous scale of the LR into descriptive categories, such as 'weak', 'moderate', and 'strong' evidence. Such a crude categorization allows a quick and easily understood characterization of the evidence for one hypothesis over another. For example, strong evidence supporting the alternative hypothesis of excess risk over the null (no risk) exists if $LR \geq k$. Similarly, strong evidence supporting the null (no risk) exists if $LR < 1/k$. The threshold values of $k = 8$ and 32 have been suggested to distinguish between weak, moderate, and strong evidence (Royall 1997). Other similar benchmarks also have been proposed in the literature.

However, these descriptive categories are designed for inductive inference to measure the strength of statistical evidence in a single experiment. They do not guarantee the error control in the long-run behavior from repeated analyses which usually refers to type I error in the frequentist paradigm. Thresholds and boundaries similar to k are also available for some of the specific methods of section 8.3 with the control of type I error in mind. With repeated analyses over an accumulating body of data, error control is an important part of the decision-theoretic framework (Goodman 1999). As a result, study designs need to be evaluated in terms of their operational characteristics such as the frequency with which a particular study design will produce misleading or weak evidence (Blume 2002). In this sense, the likelihood approach still can use the LR to represent the strength of the evidence in the data, but also provides controllable probabilities of observing misleading or weak evidence which is analogous to the type I and type II error rates of hypothesis testing. This connection between inference in individual experiments and the number of errors we make over time is not found in the frequentist paradigm. It is found only in properly assessing the strength of evidence from an experiment with a likelihood paradigm and uniting this with a synthesis of all the other scientific information that bears on the question at hand.

8.2.4 A Universal Bound

Dynamic safety monitoring is a continuous process that requires continuous examination of the clinical trial(s) data as the trial(s) progresses. In the case of multiple looks at the data, the interpretation of statistical evidence is unaffected because the LR does not directly depend on the experimental design. In terms of repeated looks at accumulating data, the cumulative probability of observing some misleading data increases with each look. This is similar to a study design that checks the data as observations are collected and stops the study as soon as a significant result is attained. The statistical consequence of hypothesis testing is that a large number of looks may substantially inflate the probability of erroneously obtaining a significant result when the null hypothesis is true. With a likelihood approach, the probability of observing strong misleading evidence is naturally low. For any fixed sample size and any pair of probability distributions, the probability of observing misleading evidence of strength k or greater is always less than or equal to $1/k$. Moreover, the probability of observing misleading evidence increases with each look but remains bound by the universal bound of $1/k$, no matter how many looks taken (Royall 2000, Royall 1997). For safety monitoring, the potential number of looks at the data is large. The universal bound is an excellent safeguard against the effects of misleading data in cumulative reviews of safety data and a useful application of the likelihood approach in error probability.

8.3 Overview of Likelihood-Based Statistical Methods

This section covers methods that range from a simple null and alternative hypothesis about a single group to multiple groups and outcomes with compound hypotheses. Likelihood-based statistics can indicate a level of evidence or be used for full hypothesis testing. Because many decisions, not just about safety, align with a hypothesis test paradigm, the following will indicate hypotheses in question, but also describe the question itself as in the summary table of the various methods.

8.3.1 The Likelihood as a Basis for Comparison of Multiple Sets of Conditions

Likelihood-based methods rely on comparison of the likelihood function for two different sets of conditions, for example, test drug versus control drug. Before covering likelihood methods useful for sequential monitoring of safety data as they accumulate, we first introduce the likelihood principle. The concept of the likelihood principle was first introduced in 1965 (Hacking 1965) by stating the formal expression of the Law of Likelihood and using the LR for comparing two simple hypotheses, such as $H_0: \theta = \theta_0$, versus a single point alternative $H_1: \theta = \theta_1$, for a parameter θ, under the assumption that a background model is true. The evidence value of a particular result is often called the "likelihood ratio" of the two hypotheses, $LR = L(\theta_1, data)/L(\theta_0, data)$, and provides an objective measure of the strength of evidence. Focused on interpreting data as statistical evidence, the evidential paradigm uses LRs to measure the strength of statistical evidence (Blume 2002). For example, consider the case that the LR for the alternative hypothesis compared with the null hypothesis is 5. This result provides inductive evidence that the data support the alternative hypothesis five times as strongly as it does the null. Both H_0 and H_1 are "simple" hypotheses because they specify a single numeric value for the probability of observing x. A "composite" alternative hypothesis, such as $H_1: \theta > \theta_1$, specifies a set of numeric values for θ in H_1. A generalization of this approach that approximates the likelihood based on asymptotic normality has been developed to expand the likelihood paradigm. Another approach is to use the Bayesian model by specifying a prior distribution for the parameter space of θ. The prior provides a mechanism for reducing the composite hypothesis to a simple hypothesis. This new LR is often called a Bayes factor and takes the form $Pr(data | H_1)/Pr(data | H_0) = Pr(\theta | data, H_1)/Pr(\theta | data, H_0) * Pr(\theta | H_0)/Pr(\theta | H_1)$.

A likelihood ratio test (LRT) often takes advantage of the known distributional features of ratios of the likelihood function under null and alternative hypotheses – typically a type of Chi-square or F-distribution. A sequential probability ratio test is a type of LRT performed repeatedly during the accumulation of data. Why use a sequential probability ratio test for drug safety monitoring? Safety data typically accrue over time and often with no fixed end time available to provide a stopping point for a "final analysis", especially in the case of post-approval surveillance of drug safety data. The following LRT-based tests all assume that exposure is known, such as the number observations or treatment exposure time. Because safety risks can involve serious health consequences, monitoring needs a dynamic approach in which data are regularly evaluated as they accrue.

8.3.2 Wald Sequential Probability Ratio Test (SPRT)

We start with the most basic SPRT which uses only a single-valued alternative such as the simple alternative of $p = p_1$ and then cover progressively more complicated SPRT variants. First let us cover the basic terminology for the simple, single alternative hypothesis SPRT. Wald introduced the first SPRT in the context of manufacturing quality testing for addressing the sequential hypothesis testing problem (Wald 1945, Wald 1947). The test consists of a null hypothesis that the safety parameter, p, equals p_0, $H_0: p = p_0$, versus a single point alternative $H_1: p = p_1$. In safety monitoring, observations typically are events/counts, and the parameter may be a type of rate or proportion from a corresponding distribution function. As each new observation arrives, the cumulative log-likelihood ratio based on these hypotheses is updated with the new data point and tested against a "stopping boundary" which defines a threshold beyond which the accumulated data support rejection of the null hypothesis. Additionally, another threshold can also set the value beyond which the LR leads to stopping the monitoring process for the acceptance of the null hypothesis. In the context of safety monitoring, this null threshold usually has less importance than the alternative threshold. The specific threshold values chosen will depend on the desired type I and II errors, and which hypothesis defines the likelihood in the numerator of the LR. Figure 8.1 illustrates the change in cumulative log-likelihood over time for a case where the null gets rejected.

8.3.3 Maximized Sequential Probability Ratio Test (MaxSPRT)

The simple alternative hypothesis of a Wald SPRT limits its utility for safety monitoring. The simple alternative of SPRT can lead to difficulties setting up and interpreting hypotheses and test results because it may not be possible to decide on excess risk beforehand and specify a simple alternate hypothesis. For instance, with $p_0 = 0.1$ and $p_1 = 0.2$ a test result can more often accept the null than with $p_0 = 0.1$ and $p_1 = 0.12$, see Figure 8.1 in Kulldorff (2011). It shows that depending on the assumed alternative, the log of the LRT for the basic SPRT will over time drift more slowly toward the alternative or potentially even go in the opposite direction.

One variation of the SPRT resolves this issue by expanding the test to allow for a composite alternative hypothesis such as $H_1: p > p_0$. The

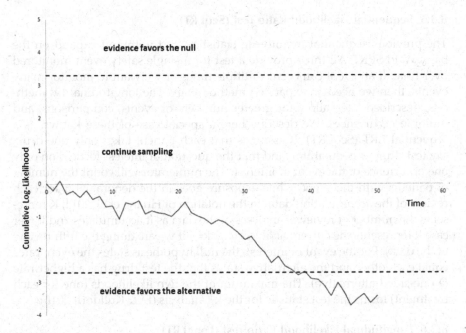

FIGURE 8.1
Cumulative log-likelihood by time for SPRT

MaxSPRT does this by calculating test statistic as the maximum likelihood under the composite alternative divided by the likelihood under the null which is worked out for the cases of Poisson and binomial data distribution in Kulldorff 2011 (see also Kulldorff & Silva 2017). The null hypothesis is rejected if the test statistic reaches the critical value before a pre-defined upper limit on the length of surveillance is reached. For the MaxSPRT, the user needs to specify the length of surveillance, defined as the expected number of events as a function of time. The main advantages of MaxSPRT over SPRT are that it 1) has a limit on the length of surveillance and 2) allows for the composite alternative.

8.3.4 Conditional Maximized Sequential Probability Ratio Test (CMaxSPRT)

In many safety monitoring applications, we may not know the expected number of events as a function of time because of limited or no data on rare events from pre-approval studies or other historical data sources. To address this lack of information, an adjustment to the SPRTs can account for information gained over time as monitoring progresses. CMaxSPRT combines historical data with accumulating data from monitoring to better define event rates over time by incorporating randomness and variability from the historical data and the surveillance population (Li & Kulldorff 2010).

8.3.5 Sequential Likelihood Ratio Test (SeqLRT)

The previous sequential monitoring statistics start with and expand on the basic Wald SPRT. All three provide a test for a single safety event monitored over time. If an event can occur multiple times or we want to monitor many events, then we need an expanded methodology. The longitudinal LR methods, described later, allow for greater numbers of events, comparisons, and multiple occurrences. We describe here a special case of these known as a sequential LRT (SeqLRT). It assumes that each patient takes only one drug, say test drug or comparator, and that the question of interest focuses on only one occurrence of the event of interest. The numerator will count the number of patients with the event, while exposure goes in the denominator for each review of the accumulating data. In the notation of Huang et al. 2014, K represents the number of reviews/analyses of the data as it accumulates and lower case k represents the current analysis (e.g. k=5 if we are doing the fifth review of the data). For the event of interest, the null hypothesis states the event rates equal each other and the alternative states that the test drug has a higher rate (a one-sided alternative). The maximum of the two likelihoods (one for each treatment) forms the test statistic for the kth analysis (Li & Kulldorff 2010).

8.3.6 Longitudinal Likelihood Ratio Test (LongLRT)

The SeqLRT is a special case of the LongLRT methods which is obtained by restricting to one comparison of drug and control for one event of interest. LongLRT methods do not restrict to a single AE per patient (either in type or recurrence) or one treatment comparison. The basic layout for the LongLRT assumes a sequence of I by J tables. For any given review of the accumulating data, the I rows represent the events under analysis, and the columns represent the J different treatments. Conceptually, one I by J table contains the number of events for each i, j combination observed up to analysis time k. A second I by J table contains information on the exposure up to the time k. The definitions of events and observation times form a key feature of the LongLRT approach. In their paper, Huang et al. (2014), they define drug exposure for each patient as the time from the start of treatment to either the event or end of treatment, whichever comes first. Importantly, if more than one event occurs during a given treatment exposure period, then the exposure will depend on the time each event occurs. For example, if a patient takes drug j for 120 days and experiences event 1 on day 15 and event 2 on day 40, then the patient contributes to the total exposure period for row 1 in the exposure table an amount of 15 days; and 40 days for event 2. If the same patient never experiences event 3 while taking drug j, then the patient contributes an exposure value of 120 days for row 3 total in the exposure tables. Similar logic applies to patients exposed to more than one treatment and to recurrences of any given event. In short, the drug exposure can be defined as 1) event time, 2) person time, or 3) exposure time.

Based on the proposed definitions for LongLRT, it can be applied to repeated tests of differences in the rates of occurrence of various safety events for a given treatment (when looking for events that rise above the background event noise) or to differences between treatments for a given safety event or recurrent safety events.

8.3.7 Sequential Generalized Likelihood Ratio Test (SGLRT)

The previous sequential monitoring statistics start with and expand on the basic Wald SPRT. The preceding tests all consider a simple null hypothesis, when in the safety setting, interest may rest in being at or below a threshold rate. The SGLRT allows for composite null and alternative hypotheses. SGLRT uses a group sequential paradigm with distributions in the exponential family of distributions, allowing the computation of stopping boundaries for both futility and rejection of the null (Lai & Shih 2004, Shih et al. 2010). The use of boundaries can aid in reducing the sample size if the difference in rates is larger than planned.

The SGLRT and LongLRT are not overlapping in their generalizability in which the SGLRT is not more general than the LongLRT in handling complicated comparisons, but it does allow for a compound null hypothesis. It can be extended to treat more complex composite null and alternate hypotheses involving multivariate parameters. For convenience, the SGLRT is placed as the end of the following table.

8.3.8 Summary of Likelihood Ratio Methods

The following table, Table 8.1, describes the key features of each test covered in this section. Each method addresses a certain type of question as indicated in the right-hand column of the summary table. This in turn corresponds to null and alternative hypotheses. Finally, each method has some assumptions about the data to which it applies, briefly described in the second column from the left.

The references for these methods are noted in the text of the section that describes each of them. Additional, in depth information of LRT methods in safety applications can be found in the text by Zalkikar, Huang, and Tiwari (2021).

8.4 Examples of Several Likelihood-Based Methods in Safety Monitoring

This section outlines three examples of LRT methods used in safety monitoring. We encourage the reader to examine the source references that will further aid in understanding important details for using them. Here, we aim merely to generate interest in how the methods can work in reality.

TABLE 8.1

LRT Methods Used for Sequential Monitoring of Safety Events. (Each row summarizes a procedure used for analyses at monitoring time k.)

Method	Data Assumptions	Null Hypothesis	Alternative Hypothesis	Question Associated with Hypotheses
SPRT	One treatment, one event per patient; no comparator	H_0: $p = p_0$ Test drug rate is p_0	H_1: $p = p_1$ Test drug rate is p_1	Does treatment have an event rate similar to historical/background rate versus a single given/assumed rate?
MaxSPRT Assumes knowledge of expected number of events as a function of time	One treatment, one event per patient; no comparator	H_0: $p = p_0$	H_1: $p > p_0$ Test drug rate is greater than p_0	Does treatment have an event rate similar to historical/background rate versus a rate greater than the background rate?
CmaxSPRT Allows updating of the expected number of events as a function of time	One treatment, one event per patient; no comparator	H_0: $p = p_0$	H_1: $p > p_0$ Test drug rate is greater than p_0	Same question as MaxSPRT but allows for updating with new information
SeqLRT	One treatment and one comparator; one event per patient	H_0: $p_1 = p_2$ Test drug and comparator have the same rate	H_1: $p_1 > p_2$ Test drug rate is greater than the comparator	Does treatment have an event rate similar to control treatment rate versus a rate greater than the control treatment rate?
LongLRT Case of differing rates across events for a given drug	One or more treatments, one or more events per patient	H_0: $p_1 = p_0$ All events have the same rate for treatment	H_1: $p_i > p_0$ Rates differ for at least one event for treatment	Do different events occur at similar rates for a given treatment?
LongLRT Case of differing rates across drugs for a given event	One or more treatments, one or more events per patient	H_0: $p_1 = p_0$ All treatments have the same rate for event	H_1: $p_j > p_0$ At least one drug has a rate greater than others, for event	Do different treatments have similar rates for a given event?
Sequential Generalized LRT Composite H_0 and H_1	One treatment or treatment vs control for one event	H_0: $p \le p_0$ Treatment has rate same or less than control	H_1: $p > p_0$ Treatment has rate greater than control	Does treatment have the same or lower event rate than control treatment rate versus a rate greater than the control rate? Control can be replaced by a historical/background rate.

8.4.1 Monitoring for Fever and Neurological Effects of a Vaccine

Randomized controlled trials of drug treatments are primarily designed to provide reliable information for regulatory approval or establishing medical standards of practice. However, for some safety questions, a single-arm study may suffice to answer the question of whether the rate of an event's occurrence is low enough to use the therapy safely in a large population, such as the case with vaccines. This can work quite effectively in the case where the event under question has a well-quantified historical rate of occurrence in the target population. In Kulldorff & Silva (2017), the MaxSPRT, under a Poisson model, is applied to vaccine safety data for monitoring fever and neurological events. The data sources come from various observational sources but are modified to mimic a prospective source. For comparison with the standard SPRT, the authors examine the time until an event is declared to have a relative risk greater than one or safely ruled to have no elevation in relative risk. As anticipated, the MaxSPRT detects more quickly the event than the SPRT; in the case of fever, the MaxSPRT rejects the null after 13 weeks of accumulated data compared to a range from 13 weeks to as long as 36 weeks for the SPRT. The range for the SPRT reflects the need to choose a single-point alternative. If the SPRT is applied with an alternative of two or greater, then it fails to detect the fever and stops monitoring because it incorrectly finds overwhelming evidence in favor of the null.

8.4.2 Rotavirus Vaccine Monitoring for a Serious Intestinal Event

This next example (Shih et al. 2010) applies an LRT method to a study based on learnings from a similar therapy that was withdrawn from the market for safety concerns. Intussusception is a rare but serious disorder of the intestinal tract which can block the passage of food or fluids (https://www.mayoclinic.org/diseases-conditions/intussusception/symptoms-causes/syc-20351452). The tetravalent rhesus–human reassortant rotavirus vaccine was withdrawn from the market in 1999 after a post-approval safety surveillance study using the VAERS database found an increase in the risk of intussusception (Murphy et al. 2001). As a result of this earlier issue, another rotavirus, a pentavalent human-bovine reassortant rotavirus vaccine (RV5), added an SGLRT approach to its pre-approval, blinded, placebo-controlled study. Some of the key features of that approach used in the study known as the rotavirus efficacy and safety (REST) study are described by Vesikari et al. (2006).

In the REST study, the null hypotheses stated that RV5 had the same or lower intussusception rate than placebo and the alternative hypothesis stated that it was greater than placebo (Heyse et al. 2008; Shih et al. 2010). Assuming a Poisson process for each treatment, the study design allowed for the calculation of stopping boundaries which effectively could either accept the null hypothesis of a neutral or favorable difference or reject it and conclude intussusception occurred more frequently in the vaccine arm of the study. A data safety monitoring board (DSMB) reviewed the data of adjudicated cases of Intussusception. The DSMB followed the pre-specified stopping boundaries and reached a conclusion that the vaccine did not have a higher event rate than placebo. See Heyse (2008) for an in-depth description of the study's statistical characteristics.

8.4.3 Example of Possible Applications of LongLRT

In this last example, the researchers aim to demonstrate testing for multiple drugs in a setting more like that of a dataset with many drugs and many events. The application of LongLRT methods by Huang (2014) demonstrates their combination of flexibility and power using pooled clinical trial data that contain safety events for osteoporosis treatments alone (drug) or in combination with a proton pump inhibitor (PPI) drug (drug +PPI); placebo is treated as a "drug", not a PPI. In one example, a composite event is followed for k=1, ..., 5 reviews of the accumulating data. An interesting point in the example is the ability of the method to incorporate repeated occurrences of the event. The LongLRT analysis finds that relative risks are higher for all drug+PPI groups than drug alone.

8.5 Comments on Other Methods and Software for LRT Methods

Although this chapter primarily covers LRT methods, the methods in section 8.3 typically require some "denominator" which usually takes the form of an exposure time of some sort. Many non-LRT methods have arisen to handle similar data for single or multiple events at a time with monitoring in a dynamic fashion.

8.5.1 Non-LR Methods for Pre- and Post-Marketing Surveillance

Methods originally intended for use on data without exposure information can sometimes still apply to data with exposures. The proportional reporting ratio and multi-item gamma Poisson shrinker are two such methods. The Bayesian confidence propagation neural network is another flexible method that can apply to data as described here. Hu (2015) developed a Bayesian method using a Poisson-Dirichlet process to handle data without exposures and applied it to the FDA's FAERS data. Finally, and importantly, the false discovery rate (Mehrotra & Heyse 2004) is used in safety surveillance and determines events of importance through the application of the multiplicity methods such as Benjamini and Hochberg (1995). For multiple looks at data, LRT methods naturally handle that multiplicity; however, they may not be set up to handle multiple endpoints. Some such as LongLRT should be able to handle both types of multiplicity.

As noted by Blume (2002), when one compares frequentist and Bayesian approaches, a common feature for both approaches is that they include more information in their interpretation of the data than what the data provide. Bayesian inference includes prior information, whereas frequentist inference includes information about the sample space. Likelihood-based approaches split the difference between the frequentist and Bayesian approaches. The approach provides a framework for presenting and evaluating LRs as measures of statistical evidence for one hypothesis over another. Although LRs figure prominently in both frequentist and Bayesian methodologies, they are neither the focus nor the endpoint of either methodology. The key difference is that, under the evidential paradigm, the measure of the strength of statistical evidence is decoupled from the probability of observing misleading evidence. As a result, controllable probabilities of observing misleading or weak evidence provide quantities analogous to the type I and type II error rates of hypothesis testing, but do not themselves represent the strength of the evidence in the data. Likelihood methods characterize the data as evidence with respect to pairs of simple statistical hypotheses. Bayesian methods use the data to update a probability distribution over the hypothesis space. Both the likelihood and Bayesian methods conform to the likelihood principle and fit together nicely.

8.5.2 A Brief Note on Software for LRT Methods

The use of the LRT methods presented in this chapter can be facilitated through various tools and software resources. These include standalone programs developed for a specific method(s) and/or statistical packages developed for performing one or more of these methods. Standalone programs can be developed using general programs such as SAS, R, and Python and are sometimes made available by the authors of the methods. An example of a standalone program can be seen in Hutson 2003 who presented a SAS program for implementing a modification of the SPRT for flagging a significant increase in the mortality rate of a treatment relative to a control. Another example of a standalone program was presented by (Nieminen 2017) who developed an R program that implements a method, the BmaxSPRT (MaxSPRT in the binomial case) method, which considers hypotheses concerning the relative incidence of AEs during specified risk and control periods. A more recent approach to implement the methods is to leverage packages that are available for performing these analyses along with visualizations. In R, these packages include the *Sequential, SPRT, MSPRT,* and *sglr* packages. The packages have different degrees of functionality and complexity. For example, the *Sequential* package includes functions to calculate exact critical values, statistical power, expected time to signal, and required sample sizes for performing exact sequential analysis which can be done for either Poisson or binomial data, for continuous or group sequential analyses, and for different types of rejection boundaries. In both R and Python, one can also leverage interactive functionality by putting packages as a backend to R Shiny and Dash in Python. As an example, for the *Sequential* package, an online web-based interface is available where a user conducts analyses implemented in this package without installing and writing any code. Interactive functionality can also be seen in the *sglr* which implements functions for computing power and boundaries for pre-licensure vaccine trials using the generalized LR tests and the *openFDA* tool which implements LRT for common events for a drug and vice versa. Additional details of the various tools and software resources are provided online at https://community.amstat.org/biop/workinggroups/safety/safety-home.

8.6 Summary and Discussion

The likelihood principle provides a framework for sequential analyses of accumulating safety data. In the case of accumulating clinical trial data, it can be applied to an individual study or a combination of studies as each completes. The methodologies can also be applied in the post-marketing stage, either with spontaneous safety reporting (for some methods such as LongLRT) or in electronic health information from the Sentinel network. The design of the studies can affect the degree of causal inference, but this does not affect the application of likelihood methods.

It is worth mentioning that the LRT has a special application as the TreeScan method for safety signal detection in the Sentinel system. The TreeScan method leverages the hierarchical structure in the AE coding system. Under the MedDRA dictionary, the system organ class and preferred term form the basic structure of the tree. The LR statistic is calculated for each branch of the tree under the null and alternative hypotheses. These can be used to scan the tree and detect those AEs that have an elevated probability to occur beyond the play of chance. More details of the TreeScan method will be provided in Chapter 11.

In this chapter, we cover six progressively more general forms of likelihood-based sequential monitoring methods. The relative strengths and weaknesses of each essentially relate to their complexity with methods, such as LongLRTs, having greater applicability and better statistical behavior, but, at the cost of less simple computation. Each method has its utility and appropriateness for a given application, but the standard SPRT perhaps has the least utility given the advances with the others.

References

Benjamini Y., and Hochberg Y., 1995, Controlling the false discovery rate - a practical and powerful approach to multiple testing. *Journal of the Royal Statistical Society. Series B (Methodological)* 57: 289–300.

Blume J.D., 2002, Likelihood methods for measuring statistical evidence. *Statistics in Medicine* 21 (17): 2563–2599.

Goodman S.N., 1999, Toward evidence-based medical statistics. 1: The p value fallacy. *Annals of Internal Medicine* 130 (12): 995–1004.

Hacking I., 1965, *Logic of statistical inference*, Cambridge University Press.

Heyse J., Kuter B., Dallas M., and Heaton P., for the REST Study Team, 2008, Evaluating the safety of a rotavirus vaccine: the REST of the story. *Clinical Trials* 5: 131–139.

Hu N., Huang L., and Tiwari R., 2015, Signal detection in FDA AERS database using Dirichlet process, *Statistics in Medicine* 34: 2725–2742.

Huang L., Zalkikar J., and Tiwari R., 2014, Likelihood ratio-based tests for longitudinal drug safety data. *Statistics in Medicine* 33: 2408–2424.

Hutson D.A., 2003, A sequential test procedure for monitoring a singular safety and efficacy outcome. *Tropical Journal of Pharmaceutical Research* 2: 197–206.

Kulldorff M., David R., Kolczak M., Lewis E., Lieu T., and Platt R., 2011, A maximized sequential probability ratio test for drug and vaccine safety surveillance. *Sequential Analysis* 30(1): 58–78.

Kulldorff M., and Silva I., 2017, Continuous post-market sequential safety surveillance with minimum events to signal. *REVSTAT* 15: 373–394.

Lai T., and Shih M., 2004, Power, sample size and adaptation considerations in the design of group sequential clinical trials. *Biometrika* 91(3): 507–528.

Li L., and Kulldorff M., 2010, A conditional maximized sequential probability ratio test for pharmacovigilance. *Statistics in Medicine* 29: 284–295.

Mehrotra D., and Heyse J., 2004, Use of the false discovery rate for evaluating clinical safety data. *Statistical Methods in Medical Research* 13: 227–38.

Murphy T.V., Gargiullo P.M., Massoudi M.S., Nelson D.B., Jumaan A.O., Okoro C.A., Zanardi L.R., Setia S., Fair E., LeBaron C.W., Wharton M., and Livengood J.R., Rotavirus Intussusception Investigation Team. 2001, Intussusception among infants given an oral rotavirus vaccine. *The New England Journal of Medicine* 344: 564–572.

Nieminen T., 2017, Vaccine safety surveillance with self-controlled study designs, Master's Thesis, University of Helsinki, https://github.com/TuomoNieminen/Thesis.

Po A.W., and Zhang W., 1998 What lessons can be learnt from withdrawal of mibefradil from the market? *Lancet* 351(9119): 1829–1830.

Royall R., 1997 *Statistical evidence: a likelihood paradigm*, Vol. 71. Chapman and Hall/CRC Press, London.

Royall R., 2000 On the probability of observing misleading statistical evidence. *Journal of the American Statisical Association* 95(451): 760–768.

Shih M., Lai T., Heyse J., and Chen J., 2010,Sequential generalized likelihood ratio tests for vaccine safety evaluation. *Statistics in Medicine* 29: 2698–2708.

Singh S., and Loke Y.K., 2012, Drug safety assessment in clinical trials: methodological challenges and opportunities. *Trials* 13(1): 138.

Strober B., Karki C., Mason M., Ning G., Holmgren S., Greenberg J., and Lebwohl M., 2018, Characterization of disease burden, comorbidities, and treatment use in a large, US-based cohort: Results from the Corrona Psoriasis Registry. *Journal of the American Academy of Dermatology* 78(2): 323–332.

Szarfman A., Machado S., and O'Neill R., 2002, Use of screening algorithms and computer systems to efficiently signal higher-than-expected combinations of drugs and events in the US FDA's spontaneous reports database. *Drug Safety* 25(6): 381–392.

Vesikari T., Matson D., Dennehy P., Van Damme P., Santosham M., Rodriguez Z., Dallas M., Heyse J., Goveia M., Black S., Shinefield H., Christie C., Ylitalo S., Itzler R., Coia M., Onorato M., Adeyi B., Marshall G., Gothefors L., Campens D., Karvonen A., Watt J., O'Brien K., Dinubile M., Clark H., BosJego J., Offit P., and Heaton P., for the (REST) Study Team. 2006, Safety and efficacy of a pentavalent human–Bovine (WC3) reassortant rotavirus vaccine. *The New England Journal of Medicine* 354: 23–33.

Vestbo J., Leather D., Bakerly N., New J., Gibson J.M., McCorkindale S., Collier S., Crawford J., Frith L., Harvey C., Svedsater H., Woodcock A., for the Salford Lung Study Investigators 2016, Effectiveness of fluticasone furoate–vilanterol for COPD in clinical practice. *The New England Journal of Medicine* 375: 1253–60.

Wald A., 1945, Sequential tests of statistical hypotheses. *Annals of Mathematical Statistics* 16: 117–186.

Wald, A., 1947, *Sequential Analysis*. New York: Wiley.

Wasserstein R.L., and Lazar N.A., 2016, The ASA's statement on p-values: Context, process, and purpose. *The American Statistician* 70(2): 129–133.

Zalkikar J., Huang L., and Tiwari R., 2021, *Signal Detection for Medical Scientists: Likelihood Ratio Based Test-Based Methodology*. Chapman & Hall/CRC Biostatistics Series, Boca Raton.

9

Meta-Analysis of Drug Safety Assessment

Rositsa B. Dimova
Center for Biologics Evaluation and Research, US Food and Drug Administration, Silver Spring, MD, USA

Sourav Santra
Sarepta Therapeutics Inc., Cambridge, MA, USA

Melvin Munsaka
AbbVie, Inc., North Chicago, IL, USA

CONTENTS

9.1 Introduction 188
9.2 Planning of a Meta-Analysis for Safety Assessments 189
9.3 Individual Participant and Aggregated Data Meta-Analysis
    of Safety 191
9.4 Meta-Analysis for Rare Events 195
    9.4.1 The Case of Rare Events 195
    9.4.2 Methods for Dealing with Rare Events 196
9.5 Multivariate Safety Endpoints 201
    9.5.1 Why Multivariate Meta-Analysis? 201
    9.5.2 Defining Multivariate Meta-Analysis 201
    9.5.3 Multivariate Random Effects Model 202
    9.5.4 Statistical Challenges in Multivariate Meta-analysis 202
9.6 Cumulative and Trial Sequential Meta-Analysis 204
    9.6.1 Cumulative Meta-Analysis 204
    9.6.2 Trial Sequential Meta-Analysis 204
9.7 Reporting of Meta-Analysis 206
9.8 Visualizing Meta-Analysis Results 206
9.9 Analysis Tools and Software for Meta-Analysis 206
9.10 Conclusions 207
References 208

Disclaimer: This manuscript reflects solely the views of the author and should not be construed to represent FDA's views or policies.

DOI: 10.1201/9780429488801-11

9.1 Introduction

Safety data are collected systematically during the pre-market drug development phase, as well as during the post-marketing phase. The goal for these collections is usually related to the assessment of the benefit-risk profile of the medical products under consideration, or for safety surveillance purposes, or for evaluations of suspected associations of the use of a specific product and certain adverse events (AEs). Due to the availability of multiple safety data sources, for example, as a result of multiple studies of safety of a product of interest being conducted, the need for data synthesis, also referred to as meta-analysis, arises. There are two major classes of data sources that are used for performing a meta-analysis. These include: (i) summarized data (such as treatment effects/risks) extracted from the published literature and (ii) subject-level data from the conducted studies, which are mainly available to those generating the data (e.g. study sponsors) or to the regulatory agencies. While meta-analysis techniques have been frequently applied for the synthesis of efficacy data, their application to safety data has not been that extensive. For example, Jacobson, Targonski, and Poland (2007) surveyed the published literature for meta-analyses in vaccinology and found that among the 134 published meta-analyses of vaccines against infectious diseases that they identified, approximately 80% evaluated efficacy, effectiveness, or immunogenicity, while only approximately 13% analyzed safety. Some of the major challenges to the conduct of meta-analysis using safety data include the lack of consistency of the definitions of the AEs across the different studies, the variability in the exposure and observation times, as well as the poor reporting, or lack of it, in the published literature (for literature based meta-analyses). For illustration, Stassijns et al. (2016) conducted a systematic review and meta-analysis of safety of newly adjuvanted vaccines in comparison to other vaccines in children and reported that the serious adverse events (SAEs) were collected for the duration of the trials, which varied from a few months up to two years. Such heterogeneity in the synthesized data may have great impact on the conclusions of the meta-analysis. Additionally, in the published literature, the reporting of safety data is usually less comprehensive than that of efficacy. Some study publications report AEs selectively, while others even omit reporting of any AEs, possibly due to low event rates or even due to publication size restrictions from the journals, which ultimately leads to reporting bias.

In 2016, the Council for International Organizations of Medical Sciences (CIOMS) Working Group X published a report entitled "Evidence Synthesis and Meta-analysis for Drug Safety," which discusses principles related to the design, analysis, reporting, and interpretation of meta-analyses of safety data.

In 2018, the U. S. Food and Drug Administration (FDA) issued a Draft Guidance for Industry "Meta-Analyses of Randomized Controlled Clinical Trials to Evaluate the Safety of Human Drugs or Biological Products Guidance for Industry." This draft guidance outlines a general framework for evaluation of the level of evidence that meta-analyses of safety provide in the FDA regulatory setting. Of note, it addresses only meta-analyses of randomized controlled trials (RCTs) due to the difficulties associated with the interpretation of meta-analyses that include non-RCTs. This guidance puts emphasis on the importance of the quality of the studies that are included in the meta-analysis. While observational studies are more prone to bias, due to both known and unknown factors, they may be of use for investigations of certain rare AEs that are less likely to be detected in RCTs. Additionally, certain events, such as some autoimmune conditions, require longer follow-up time, which is not always feasible with RCTs. Synthesizing the evidence from observational studies in vulnerable populations such as children, pregnant women, or those with serious medical conditions who are not typically studied in RCTs may also be informative. Meta-analyses of non-RCTs may also be useful for the purpose to inform the design of a new study.

The goal of this chapter is to provide an overview and discussion on the process of conducting a meta-analysis using safety data for safety assessments, some of the methods that are being used, and some difficulties that may be encountered, as well as approaches how these difficulties may be reduced or overcome. For a deeper insight into the concepts and methods that are discussed, the reader is referred to the cited references. Specific topics considered in this chapter are the following. Section 9.2 discusses the role of planning of meta-analysis for safety assessments that determines the quality of the meta-analysis study. Section 9.3 provides an overview of the two main classes of meta-analysis, namely, the individual participant data (IPD) meta-analysis and the aggregated data meta-analysis. Section 9.4 discusses the most frequently used approaches for meta-analysis of rare safety endpoints. In Section 9.5, we provide an overview of multivariate meta-analysis approaches. Section 9.6 discusses cumulative and trial sequential methods. Section 9.7 discusses the importance of reporting the methods, conduct, and results of meta-analysis studies according to established principles for transparency. Sections 9.8 and 9.9 briefly highlight some visualization and analysis tools and software for meta-analysis, respectively.

9.2 Planning of a Meta-Analysis for Safety Assessments

As with original clinical trials, the planning procedure of a meta-analysis plays a significant role in its conduct, the final results, and conclusions. This procedure requires the development of a study protocol and statistical analysis plan prior to initiation of the meta-analysis. An important component of the meta-analysis planning procedure is the specification of the outcomes of interest and the respective estimands. The safety outcomes collected in the individual studies may be different for different therapeutic areas and medical product classes. In contrast to efficacy endpoints, which are typically pre-defined prior to initiation of the study, most safety events, especially those that are unexpected, are not pre-specified in the clinical trials and may not be known a priori. In general, the safety outcomes that are collected have several attributes. These include:

(i)   the nature of their collection (active or passive);

(ii) the timing of collection with respect to exposure;

(iii) the duration of collection (follow-up);

(iv) the specific definition of the safety event, including grading.

Each of these attributes can be a source of heterogeneity between the studies considered for inclusion in the meta-analysis. Therefore, the inclusion criteria need to specify appropriate values or a range of values for each of these attributes. A very wide range of values for an attribute, while increasing the sample size, may result in inclusion of very heterogeneous studies, which may not be synthesizable. For literature-based meta-analysis, it may be challenging to determine all of the attributes of the safety outcomes of interest for some studies, which may result in exclusion of these studies. In this respect, various sensitivity analyses may be useful.

Regarding the first attribute specified above, the safety events are referred to as solicited (actively collected) or unsolicited (passively collected). In vaccine studies, the solicited events that are typically collected are the local injection site reactions (such as pain, swelling, and redness) and systemic reactions (such as fever, headache, and malaise), and are usually collected for up to seven days after each vaccine dose exposure. The unsolicited events are not pre-defined and are usually collected for a pre-specified period of time, for example, throughout the duration of the trial. Another safety outcome, experience of at least one SAE, refers to a category of events and is always collected in studies intended for regulatory purposes. The SAEs are usually defined as life-threatening events or events that result in persistent disability, hospital admission, or death and are collected throughout the duration of the trial. Some trials may also collect adverse events of special interest, for which prior information suggests that a potential risk may exist.

The Addendum R1 to the International Council on Harmonization (ICH) E9 Guideline introduces the concept of estimand, that is "a precise description of the treatment effect reflecting the clinical question posed by the trial objective." The estimand for a study objective is described in the addendum to include the following components: (i) the treatment(s) of interest; (ii) the population; (iii) the variable (endpoint) that corresponds to the objective; (iv) the specification of how to account for intercurrent events; and (v) the population-level summary for the variable. The intercurrent events are specified as "events occurring after treatment initiation that affect either the interpretation or the existence of the measurements associated with the clinical question of interest." Examples of these events are the use of rescue medication, discontinuation of treatment, and death. The addendum describes five strategies to account for intercurrent events, namely, treatment policy strategy; composite strategy; hypothetical strategy; principal stratum strategy; and while on treatment strategy. The strategy, in general, would depend on the trial objective and on the medical product that is evaluated. Thus, the starting point for defining an estimand is the scientific question that is derived from the targeted study objective. While the ICH E9 (R1) guideline refers to individual studies, the same principles are applicable to the planning of meta-analyses. Estimands for safety assessments are discussed in detail in Chapter 14.

An additional point for consideration, when planning a meta-analysis, is the possibility for some of the individual studies, that are assessed for inclusion, to have been affected by unplanned disruptions. Examples of such disruptions are extreme events, such as the global COVID-19 pandemic. These may affect the overall study conduct and consequently the quality of the data collected during such disruptions. Specifically, the effect of such events on the studies may manifest as modifications of the initially planned study procedures (e.g. televisits instead of in-person visits), intercurrent events, missing data, poor or lack of ascertainment of the efficacy, and safety outcomes. Subsequently, these may potentially lead to heterogeneity between the individual studies' results. Therefore, such considerations, if applicable, are important and need to be addressed.

9.3 Individual Participant and Aggregated Data Meta-Analysis of Safety

The most widely used form of meta-analysis synthesizes summarized (aggregated) results (e.g. treatment effects) extracted from published studies. While this form of meta-analysis is very accessible, it may be subject to publication bias, poor and limited reporting of the data of interest, and limited ability to assess for treatment effect interactions. These difficulties are especially evident in the literature-based meta-analyses of safety data due to primarily selective or poor reporting of AEs in the published literature. In addition, in comparison to efficacy data where studies with significant results tend to be more likely to be submitted and published, the publication bias regarding safety data may be in the opposite direction, i.e. significant safety findings may be under-reported in the literature. An alternative approach, which has gained increasing popularity in recent years, is the IPD meta-analysis. This form of meta-analysis synthesizes subject-level data from multiple data sources (studies). Thus, access to individual subjects' data is required for this type of meta-analysis. The FDA, as a regulatory agency, has access to subject-level data submitted by pharmaceutical companies as part of the drug approval application process. However, this is often not the case for other parties conducting research.

The 2018 FDA Draft Guidance for Industry on meta-analyses intended for evaluation of safety of medical products discusses the level of evidence a meta-analysis of safety data provides in the regulatory setting. Accordingly, the highest level of evidence is assigned to a meta-analysis that "is prospectively planned prior to the conduct of the trials to be included, and where the component trials are designed with the meta-analysis objectives in mind. The trials have well-ascertained outcomes and exposure periods, and subject-level data are available for analysis." The draft guidance recognizes that this level may be hard to achieve, nevertheless it considers it as a useful "benchmark" for evaluation of the quality of meta-analysis studies. The next level in the hierarchy of evidence from a meta-analysis of safety, described in the draft guidance, is "a prospectively planned meta-analysis based on existing trials that were designed for other purposes but for which the quality of the data and the ascertainment of outcomes and exposure are adequate to support the planned analysis." This level of evidence also requires that the meta-analysis study plan and the decisions for inclusion of the screened studies be made without knowledge of the study outcome. Finally, the lowest level of evidence, specified in the draft guidance, corresponds to meta-analyses "for which prospective planning did not occur, or is in doubt, study outcomes and trial inclusion decisions were made with outcome data in hand, and one or more of the important quality factors is in question." The draft guidance also specifies that meta-analyses based only on study-level (aggregated) summary data are considered of lower evidence level than meta-analyses based on subject-level data. Nevertheless, it recognizes that the aggregated data meta-analyses may be appropriate in situations in which the outcome is "well-ascertained" (such as mortality) or for hypothesis generation.

The majority of the safety outcomes represent occurrence of AEs and are usually represented through binary variables. Thus, the treatment effects on safety are mainly expressed as a risk difference (RD), risk ratio (RR), odds ratio (OR), or hazard ratio (HR). A typical aggregated data random effects meta-analysis model is expressed as follows (e.g. Hardy and Thompson 1996):

$$\hat{\theta}_i = \theta_i + \varepsilon_i, \ i = 1,\ldots,m$$

$$\theta_i \sim N\left(\theta, \tau^2\right)$$

$$\varepsilon_i \sim N\left(0, \sigma_i^2\right), \tag{9.1}$$

where $\hat{\theta}_i$ is the observed treatment effect in the i-th study (e.g. the log OR), θ_i is the i-th study mean effect, σ_i^2 is the within study variance, θ is the overall treatment effect, and τ^2 is the between study variance (the heterogeneity parameter). The population parameter θ can then be estimated as the weighted average

$$\hat{\theta} = \frac{\sum_{i=1}^{m} w_i \hat{\theta}_i}{\sum_{i=1}^{m} w_i}, \tag{9.2}$$

where $w_i = 1/(\hat{\sigma}_i^2 + \hat{\tau}^2)\hat{\sigma}_i^2$ is an estimate of the within study variance, and $\hat{\tau}^2$ is an estimate of the between study variance, for example, based on the REML or the DerSimonian-Laird estimators (DerSimonian and Laird 1986).

When $\tau^2 = 0$, the model (9.1) reduces to a fixed effects model. In addition to these inverse variance weight approaches, other fixed effects approaches for binary outcomes include the Mantel–Haenszel method (Mantel and Haenszel 1959; Greenland and Robins 1985) and the Peto method (Yusuf et al. 1985). These are especially useful for sparse data (rare events). We discuss methods for rare events in the next section.

Often, in meta-analyses, when heterogeneous treatment effects are observed, an attempt is made to explain the differences or to investigate for subgroup effects based on study-level variables. Respectively, the model (9.1) is extended into a meta-regression model by including study-level covariates, such as study characteristics (e.g. drug dose) or aggregated subjects' characteristics (e.g. proportion of male subjects or mean age). However, it has been shown (Lambert et al. 2002) that the meta-regression approach has low power to detect interactions and is prone to aggregation bias (ecological fallacy).

The IPD meta-analysis approaches, on the other hand, have a number of advantages over the traditional aggregated data meta-analyses. These include the ability to adjust for subject-level characteristics (such as demographics, baseline-level variables, etc.), to investigate for subject-level interactions, as well as the possibility for standardization of the outcome definitions across the different studies, subgroup analyses, etc. Nevertheless, in some situations, the IPD meta-analysis can still be challenging to conduct, especially in the setting of analysis of safety, as discussed in the previous section.

In general, there are two approaches to IPD meta-analysis. These are the two-stage IPD meta-analysis and the one-stage IPD meta-analysis (Debray et al. 2015; Stewart et al. 2012).

The two-stage IPD meta-analysis model includes the following two steps. First, study-specific treatment effects are produced using the IPD from each study. Second, the study-specific treatment effects are then combined using the traditional methods for aggregated data meta-analysis described above. Although this approach is more easily interpretable, when there are a few studies or a few participants per study, it may be subjected to difficulties in estimating the pooled effects, standard errors, and the between study heterogeneity (Debray et al. 2015). Another disadvantage of the two-stage approach is its lower power in detecting treatment-covariate interactions (Simmonds and Higgins 2007).

In the one-stage IPD approach, all data are analyzed simultaneously using a single statistical model that takes into account the clustered structure of the data. Typically, the one-stage IPD meta-analysis model is expressed as a hierarchical model, such as the generalized linear mixed effects model (e.g. Burke, Ensor, and Riley 2017). An example of a one-stage IPD model for a binary outcome is the following logistic regression random effects model.

$$
\begin{aligned}
\text{logit}\left(\pi_{ij}\right) &= \alpha_i + \theta_i x_{ij} + \beta_i z_{ij} \\
\theta_i &= \theta + u_i; \\
u_i &\sim N\left(0, \tau^2\right), \; i = 1, \ldots, m, \; j = 1, \ldots, n_i,
\end{aligned}
\tag{9.3}
$$

where π_{ij} is the probability of subject j from study i to experience an event, α_i is the study effect, here assumed fixed but estimated for each study, x_{ij} is the treatment/control status, θ_i is the treatment effect, z_{ij} is a baseline covariate, and β_i is the covariate effect. Here, the treatment effect θ_i is assumed random, but it can also be assumed fixed, i.e. $\theta_i = \theta$ for all i. This model can be further extended by assuming a random effect for the intercept, or for the covariate, or by including a treatment/covariate interaction.

Although the one-stage approach offers more flexibility than the two-stage approach, it may be subjected to additional computational complexity. Stewart et al. (2012) conducted a comparison of the two IPD meta-analysis approaches and concluded that they produce similar results. Burke, Ensor, and Riley (2017) noted that in some situations differences in the results between the two methods may arise. They pointed out that discrepancies arise mainly due to differences in the modeling assumptions and estimation methods.

As an example, let us consider the meta-analysis of suicidal events in subjects treated with antidepressants. An IPD meta-analysis of safety of a class of antidepressant drugs was described in the FDA draft guidance on meta-analyses of RCTs for evaluation of safety (FDA 2018). In that IPD (two-stage) meta-analysis, the FDA evaluated the risk of suicidal events in adults treated with antidepressants. The study was prospectively planned on completed RCTs (retrospective data) and was triggered by a prior meta-analysis, conducted by the FDA, that investigated the risk of suicidal events in pediatric subjects treated with antidepressants (Hammad, Laughren, and Racoosin 2006). The study in pediatric subjects showed an association between suicidal behavior and ideation and use of antidepressants, and, based on these results, the FDA added a boxed warning to the labels of all antidepressants on their use in pediatric subjects. For the meta-analysis in adults, the FDA requested patient-level data from RCTs from nine manufacturers of antidepressants. As a result, data from 372 trials were included in the meta-analysis. Because of the availability of patient-level data, it was possible to define the outcome uniformly across the studies and for adjudication. The definition of the outcome required that the event occurred while the subjects were on randomized treatment or within one day of stopping treatment. The FDA conducted various sensitivity and subgroup analyses. Subsequently, as a result of this analysis, the FDA updated the warning on all antidepressants labeling to include young adults in addition to pediatric subjects.

In addition to IPD meta-analyses and aggregated data meta-analyses, there also have been developed approaches that combine both types of data in situations in which subject-level data are available only for a part of the included studies (Sutton, Kendrick, and Coupland 2008; Riley and Steyerberg 2010). Wallach et al. (2020) conducted a meta-analysis study of the risk of cardiovascular (CV) events following treatment with Rosiglitazone using IPD. They inferred that although an increased risk of myocardial infarctions (MIs) was observed across analyses, the strength of the evidence varied. They also noted that the effect estimates were attenuated when summary level data were used in addition to the IPD. They further concluded that because more MIs and fewer CV-related deaths were reported in the IPD than in the summary level data available in literature publications, sharing of IPD might be necessary when performing meta-analyses focused on safety.

9.4 Meta-Analysis for Rare Events

9.4.1 The Case of Rare Events

In order to fully understand the risk associated with the use of a medical product, the occurrences of rare SAEs may be of interest. These are events that occur at low frequency rates (such as <1/1,000). Such assessments that intend to investigate a possible increased risk for occurrence of an event of interest require large clinical trials. These may not always be feasible. Therefore, as an alternative, meta-analysis techniques may be of use in order to combine the data from multiple independent studies (data sources). There are certain challenges in the application of meta-analysis techniques when the endpoint of interest is a rare event, specifically, when one or more of the trials, that are intended for synthesis, report no events in the treatment and/or control arms (zero-events trials).

Most safety outcomes are represented by binary variables and as noted previously, the respective measures of risk are thus expressed as a RD, RR, OR, or HR. When one or both arms in a study have zero events, while the RD can be defined, the relative risk measures and their variances cannot. Several approaches have been proposed and used in the literature to address this issue.

One of the most widely discussed examples in the literature related to meta-analysis of rare safety events is the study by Nissen and Wolski (2007) that assessed the risk for MI and CV mortality following the use of the diabetes drug Rosiglitazone (see also the previous section). This was a study-level literature-based meta-analysis, which identified 48 studies as satisfying its inclusion criteria. The effect measure in this meta-analysis was the OR, and the Peto method was used for data synthesis. Of the 48 studies, six did not report any MIs or deaths from CV causes and were thus excluded by the authors for the reason that the effect measure could not be calculated. Studies that had no events in either group (see Table 3 in Nissen and Wolski 2007) were also excluded. Based on their analysis, Nissen and Wolski (2007) showed a significant increase in the risk of MI and a "borderline significant" increase in the risk of death from CV causes. Later, the same authors (Nissen and Wolski 2010) updated their analysis by including additional studies. That analysis demonstrated an increased risk for MI, but not for CV. Multiple studies have since then been published, re-analyzing the same data but using different analytical approaches (e.g. Shuster, Guo, and Skyler 2012; Böhning, Mylona, and Kimber 2015; Wallach et al. 2020).

9.4.2 Methods for Dealing with Rare Events

In this section, we discuss the most commonly used frequentist methods for meta-analysis that include studies with rare events. Tables 9.1 and 9.2 provide brief summaries of the key characteristics of these methods.

One of the most frequently used approaches intended to account for zero-events studies is the continuity correction approach. Specifically, the constant continuity correction method consists of the procedure of adding a small constant, k, for example, $k = 0.5$, to the events' counts for studies with zero events. The transformed data are then analyzed using traditional methods for meta-analysis. In addition to the constant continuity correction approach, other correction approaches have been proposed in the literature. Such approaches, for

TABLE 9.1

Summary of Frequentist Approaches Not Requiring Continuity Corrections Used in Meta-Analyses of Relative Risk Measures in the Case of Rare Events

Approaches	Performance[a]
1. **Peto** (e.g. Sweeting et al., 2004, Bradburn et al., 2007)	Performance was good for balanced studies when event rates were < 1% and the treatment effect was small to moderate. The bias increased when the imbalance in the sample sizes between the compared arms increased. This method outperformed the MH method (with no or with 0.5 constant continuity correction) for the balanced sample size case.
2. **Mantel-Haenszel** (e.g. Sweeting et al., 2004, Bradburn et al., 2007)	This method outperformed the Peto method when there was imbalance between the treatment arms sample sizes and for larger event rates and treatment effects.
3. **Logistic Regression** (e.g. Sweeting et al., 2004, Bradburn et al., 2007)	This method showed good performance for balanced studies. Its performance was similar to that of the MH method with no continuity correction when there was imbalance between the treatment arms.
4. **Generalized linear mixed effects model** (e.g. Stijnen et al., 2010, Böhning et al., 2015)	This method uses an exact likelihood (binomial, Poisson, etc.) for the event outcome and allows for inclusion of single zero and double zero studies and random effects in the model. Stijnen et al. (2010) showed better performance for various GLMM models versus the traditional inverse variance weighted average with constant continuity correction approach for OR and RR effect measures. Böhning et al. (2015) used a Poisson regression model with random effects and showed, using a real data example, that the results were similar to those obtained with the MH (without continuity correction) approach.

[a] Performance as assessed in the cited publications through simulations based on the bias and the coverage of the 95% CIs of the treatment effect.

example, are the treatment arm correction approach and the empirical correction approach (Sweeting, Sutton, and Lambert 2004). The treatment arm correction approach uses a function of the reciprocal of the opposite arm sample size to correct for zero events, while the empirical correction approach uses the estimated pooled effect size from the rest of the studies (non-zero events studies) included in the meta-analysis. The downside of the continuity correction approaches is that the addition of a constant to the data may affect the results and respectively the conclusions of the meta-analysis. Other popular approaches, which do not require the use of a continuity correction, are the classical Mantel–Haenszel (MH) and the Peto methods, although in some cases continuity corrections have been used with the MH method. Other methods, which also do not require continuity correction for zero-events studies, use statistical models, such as generalized mixed effects models, to model the observed event counts from each study intended for synthesis.

TABLE 9.2

Summary of Continuity Correction Approaches Used in Meta-Analyses of Relative Risk Measures in the Case of Rare Events

1. **Constant Continuity Correction**: Adds a small constant k (e.g. $k = 0.5$) to each cell of the 2x2 contingency tables for the studies with zero events. References: e.g. Sweeting et al. (2004); Bradburn et al. (2007).

Applied to:	Performance:[a]
• Inverse variance weighted average; DerSimonian-Laird	This approach produced higher bias than other approaches (e.g. Peto and Mantel-Haenszel) in most situations that were studied in the referenced publications.
• Mantel-Haenszel	Performance varied when different constants were used and for different levels of imbalance between the sample sizes of the compared arms. Overall, this approach outperformed the inverse variance weighted average method in the referenced publications.

2. **Treatment Arm Correction**: Adds a factor of the reciprocal of the sample size of the opposite treatment arm to the cells. For example, for a study with sample sizes of the treatment and control arms of $n_1 = 200$ and $n_2 = 100$, the treatment arm would be adjusted by a constant $c_1 = k/100$, and the control arm would be adjusted by a constant $c_2 = k/200$. Reference: Sweeting et al. (2004).

Applied to:	Performance:[a]
• Inverse variance weighted average	This approach resulted in poor performance. The bias was substantially higher than that of the MH method using the same continuity correction.
• Mantel–Haenszel	Showed better performance than the inverse variance weighted average approach and the MH with 0.5 continuity correction approach.

3. **Empirical Correction**: Adds continuity corrections, derived from the pooled treatment effect of the studies without zero events, to the cells of the 2x2 contingency tables for the studies with zero events. Reference: Sweeting et al. (2004).

Applied to:	Performance:[a]
• Inverse variance weighted average	This approach resulted in poor performance. The bias was substantially higher than that of the MH method using the same continuity correction.
• Mantel–Haenszel	The performance of this approach was comparable to that of the Peto method in the balanced sample size case. In the unbalanced sample size case, this approach outperformed the Peto method and the MH with constant continuity correction method and was comparable to the MH with treatment arm correction method.

[a] Performance as assessed in the cited publications through simulations based on the bias and the coverage of the 95% CIs of the treatment effect.

A number of publications in the literature examined and compared the performance of the most popular meta-analysis methods for rare events (e.g. Sutton et al. 2002; Sweeting, Sutton, and Lambert 2004; Bradburn et al. 2007; Bhaumik et al. 2012; Shuster, Guo, and Skyler 2012). Sweeting, Sutton, and Lambert (2004) compared through simulations the performance of several classical meta-analysis methods for combining ORs using different continuity corrections. They considered constant continuity corrections (0.5, 0.05, and 0.005), alternative continuity corrections (treatment arm correction and empirical correction), or no continuity correction and various group imbalance ratios (ratio of the numbers of subjects assigned to the treatment and to the control groups, e.g. 1:1, 2:1, 3:1, etc.). They concluded that the inverse variance method with any continuity correction performed consistently poor. They also showed that the Peto method was almost unbiased when the sample sizes of the compared arms were balanced and that the bias increased when the imbalance of the sample sizes increased. Additionally, they concluded that the MH method with the alternative continuity corrections (treatment arm correction and empirical correction) and the logistic regression model performed consistently well.

Similarly, based on a comprehensive simulation study of sparse data, under the assumption of a common treatment effect for all studies (a fixed effect model), Bradburn et al. (2007) showed that the bias in the estimated treatment effect was the greatest for the inverse variance and for the DerSimonian and Laird OR and RD methods and for the MH OR method with a 0.5 zero-cell continuity correction. They also showed that for events with rates <1%, the Peto one-step OR method (Yusuf et al. 1985) had the best performance when the within trials groups' sample sizes were balanced and the treatment effects were not very large. In other situations, they showed that the MH OR method without continuity correction, the exact method, and the logistic regression model outperformed the Peto method. Bradburn et al. (2007) also showed that the RD methods resulted in wide confidence intervals (CIs) and had low statistical power when the events were rare.

Bhaumik et al. (2012) discussed moment-based methods for meta-analysis of sparse data when there is a heterogeneity in the treatment effect across the studies. Through a simulation study, they compared the performance of the simple average (unweighted) with a constant continuity correction of 0.5 estimate of the overall OR to those estimates obtained through the MH, the DerSimonian, and Laird methods, or through the inverse variance with a constant continuity correction of 0.5 methods. They observed that for sparse data, when there is a higher between-study heterogeneity of the treatment effect, the simple average approach produced less biased estimates of the overall treatment effect than the other three approaches. While the impact of a higher level of between-study heterogeneity was minimal for the simple average estimate of the treatment effect, it was much more pronounced for the MH, the DerSimonian and Laird, and the inverse variance approaches.

In the meta-analyses of rare AEs, it is not uncommon to encounter studies with zero events in both arms that are compared (double-zero studies). These are usually handled through exclusion of such studies from the synthesis. An alternative approach involves the use of models for count data such as generalized linear mixed effects models. These models allow for inclusion of both single- and double-zero studies without the need for exclusion or continuity correction (e.g. Stijnen, Hamza, and Ozdemir 2010; Böhning, Mylona, and Kimber 2015). Böhning, Mylona, and Kimber (2015) considered Poisson regression models with random effects and zero-inflated mixed effects models as methods for meta-analysis. They investigated the effect of exclusion of studies with zero events in either or both arms on the treatment effect estimates. The authors re-analyzed data from a published meta-analysis which assessed the risk for MI and CV mortality following the use of the diabetes drug Rosiglitazone (Nissen and Wolski 2010), discussed above. They found a borderline significant risk for MI, but not for CV death, using the Poisson regression model when including both single-zero and double-zero studies. They noted that the effect of exclusion of the double-zero studies from the analysis on the estimate of the RR and on the standard error, when using the Poisson model, was negligible. Their results were similar to the results obtained using the MH approach, which is unaffected by double-zero studies. The effect of exclusion of single-zero studies from the analysis was more pronounced and led to different conclusions. An advantage of the use of generalized linear mixed effects models is that they allow for inclusion of additional variables and for random effects in the model.

Overall, the performance of the various meta-analysis approaches for rare events described in the literature varies. The main factors that affect the performance of the different approaches are: the risk measure that is used; the balance between the sample sizes of the treatment arms that are being compared; the between-study heterogeneity; and the risk size (e.g. Sweeting, Sutton, and Lambert 2004; Bradburn et al. 2007). Although, in the reviewed literature, there was no single method that performed consistently better than the rest in all cases, the inverse variance weighted average and the DerSimonian and Laird approaches, using a constant continuity correction, were shown to demonstrate higher levels of bias than the rest of the methods. The 2018 FDA Draft Guidance for Industry on meta-analyses of RCTs intended to evaluate safety of drugs or biologics, while not recommending particular meta-analysis methods as a choice when dealing with sparse data, recommends against the use of continuity corrections. Thus, when conducting a meta-analysis to assess the risk for rare AEs associated with a medical product, it may be useful to additionally provide sensitivity analyses using several alternative approaches.

9.5 Multivariate Safety Endpoints

9.5.1 Why Multivariate Meta-Analysis?

In many applications of meta-analysis, estimates of a single parameter of interest are combined across several studies. Many research settings, on the other hand, typically evaluate the impact of a treatment on more than one parameter or outcome of interest. For example, in hypertension trials, there may be interest in looking at both systolic and diastolic blood pressure simultaneously. Pain and nausea may be of interest in trials for migraine. Even for these settings, meta-analyses tend to be performed separately on each outcome of interest to obtain summary estimates of the treatment effect on each outcome separately. However, it may be desirable to combine two or more outcomes of interest in a single meta-analysis. Multivariate meta-analysis has a variety of potential applications when multiple trials are available, including, modeling the outcome separately in each arm of a clinical trial, exploring treatment effects simultaneously on several clinical outcomes, and combining trials comparing more than one treatment. Despite its potential value, multivariate meta-analysis has not been widely used, and researchers still prefer to conduct univariate meta-analyses. There are a variety of reasons for this, including traditional practice, lack of available software, computational complexity, challenges in understanding the correlation structure of the study effects, and underestimation of the impact of ignoring correlation. As outcomes measured on the same population tend to be correlated, in applying meta-analysis to each outcome separately, any possible correlation structure is ignored. Riley (2009) showed that separate meta-analyses of correlated outcomes can lead to overestimated variance of the summary effect size and biased estimates. Additionally, independent testing of treatment effects on multiple outcomes, which are indicators of the general effectiveness of an intervention, increases the chances of finding spuriously significant treatment effects, and adjustment for multiple comparisons might be needed. In these situations, multivariate meta-analysis methods can be used for joint synthesis of effect estimates, accounting for within-study and between-study correlations of the outcomes.

9.5.2 Defining Multivariate Meta-Analysis

Multivariate meta-analysis is an expansion of a single parameter meta-analysis which combines simultaneously the estimates of several related parameters. It allows the joint synthesis of parameters of interest and accounts for their within-study and between-study correlation. It may also provide a solution to the multiplicity problem by summarizing simultaneously all outcomes of interest, instead of conducting many separate univariate meta-analyses. The data required for conducting a multivariate meta-analysis are the IPD for each study or study-specific effect sizes and their variance–covariance matrices. Thus, multivariate meta-analysis can help obtain estimates for multiple effects in a single modeling framework, as well as describe the relationship between the effects. The resulting estimates, then, may exhibit better statistical properties than if univariate meta-analysis had been used. The utilization of the correlation structure may contribute to improved estimates of the between-study variance, as well as of the pooled estimates. Incorporating the correlations among the outcomes helps to borrow strength across outcomes. When data of some of the outcomes for some of the studies are missing, under the missing at random assumption, a multivariate meta-analysis thus borrows strength across outcomes and eventually may help to reduce the impact of outcome reporting bias (Riley et al. 2015). Jackson, Riley, and White (2011) mentioned a variety of outcome types, including continuous, longitudinal, binary, and survival, that have been used in multivariate meta-analysis applications. Depending on the data availability (IPD or summary), and on the computational complexity, a one-stage or a two-stage method of analysis may be adopted. In a two-stage approach, first, the necessary parameter estimates, including covariances or correlations, are obtained, either from study data or from publications. Then, these estimates are combined in the second stage into the multivariate model.

9.5.3 Multivariate Random Effects Model

In a univariate meta-analysis, the fixed effects model works under the assumption of homogeneity of the studies' treatment effect. However, in many circumstances, these homogeneity assumptions are not satisfied. The random effects model in univariate meta-analysis is a very popular modeling strategy to address the underlying heterogeneity across studies. van Houwelingen, Arends, and Stijnen (2002) presented a bivariate random-effects meta-analysis (BRMA) considering two correlated endpoints. This model requires the summary measures and their standard errors for both endpoints for each study; in addition, the within-study covariances (or correlations) between the two endpoints are assumed to be known. Studies with one missing endpoint can be included in the analysis under the missing at random assumption. Multivariate random effects meta-analysis may be utilized for synthesizing data from studies reporting multiple correlated outcomes. It is a generalization of the standard univariate model, or the BRMA.

9.5.4 Statistical Challenges in Multivariate Meta-analysis

Despite their advantages, multivariate meta-analysis methods come at the price of making more assumptions in comparison to univariate meta-analysis. They are often based on the multivariate normality assumption which is hard to verify. Moreover, an implicit assumption in the multivariate case is that the effects have a linear relationship among studies. Multivariate meta-analysis methods are also often difficult to apply, primarily due to the difficulty in the estimation of the between-study correlation data structure complexities, for example, categorical data adding to the complexity. Furthermore, obtaining and estimating within-study and between-study correlations is often difficult. The information gained by utilizing correlation is often small and may not change clinical conclusions. The method also assumes that the outcomes are missing at random, which may not hold when there is selective outcome reporting. Thus, simplification of the assumptions may be needed to deal with a large number of unknown variance parameters.

As an example of an application of multivariate analysis, we discuss an application considered by Riley et al. (2015) based on IPD from hypertension trials. The INdividual Data ANalysis of Antihypertensive intervention trials (INDANA) data set consists of IPD from ten randomized control trials. Participants received either active anti-hypertensive drugs (Treatment), or placebo, or no treatment (Control). The effects of the treatment with anti-hypertensive drugs were assessed through a meta-analysis of the four end-points, including systolic blood pressure (SBP), diastolic blood pressure (DBP), risk of cardiovascular disease (CVD), and risk of stroke. It should be noted that SBP and DBP were studied as efficacy endpoints in this application. In addition, these endpoints can be analyzed either as continuous variables or as binary endpoints by dichotomizing as appropriate. Riley et al. (2015) presented three different bivariate meta-analyses; first, taking SBP and DBP as two continuous variables, second, taking SBP and DBP as two binary variables, and third, taking CVD and stroke as survival endpoints. In the two-stage approach, first each outcome was modeled for each study using IPD to produce the study-specific treatment effect estimates and their variances. Within-study correlations were obtained using bootstrapping (e.g. Efron and Tibshirani 1993), where one takes a number of bootstrap samples, gets treatment effect estimates and their variances for each bootstrap sample for each outcome, and then calculates the observed correlation between the treatment effect estimates. Alternatively, one may directly model outcomes jointly within each trial, when possible. For example, one can use a bivariate normal model for SBP and DBP, when taken as continuous variables, and directly obtain the within trial correlations between the outcomes. Once the treatment effect estimates, their variances, and within study correlations are obtained for each study, then, in the second stage, these estimates are used in the final multivariate model. Riley et al. (2015) also presented multivariate meta-analysis of all four variables, treating SBP and DBP as continuous variables and CVD and stroke as survival outcomes.

Copas et al. (2018) reviewed the practical value of multivariate approaches in meta-analysis using the above example. They concluded that under certain assumptions the multivariate meta-analysis approach brought little or no improvement over univariate methods. Thus, one needs to pay close attention to data availability and assumptions before performing the complex sequences of analyses and carefully think through the need to pursue a multivariate meta-analysis.

9.6 Cumulative and Trial Sequential Meta-Analysis

9.6.1 Cumulative Meta-Analysis

Cumulative meta-analysis (CMA) is performed by adding studies in chronological order one at a time, until all relevant studies have been included in the analysis. It can reveal whether there is consistency in the results of consecutive studies and can indicate the point at which no further studies are required because the definitive conclusion is evident. It also has the advantage of revealing uneven irregular changes in effect sizes as well as multiple shifts in opposite directions. It can also be used to evaluate the extent to which further studies are needed to confirm or refute a hypothesis. In general, no adjustment is made for repeated testing of the null hypothesis of treatment equivalence on cumulative data. In most practical applications, CMA is used to display the effect size estimate over time. If conducted prospectively, CMAs can be updated every time a new study is completed. Performing a new meta-analysis, whenever the results of a new trial are available, permits the evaluation of study of trends and makes it possible to determine when a treatment appears to have a significant effect, or otherwise.

CMA, along with other standard sequential methods and sequential meta-analysis, may utilize quality control charts and a penalized z-test for monitoring effects. Within the context of CMA, these methods are only effective when monitoring fixed effects of meta-analysis. For random effects, the analysis incorporates the heterogeneity variance, for which estimation gets complicated. Dogo (2016) proposed the use of a truncated CUSUM-type test for sequential monitoring in a random-effects CMA and explored the effect of accumulating evidence. He showed via a simulation that there are challenges with the type I error control. He also noted that the bias associated with accumulating evidence (sequential decision bias) and sequential design bias are non-negligible and increase with increases in heterogeneity.

9.6.2 Trial Sequential Meta-Analysis

Sequential meta-analysis (SMA) methods involve the use of formal group sequential boundaries to monitor accumulative evidence. They were proposed by Pogue and Yusuf (1997) to address the question of inflated type I error in CMA. The crossing of the monitoring boundaries of group sequential methods can indicate a significant change in the cumulative effect and may be used to stop a meta-analysis when there is sufficient evidence of effect based on the pre-specified significance level and power, see Higgins, Whitehead, and Simmonds (2011). There are several group sequential designs which can maintain the overall significance level. The method by Lan and DeMets (1983) for alpha spending is more flexible in the sense that it does not require the number or the times of interim analyses to be specified in advance. Since this type of meta-analysis is a continuous process and the number of interim analyses is not known in advance, Pogue and Yusuf (1997) suggested the use of Lan and DeMets (1983) method for SMA. A key issue in conducting a SMA is the calculation of the optimum information size (OIS) needed to define the monitoring boundaries. The OIS is the amount of information needed to detect a significant treatment effect had a well-designed trial been planned. It is a function of the maximum sample size needed to achieve the power requirement for the test at the given significance level.

There are many discussions in the literature on considerations and methods used in cumulative and trial sequential meta-analysis; for example, Hu, Cappelleri, and Lan (2007) applied the law of iterated logarithm to control the type I error in CMA of binary outcomes; Higgins, Whitehead, and Simmonds (2011) discussed the use of sequential methods for meta-analyses that incorporate random effects to allow for heterogeneity across studies; Shuster and Neu (2013) applied the Pocock approach to sequential meta-analysis in the role of probiotics in preventing necrotizing enterocolitis in preterm infants; Miladinovic and Djulbegovic (2013) discussed an implementation of trial sequential meta-analysis in STATA using the Lan–DeMets α-spending function approach, and Shah and Smith (2020) provided a checklist on points to consider when interpreting the results of a trial sequential meta-analysis.

Some applications of cumulative and sequential trial meta-analysis can be found in the literature. For example, Lau et al. (1992) applied CMA to assess the use of intravenous streptokinase as thrombolytic therapy for acute infarction. Juni et al. (2004) used CMA in assessing the risk of CV events in patients receiving rofecoxib. Imberger et al. (2014) examined the effect of anesthesia with nitrous oxide on mortality or CV morbidity. Shuster and Neu (2013) used the Pocock approach to sequential meta-analysis on the role of probiotics in preventing necrotizing enterocolitis in preterm infants. Mihara et al. (2016) examined the effects of steroids on quality of recovery and AEs after general anesthesia using both meta-analysis and trial sequential analysis. Xing et al. (2018) used trial sequential meta-analysis to study the effects of thrombolysis on outcomes of patients with deep venous thrombosis. A useful way to visualize the results of a CMA is through the use of a cumulative forest plot (See Chapter 15).

9.7 Reporting of Meta-Analysis

Appropriate reporting of the components of a meta-analysis, which include the planning stage, the conduct stage, and the final results, is of great importance. The way these components are reported may affect the interpretation of the results by the readers and the recommendations that are made based on them. There are several guidelines and checklists on the meta-analysis items that should be reported. In 2009, the Preferred Reporting Items for Systematic Reviews and Meta-Analyses (PRISMA) Statement (Liberati et al. 2009), that includes a 27-item checklist and a flow-diagram, was published. Later, additional PRISMA statements were published, such as for the reporting of protocols of systematic reviews and meta-analyses (Moher et al. 2015), for the reporting of IPD meta-analyses (Stewart et al. 2015), of network meta-analyses (Hutton et al. 2015), of systematic reviews and meta-analyses of harms (Zorzela et al. 2016). The CIOMS Working Group X (2016) also published a checklist for the reporting of meta-analyses of safety assessments.

9.8 Visualizing Meta-Analysis Results

Graphs are an essential tool for exploring, presenting, conveying, and communicating the results of a meta-analysis of multiple studies and also serve to supplement the summary statistics used in the meta-analysis. Some graphs are intended to address or assess questions regarding the study selection, such as heterogeneity, data quality and reliability, publication bias, influential or outlier studies (influence diagnostics), and robustness of the results. Other graphs are intended to assess treatment effects and consistency of treatment effects across studies and subgroups, as well as trends in accruing data from cumulative or trial sequential meta-analysis. Details of the graphs used in meta-analysis and their interpretation can be seen for example in the studies by Anzures-Cabrera and Higgins (2010), Baujat et al. (2002), Kossmeier, Tran, and Voracek (2020), Reid (2006), Sterne and Egger (2001), Thomson and Thomas (2013), Thomson and Thomas (2017), and Westgate (2019).

9.9 Analysis Tools and Software for Meta-Analysis

The increased use of meta-analysis in the analysis of safety data and other areas has also been accompanied by an increase in the available software and tools for performing meta-analyses. These come with different degrees of complexity, functionality, and ease of use. Additionally, new tools continue to evolve, with some being more general, while others may focus on a specific analysis or methodology. For discussion purposes, we will classify them into the following contexts: commercial general tools with functionality for performing meta-analysis, commercial software tools built specifically for performing meta-analysis, open-source general tools with functionality for performing meta-analysis, and open-source tools specifically built for meta-analysis. General commercial tools include SAS (see for example, Huffcuyt, Arthur, and Bennett 1993; Arthur, Bennett, and Huffcutt 2001; and Lane, Galwey, and Quartey 2018; Senn et al. 2011; Hertzmark and Spiegelman 2017), STATA (e.g. Palmer and Sterne 2015; STATA 2019), and SPSS (e.g. Lipsey and Wilson 2000). General purpose open-source software tools include R (see for example, Dewey 2021; Viechtbauer 2010, 2019; Hamilton, Aydin, and Mizumoto 2017) and Python (e.g. Hongyong 2020, 2021; Yarkoni, Salo, and Nichols 2020). The last set of software tools are those that are specifically for the sole purpose of meta-analysis. They can be both commercial and open-source. They include Comprehensive Meta-Analysis software, CMA 2019; EasyMA, EpiMeta, Mix, Meta Analyst, Meta Stat, MIX, RevMan, and MetaWin, see Bax et al. (2007).

9.10 Conclusions

Meta-analysis techniques have been extensively implemented for the synthesis of efficacy endpoints, while less frequently for safety data due to a number of difficulties. For example, Dimova, Egelebo, and Izurieta (2020) surveyed the published literature of meta-analyses of vaccine safety to assess the utilization of the technique of meta-analysis in the area of vaccine safety and to describe their methodological characteristics. They noted that most of the meta-analyses, that were identified, were literature-based and that the most frequently meta-analyzed safety outcomes were the local and systemic reactions and the SAEs after vaccine administration. They also pointed out that the most often reported challenges to the conduct of meta-analysis were the differences in the definitions of the safety outcomes across the different studies and the poor reporting of safety results in the published literature. As discussed in this chapter, the safety endpoints have several attributes that describe them (Section 9.2) and that may drive heterogeneity between the studies included in a meta-analysis. As discussed also, while some safety events, such as certain AEs of special interest, are pre-specified prior to the initiation of the trial, most AEs are not pre-specified. This, and the fact that often many different AEs are observed throughout the duration of a trial, may lead to various levels of reporting (or lack of reporting) of the studies' safety results in the literature publications.

These difficulties may be reduced through the utilization of standardized safety outcomes and reporting, and through prospectively conducted IPD meta-analyses (as discussed in the FDA 2018 Draft Guidance "Meta-Analyses of Randomized Controlled Clinical Trials to Evaluate the Safety of Human Drugs or Biological Products"). The Clinical Data Interchange Standards Consortium's (CDISC) Study Data Tabulation Model (SDTM), which represents a standardized framework for organization of the study data, including common data elements and standard terminology (https://www.cdisc.org/), may aid in the standardized collection and organization of safety data across different studies and thus in the conduct of meta-analyses.

References

Anzures-Cabrera, J., and J. P. Higgins. 2010. "Graphical Displays for Meta-Analysis: An Overview with Suggestions for Practice." *Research Synthesis Methods* **11**: 66–80.

Arthur, W., W. Bennett, and A. I. Huffcutt. 2001. *Conducting Meta-Analysis Using Sas.* Lawrence Erlbaum Associates Publishers.

Baujat, B., C. Mahe, J-P. Pignon, and C. Hill. 2002. "A Graphical Method for Exploring Heterogeneity in Meta-Analyses: Application to a Meta-Analysis of 65 Trials." *Statistics in Medicine* **21**: 2641–52.

Bax, L., L-M. Yu, N. Ikeda, and K. G. M. Moons. 2007. "A Systematic Comparison of Software Dedicated to Meta-Analysis of Causal Studies." *BMC Medical Research Methodology* **7**: 40.

Bhaumik, D. K., A. Amatya, S-L. T. Normand, J. Greenhouse, E. Kaizar, B. Neelon, and R. D. Gibbons. 2012. "Meta-Analysis of Rare Binary Adverse Event Data." *Journal of the American Statistical Association* **107**: 555–67.

Böhning, D., K. Mylona, and A. Kimber. 2015. "Meta-Analysis of Clinical Trials with Rare Events." *Biometrical Journal* **57**: 633–48.

Bradburn, M. J., J. J. Deeks., J. A. Berlin, and L. A. Russell. 2007. "Much Ado About Nothing: A Comparison of the Performance of Meta-Analytical Methods with Rare Events." *Statistics in Medicine* **26**: 53–77.

Burke, D. L., J. Ensor, and R. D. Riley. 2017. "Meta-Analysis Using Individual Participant Data: One-Stage and Two-Stage Approaches, and Why They May Differ." *Statistics in Medicine* **36**: 855–75.

CIOMS Working Group X. 2016. *Evidence Synthesis and Meta-Analysis for Drug Safety: Report of Cioms Working Group X. Geneva*, Switzarland: Council for International Organizations of Medical Sciences (CIOMS).

Copas, John B., Dan Jackson, Ian R. White, and Richard D. Riley. 2018. "The Role of Secondary Outcomes in Multivariate Meta-Analysis." *Journal of the Royal Statistical Society: Series C (Applied Statistics)* 67 (5): 1177–1205.

Debray, T. P. A., K. G. M. Moons, G. van Valkenhoef, O. Efthimiou, N. Hummel, R. H. H. Groenwold, and J. B. Reitsma. 2015. "Get Real in Individual Participant Data

(Ipd) Meta-Analysis: A Review of the Methodology." *Research Synthesis Methods* **6**: 293–309.

DerSimonian, R., and N. Laird. 1986. "Meta-Analysis in Clinical Trials." *Controlled Clinical Trials* **7**: 177–88.

Dewey, M. 2021. "CRAN Task View: Meta-Analysis." https://CRAN.R-project.org/view=MetaAnalysis.

Dimova, R. B., C. C. Egelebo, and H. S. Izurieta. 2020. "Systematic Review of Published Meta-Analyses of Vaccine Safety." *Statistics in Biopharmaceutical Research* **12**: 293–302.

Dogo, S. H. 2016. "Some Statistical Problems in Sequential Meta-Analysis." PhD Thesis, University University of East Anglia, https://ueaeprints.uea.ac.uk/id/eprint/59244/1/PhD_Thesis_of_S_H_Dogo.pdf.

Efron, B., and R. Tibshirani. 1993. *An Introduction to the Bootstrap*. Chapman & Hall/CRC.

Food and Drug Administration. 2018. "Draft Guidance: Meta-Analyses of Randomized Con- trolled Clinical Trials to Evaluatethe Safety of Human Drugs or Biological Products." Available at https://www.fda.gov/media/117976/download.

Greenland, S., and J. M. Robins. 1985. "Estimation of a Common Effect Parameter from Sparse Follow-up Data." *Biometrics* **41**: 55–68.

Hamilton, W. K., B. Aydin, and A. Mizumoto. 2017. "MAVIS." http://kylehamilton.net/shiny/MAVIS/.

Hammad, T. A., T. Laughren, and J. Racoosin. 2006. "Suicidality in Pediatric Patients Treated with Antidepressant Drugs." *Archives of General Psychiatry* **63**: 332–39.

Hardy, R. J., and S. G. Thompson. 1996. "A Likelihood Approach to Meta-Analysis with Random Effects." *Statistics in Medicine* **15**: 619–29.

Hertzmark, E., and D. Spiegelman. 2017. "The Sas Metaanal Macro." https://cdn1.sph.harvard.edu/wp-content/uploads/sites/271/2012/08/metaanal.pdf.

Higgins, J. P. T., A. Whitehead, and M. Simmonds. 2011. "Sequential Methods for Random-Effects Meta-Analysis." *Statistics in Medicine* **30**: 903–21.

Hongyong, D. 2020. "PythonMeta." https://pypi.org/project/PythonMeta/.

Hongyong, D. 2021. "Pymeta." https://www.pymeta.com/.

van Houwelingen, H. C., L. R. Arends, and T. Stijnen. 2002. "Advanced Methods in Meta-Analysis: Multivariate Approach and Meta-Regression." *Statistics in Medicine* **21**: 589–624.

Hu, M., J. C. Cappelleri, and K. K. G. Lan. 2007. "Applying the Law of Iterated Logarithm to Control Type I Error in Cumulative Meta-Analysis of Binary Outcomes." *Clinical Trials* **4**: 329–40.

Huffcuyt, A. I., W. Arthur, and W. Bennett. 1993. "Conducting Meta-Analysis Using the Proc Means Procedure in Sas." *Educational and Psychological Measurement* **53**: 119–31.

Hutton, B., G. Salanti, D. M. Caldwell, A. Chaimani, C. H. Schmid, C. Cameron, J. P. A. Ioannidis, et al. 2015. "The Prisma Extension Statement for Reporting of Systematic Reviews Incorporating Network Meta-Analyses of Health Care Interventions: Checklist and Explanations." *Annals of Internal Medicine* **162**: 777–84.

Imberger, G., A. Orr, K. Thorlund, J. Wetterslev, P. Myles, and A. M. Moller. 2014. "Does Anaesthesia with Nitrous Oxide Affect Mortality or Cardiovascular Morbidity? A Systematic Review with Meta-Analysis and Trial Sequential Analysis." *British Journal of Anaesthesia* 112: 410–26.

Jackson, D., R. Riley, and I. R. White. 2011. "Multivariate Meta-Analysis: Potential and Promise." *Statistics in Medicine* 30: 2481–98.

Jacobson, R. M., P. V. Targonski, and G. A. Poland. 2007. "Meta-Analyses in Vaccinology." *Vaccine* 25: 3153–9.

Juni, P., L. Nartey, S. Reichenbach, R. Sterchi, P. A. Dieppe, and M. Egger. 2004. "Risk of Cardiovascular Events and Rofecoxib: Cumulative Meta-Analysis." *Lancet* 364: 2021–9.

Kossmeier, M., U. S. Tran, and M. Voracek. 2020. "Charting the Landscape of Graphical Displays for Meta-Analysis and Systematic Reviews: A Comprehensive Review, Taxonomy, and Feature Analysis." *BMC Medical Research Methodology* 20: 26.

Lambert, P. C., A. J. Sutton, K. R. Abrams, and D. R. Jones. 2002. "A Comparison of Summary Patient-Level Covariates in Meta-Regression with Individual Patient Data Meta-Analysis." *Journal of Clinical Epidemiology* 55: 86–94.

Lan, K. K. G., and D. L. DeMets. 1983. "Discrete Sequential Boundaries for Clinical Trials." *Biometrika* 70: 659–63.

Lane, P., N. Galwey, and G. Quartey. 2018. "Methods for Meta-Analysis in the Pharmaceutical Industry." Cochrane Scientific Committee, https://www.psiweb.org/docs/default-source/default-document-library/meta-analysis-course-notes.pdf?sfvrsn=38ffdedb_0.

Lau, J., E. M. Antman, J. Jimenez-Silva, B. Kupelnick, F. Mosteller, and T. C. Chalmers. 1992. "Cumulative Meta-Analysis of Therapeutic Trials for Myocardial Infarction." *New England Journal of Medicine* 327: 248–54.

Liberati, A., D. G. Altman, J. Tetzlaff, C. Mulrow, P. C. Gøtzsche, J. P. A. Ioannidis, M. Clarke, P. J. Devereaux, J. Kleijnen, and D. Moher. 2009. "The Prisma Statement for Reporting Systematic Reviews and Meta-Analyses of Studies That Evaluate Healthcare Interventions: Explanation and Elaboration." *BMJ* 339: b2700.

Lipsey, M. W., and D. B. Wilson. 2000. *Practical Meta-Analysis*. Sage Publications.

Mantel, N., and W. Haenszel. 1959. "Statistical Aspects of the Analysis of Data from Retrospective Studies of Disease." *Journal of the National Cancer Institute* 22: 719–48.

Mihara, T., T. Ishii, K. Ka, and T. Goto. 2016. "Effects of Steroids on Quality of Recovery and Adverse Events After General Anesthesia: Meta-Analysis and Trial Sequential Analysis of Randomized Clinical Trials." *PLOS ONE* 11: 1–14.

Miladinovic, B., and B. Djulbegovic. 2013. "Trial Sequential Boundaries for Cumulative Meta-Analyses." *The Stata Journal* 13: 77–91.

Moher, D., L. Shamseer, M. Clarke, D. Ghersi, A. Liberati, M. Petticrew, P. Shekelle, and L. A. Stewart. 2015. "Preferred Reporting Items for Systematic Review and Meta-Analysis Protocols (Prisma-P) 2015 Statement." *Systematic Reviews* 4: 1.

Nissen, S. E., and K. Wolski. 2007. "Effect of Rosiglitazone on the Risk of Myocardial Infarction and Death from Cardiovascular Causes." *Archives of Internal Medicine* 356: 2457–71.

Nissen, S. E., and K. Wolski. 2010. "Rosiglitazone Revisited: An Updated Meta-Analysis of Risk for Myocardial Infarction and Cardiovascular Mortality." *Archives of Internal Medicine* 170: 1191–1201.

Palmer, T. M., and J. A. C. Sterne. 2015. *Meta-Analysis in Stata: An Updated Collection from the Stata Journal.* Stenhouse Publishers.

Pogue, J. M., and S. Yusuf. 1997. "Cumulating Evidence from Randomized Trials: Utilizing Sequential Monitoring Boundaries for Cumulative Meta-Analysis." *Controlled Clinical Trials* 18: 580–93.

Reid, R. 2006. "Interpreting and Understanding Meta-Analysis Graphs - a Practical Guide." *Australian Family Physician* 35: 635–38.

Riley, R. D. 2009. "Multivariate Meta-Analysis: The Effect of Ignoring Within-Study Correlation." *Journal of the Royal Statistical Society: Series A (Statistics in Society)* 172: 789–811.

Riley, R. D., M. J. Price, D. Jackson, M. Wardle, F. Gueyffier, J. Wang, J. A. Staessen, and I. R. White. 2015. "Multivariate Meta-Analysis Using Individual Participant Data." *Research Synthesis Methods* 6: 157–74.

Riley, R. D., and E. W. Steyerberg. 2010. "Meta-Analysis of a Binary Outcome Using Individual Participant Data and Aggregate Data." *Research Synthesis Methods* 1: 2–19.

Senn, S., J. Weir, T. A. Hua, C. Berlin, M. Branson, and E. Glimm. 2011. "Creating a Suite of Macros for Meta-Analysis in Sas: A Case Study in Collaboration." *Statistics and Probability Letters* 81: 842–51.

Shah, A., and A. F. Smith. 2020. "Trial Sequential Analysis: Adding a New Dimension to Meta-Analysis." *Anaesthesia* 75: 15–20.

Shuster, J., and J. Neu. 2013. "A Pocock Approach to Sequential Meta-Analysis of Clinical Trials." *Research Synthesis Methods* 4: 269–79.

Shuster, J., J. D. Guo, and J. S. Skyler. 2012. "Meta-Analysis of Safety for Low Event-Rate Binomial Trials." *Research Synthesis Methods* 3: 30–50.

Simmonds, M. C., and J. P. T. Higgins. 2007. "Covariate Heterogeneity in Meta-Analysis: Criteria for Deciding Between Meta-Regression and Individual Patient Data." *Statistics in Medicine* 26: 2982–99.

Stassijns, J., K. Bollaerts, M. Baay, and T. Verstraeten. 2016. "A Systematic Review and Meta-Analysis on the Safety of Newly Adjuvanted Vaccines Among Children." *Vaccine* 34: 714–22.

STATA. 2019. "Meta-Analysis Reference Manual." https://www.stata.com/bookstore/meta-analysis-reference-manual/.

Sterne, J. A., and M. Egger. 2001. "Funnel Plots for Detecting Bias in Meta-Analysis: Guidelines on Choice of Axis." *Journal of Clinical Epidemiology* 54: 1046–55.

Stewart, G. B., D. G. Altman, L. M. Askie, L. Duley, M. C. Simmonds, and L. A. Stewart. 2012. "Statistical Analysis of Individual Participant Data Meta-Analyses: A Comparison of Methods and Recommendations for Practice." *PLoS One* 7: e46042.

Stewart, L. A., M. Clarke, M. Rovers, R. D. Riley, M. Simmonds, G. Stewart, and J. F. Tierney. 2015. "Preferred Reporting Items for a Systematic Review and Meta-Analysis of Individual Participant Data: The Prisma-Ipd Statement." *JAMA* 313: 1657–65.

Stijnen, T., T. H. Hamza, and P. Ozdemir. 2010. "Random Effects Meta-Analysis of Event Outcome in the Framework of the Generalized Linear Mixed Model with Applications in Sparse Data." *Statistics in Medicine* 29: 3046–67.

Sutton, A. J., N. J. Cooper, P. C. Lambert, D. R. Jones, K. R. Abrams, and M. J. Sweeting. 2002. "Meta- Analysis of Rare and Adverse Event Data." *Expert Review of Pharmacoeconomics & Outcomes Research* 2: 367–79.

Sutton, A. J., D. Kendrick, and C. A. C. Coupland. 2008. "Meta-Analysis of Individual- and Aggregate-Level Data." *Statistics in Medicine* **27**: 651–69.

Sweeting, M. J., A. J. Sutton, and P. C. Lambert. 2004. "What to Add to Nothing? Use and Avoidance of Continuity Corrections in Meta-Analysis of Sparse Data." *Statistics in Medicine* **23**: 1351–75.

Thomson, H. J., and S. Thomas. 2013. "The Effect Direction Plot: Visual Display of Non-Standardised Effects Across Multiple Outcome Domains." *Research Synthesis Methods* **4**: 95–101.

Thomson, H. J., and S. Thomas. 2017. "The Effect Direction Plot: Visual Display of Non-Standardised Effects Across Multiple Outcome Domains." *Research Synthesis Methods* **8**: 387–87.

Viechtbauer, W. 2010. "Conducting Meta-Analyses in R with the Metafor Package." *Journal of Statistical Software* **36**: 1–48.

Viechtbauer, W. 2019. "Metafor: A Meta-Analysis Package for R." https://wviechtb. github.io/metafor/index.html.

Wallach, J. D., K. Wang, A. D. Zhang, D. Cheng, H. K. G. Nardini, H. Lin, M. B. Bracken, M. Desai, H. M. Krumholz, and J. S. Ross. 2020. "Updating Insights into Rosiglitazone and Cardiovascular Risk Through Shared Data: Individual Patient and Summary Level Meta-Analyses." *BMJ* **368**: l7078.

Westgate, M. J. 2019. "Revtools: An R Package to Support Article Screening for Evidence Synthesis." *Res Syn Meth* 10: 606– 614.

Xing, Z., L. Tang, Z. Zhu, and X. Hu. 2018. "Effects of Thrombolysis on Outcomes of Patients with Deep Venous Thrombosis: An Updated Meta-Analysis." *PLOS ONE* **13**(9): e0204594.

Yarkoni, T., T. Salo, and T. Nichols. 2020. "PyMARE." https://pymare.readthedocs. io/en/latest/.

Yusuf, S., R. Peto, J. Lewis, R. Collins, and P. Sleight. 1985. "Beta Blockade During and After Myocardial Infarction: An Overview of the Randomized Trials." *Progress in Cardiovascular Diseases* **27**: 335–71.

Zorzela, L., Y. K. Loke, J. P. Ioannidis, S. Golder, P. Santaguida, D. G. Altman, D. Moher, and S. Vohra. 2016. "PRISMA Harms Checklist: Improving Harms Reporting in Systematic Reviews." *BMJ* **352**:i157. Erratum in: BMJ. 2016; 353:i2229.

Part C

Design and Analysis Considerations in Randomized Controlled Trials and Real-World Evidence for Safety Decision-Making

10

Pragmatic Trials with Design Considerations for Cardiovascular Outcome Trials

Li-An Lin
Merck & Co, Inc., Kenilworth, NJ, USA

Ed Whalen
Pzifer Inc., New York, NY, USA

Melvin Munsaka
AbbVie, Inc., North Chicago, IL, USA

William Wang
Merck & Co, Inc, Kenilworth, NJ, USA

CONTENTS

10.1 Introduction 215
10.2 Design Considerations 217
    10.2.1 Randomization in a Real-World Setting 218
    10.2.2 Blinding the Treatment Interventions 218
    10.2.3 Selecting a Comparator..219
    10.2.4 Choice of Endpoints and Data Collection 220
    10.2.5 Safety Evaluation and Monitoring 220
    10.2.6 Design Tools...221
10.3 CVOTs: From RCT to Pragmatic Trials 222
    10.3.1 Regulatory Guidance..222
    10.3.2 CVOT Pragmatic Trial 224
10.4 Discussion 228
References..229

10.1 Introduction

Randomized clinical trials (RCTs) are often considered the gold standard for establishing treatment efficacy of an intervention and help to advance the practice of evidence-based medicine. Nonetheless, because of the stringent

DOI: 10.1201/9780429488801-13

inclusion and exclusion criteria used to select study participants, a RCT may not reflect the real-world effectiveness and safety of an intervention. In general, standard RCTs are designed to optimize internal validity to efficiently demonstrate the efficacy of an intervention in a highly selected patient group. However, such designs have notable limitations, including a lack of generalizability and external validity, which may limit their direct applicability to routine clinical practice [1, 2].

Observational studies have been proposed as alternative means of testing the effectiveness of an intervention [3]. They are conducted in patient populations that have a wider range of demographic clinical characteristics and can be used to monitor the effects of therapies beyond the RCT setting. The availability of large, high-quality databases offers the potential to bring observational research closer to evidence-based medicine. Well-designed observational studies can offer superior external validity and provide a unique opportunity to evaluate the uptake of new treatments and their outcomes in routine practice [4]. However, observational studies are prone to three broad types of bias, including selection bias, information bias, and confounding. With appropriate methods of adjustment, observational studies can be useful especially when ethical, practical, or financial reasons preclude randomized studies from being conducted [1].

Both the traditional RCTs and observational studies have limitations in providing real-world evidence on the effectiveness and safety of treatment options [5, 6].Pragmatic trials, as first introduced by Schwartz and Lellouch [7], offer the opportunity to combine the real-world nature of an observational study with the scientific rigor of a randomized trial and thereby give better answers to questions that are relevant to general clinical practice [8]. A pragmatic trial offers gains in efficient execution of studies with an appropriate balance between internal and external validity, where randomization removes differences in both unknown and known confounding that lead to prognostic incomparability between patient groups, while study design and execution in real-world patients with routine clinical practice yield results that are likely more relevant than the isolated effect of a medicine.

The pragmatic trial aims to understand the real-world benefit of an intervention by incorporating design elements that allow for greater generalizability and clinical applicability of study results. For this reason, we may not see many pre-marketing regulatory trials that can be classified as pragmatic trials, but expect the pragmatic trial concept to be applied more often to post-marketing studies, possibly incorporated into future post-marketing regulatory requirements. Moreover, many post-marketing studies conducted under a regulatory safety reporting regimen (such as REMS, RMP, etc.) may lead to an integrated evaluation of benefits and risks under real-world conditions. The design of a pragmatic trial may receive input from both regulatory and health technology agencies to satisfy the joint need for early detection of post-marketing safety signals and to inform coverage decisions of new drugs [9].

The advent of a modern electronic health record (EHR) system makes it feasible and relatively inexpensive to conduct studies in the context of routine clinical practice [10]. However, the regulatory requirements for good clinical practice in the collection of adverse events (AEs) and the monitoring of trial conduct may impact routine clinical care [11]. This, in turn, would mean that data are not really representative of the real-life clinical situation, and the generalizability of trial results may be compromised [12]. Regulatory agencies have recognized the need for developing more proportionate approaches to the collection of safety data and the monitoring of trial conduct. Several regulatory guidances have been published to facilitate the risk-based monitoring of trial conduct [13–15].

In this book chapter, we focus on the design considerations of a pragmatic trial. In particular, we will outline important considerations in trial design that can help to mitigate the risks to internal/external validity and conversely help to identify design options that are more suited to investigation using such a design. We then illustrate these considerations in-depth concerning an application to cardiovascular outcome trial (CVOT), which examines whether the new anti-diabetic therapy increases cardiovascular (CV) risks. The regulatory guidance for conducting a CVOT will also be discussed regarding how to potentially fulfill regulatory requirements with a pragmatic trial design.

10.2 Design Considerations

The objective of a pragmatic trial is to capture the true effect of a treatment strategy in a real-world setting. If well-designed and executed, a pragmatic trial can have the strength of an RCT for internal validity and the ability to provide valid relative effectiveness and safety information among all the available interventions in current clinical practice [7, 16].

Although the pragmatic trial can be viewed as a simple clinical trial design with increased generalizability, several operational and statistical challenges arise, such as impacts on enrollment, larger sample sizes (e.g., within the framework of cluster randomization), smaller treatment effects and higher variability in the heterogeneous populations due to minimal inclusion/exclusion criteria, usual care settings with fewer and irregularly timed visits (which may render an interim analysis infeasible and reduce adherence), and biases from using an open-label design. The design elements with operational and/or statistical challenges for the pragmatic trial have been discussed widely in the literature [7, 17–20]. Key features of pragmatic trial design elements include the following: (1) randomizing trial participants to receive treatments in usual care, (2) blinding relevant parties to mitigate potential biases, (3) an alternative comparator treatment among many available treatments rather than a placebo, allowing flexible concomitant medication, complementary care, and compliance, rather than adherence to a strict protocol, (4) clinically meaningful outcomes that are routinely collected through existing channels, rather than a surrogate outcome, and (5) less frequent safety data collection and monitoring. Combining these design elements with a trial plan that does not alter the routine care received by patients will support the generalization of trial results to the appropriate target patient group in clinical practice.

10.2.1 Randomization in a Real-World Setting

The key purpose of randomization is to balance known/unknown confounding factors among the study groups and avoid biases associated with the allocation of treatment to trial participants (such as channeling bias). Although some confounding can be addressed in the study data analyses, it is difficult to be sure that all of it is taken care of. Thus, randomization remains an attractive feature for pragmatic trials. However, randomization alters routine practice by taking away the clinician's control of who gets what intervention based on their expert assessment and knowledge, turning instead to some form of random assignment in the hope that determinants of outcome other than the interventions themselves will be balanced across the groups. It causes ethical concerns since the clinician may believe certain interventions are better than the others for a particular patient. Novel designs, such as Bayesian adaptive methods, may help in mitigating clinicians' concerns [21]. For instance, randomizing more patients to the study group(s) with a greater likelihood of success may make randomization more readily acceptable when clinicians have reservations about patients' benefits. Another commonly used approach is cluster (or group) randomization where clinic sites or physicians are the units of randomization. Cluster randomization can help to avoid 'contamination' of the treatment strategies, to keep usual care as real world as possible [17, 20]. If it turns out that clinicians are not willing to randomize, then it may be more productive to study the basis for their belief [18]. More complex randomization strategies may be needed to minimize bias [19]. For example, randomization could be made more general, by randomizing patients to a treatment strategy and allowing physicians to determine the specific treatment to be used [22].

10.2.2 Blinding the Treatment Interventions

Blinding of treatments and their comparators is routinely implemented in classical RCTs to isolate the effects of the treatment and increase internal validity by mitigating potential biases associated with health care providers, patients, data collectors, outcome assessors (e.g., adjudication committees), data analysts, and others (e.g., governance/scientific committee) that may occur with knowledge of the treatment assignment [23]. However, blinding of treatment assignments is not only complex, but it hinders how generalizable the findings are to the real-world clinical setting [18]. All the steps are taken to blind the treatment intervention increase patients' awareness of being in a study and may amplify the Hawthorne effect. Moreover, the pragmatic trial assesses comparators that are available in routine clinical practice, not masked placebos. Thus, it may not be ethical or even feasible to blind all parties in a pragmatic trial. The decision on whom to blind should be based on the likelihood that one or more of these parties have or may develop opinions about the effectiveness and/ or safety of the intervention being investigated relative to the comparators and that these opinions may lead to systematic bias through conscious or unconscious mechanisms [24]. For example, an outcome adjudication committee may need to be blinded when outcomes are subjective, and bias can be introduced by differential assessments of outcomes. When a clinical event (such as all-cause mortality) has little subjectivity, the blinding is likely unnecessary in most cases.

10.2.3 Selecting a Comparator

Pragmatic trials deliver real-world evidence about the added value of new medications compared with usual care and inform decision-making earlier in development [8]. The use of a placebo is inappropriate in a pragmatic trial since the placebo would not be used in routine care. The selection of comparators and implementation of treatment strategies should resemble clinical practice as closely as possible, including existing treatment options, dosing, co-medication, treatment sequences, and the supply and reimbursement of drugs. There may be more than one treatment regimen in usual care. A single specific treatment regimen can provide clearer results as to whether the new treatment is more or less effective than each comparator. However, using multiple usual-care treatment arms will increase the complexity of the study. Usual care combining multiple options per physician choice is simpler and cheaper than multiple arms for each specific regimen and more useful for health care providers in deciding between treatment options for their patients. It is difficult to combine multiple treatment options since a new treatment may be more effective than one usual care treatment but less effective than another. Another challenge is that the variability of usual care across centers/countries and over time makes it difficult to address treatment effect heterogeneity within the usual care comparators and complicates comparator selection. Using clinical guidelines to define usual care can help standardize comparator treatments; however, this may decrease the applicability of the results to real-life settings [8].

10.2.4 Choice of Endpoints and Data Collection

The endpoints for an explanatory trial have traditionally focused primarily on physiologic or other such biologic outcomes that are directly related to the intervention's mode of action and, therefore, are most likely to produce larger, easier-to-detect effects [18]. In the real-world setting, patients and physicians are more interested in the outcomes that are of direct interest to patients and reflect the comparative treatment effects encountered in patients in real-life clinical practice to guide treatment decisions and clinical practice. However, these outcomes may be more difficult/subjective to measure, and the resulting noise may mask the effect of the intervention [18]. In the choice of endpoints, one consideration is whether the outcomes are in line with routine care and easily assessable (e.g., morbidity, mortality, functional status, well-being, and resource utilization) [19]. Methods that decrease bias or enhance the precision of trial results, such as standardization and blinding of outcome assessment, should be considered when a high risk of bias or high variability is expected [25]. Another consideration is that the outcome measurements should not interfere with real-world clinical practice to ensure the generalizability of trial results, and the use of data that are recorded in (electronic) medical files during routine clinical practice should be prioritized [25]. Many algorithms have been created to define outcomes based on data that have been routinely recorded in EHRs [26]. However, many outcomes, such as ambiguous or less severe conditions that are not medically attended to, are unlikely to be captured in an EHR. Although using a dedicated electronic case report form (eCRF) may lead to the inclusion of a nonrepresentative sample of patients, physicians, and practices and therefore affect the generalizability of the trial results, it may be necessary to fully capture the key outcomes. In this case, a hybrid approach (with both EHR and the eCRF) can be adapted where the eCRF must be kept as simple as possible to reduce the burden for practitioners and minimize the influence on routine clinical practice [27].

10.2.5 Safety Evaluation and Monitoring

Pragmatic trials differ from explanatory trials in several respects related to design features. These features of pragmatic trials may influence data and safety monitoring but, nevertheless, investigators and sponsors still have the same fundamental ethical obligations to monitor the safety of trial participants, the risks of study treatments, the integrity of trial data, and the possibility that continuing a trial may not yield a definitive result [28]. Current ICH guidelines state that all serious AEs and all drug-related events must be collected and reported to the study sponsor within 24 hours of the investigator becoming aware of an event unless directed otherwise in the protocol in an interventional trial [29]. The data that are collected in routine visits often occur too infrequently to capture sufficient safety data. Data extraction may also not be in real-time which causes time gaps between data entry and information becoming available to the sponsor [27]. This can adversely influence the validity of the study and raise safety issues for pragmatic trials that rely on EHRs for the collection of safety data [30]. Regulatory agencies have recognized the need for developing more proportionate approaches to the collection of safety data and the monitoring of trial conduct for pragmatic trials and other post-approval studies. ICH E19 provides guidance on an optimized approach to safety data collection in some late-stage pre-approval or post-approval studies when the safety profile of a drug is already sufficiently characterized [15]. The ICH is also proposing to update the existing ICH E2D Guideline to incorporate pragmatic and/or risk-based approaches to the management of safety information from existing and new data sources, to ensure that global data can be leveraged to optimize patient safety [31]. Depending on specific questions in the trial and knowledge of the safety profile of the medicine or medicine class in the patient population of interest, a risk-based approach may exempt the collection of certain non-drug related, non-serious AEs, or even serious events not related to treatment [11, 13, 14, 32]. In line with regulatory guidance, different available options to support the timely collection and reporting of safety data with sufficient details while reducing study-related follow-up visits have been discussed in the literature [11]. For example, adding telecommunication between patients and care providers in a defined time frame without physician interaction (as used in the Salford lung studies) might mitigate the risk [33].

10.2.6 Design Tools

When designing a trial, the scientific value and the operational feasibility need to be implicitly balanced for all relevant design considerations [17]. The designs can differ considerably with respect to whether they are conducted under highly controlled and defined conditions (explanatory) or broadly reflect real-world circumstances (pragmatic). The terms explanatory and pragmatic are the two extremes of a continuum. Trials are hardly ever either purely explanatory or purely pragmatic for all domains. A pragmatic trial may have some aspects of design that are more pragmatic and some that are more explanatory. Depending on the operational feasibility and the research questions that the trial is seeking to answer, certain design features can be more pragmatic while others can be more explanatory. Tools and guidances have been developed to make design features and their implications more transparent and allow careful consideration of the possible design options and their consequences. Currently, the most widely used tools are PRECIS (PRagmatic Explanatory Continuum Indicator Summary) and its variations. To make the design decisions in line with the intended explanatory or pragmatic nature of a trial, the PRECIS-2 (updated and improved from the original PRECIS tool) describes nine domains that distinguish pragmatic studies from explanatory trials: eligibility criteria, recruitment, setting, organization, flexibility in delivery, flexibility in adherence, follow-up, primary outcome, and primary analysis [34, 35]. The steps to use in the tool include: (1) be clear about the research question and intended purpose; (2) consider the design choices for each of the PRECIS-2 domains; (3) score raters' choices from 1(explanatory) to 5(very pragmatic) for each domain; and (4) review PRECIS score and compare it to study purpose. By laying out the nine domains and their scores into a wheel, the result (displayed via radar graphs) can present how pragmatic the trial is and how the design choices can satisfy the purposes of the study.

Other tools have also been developed to facilitate pragmatic trial design. ASPECT-R (A Study Pragmatic-Explanatory Characterization Tool-Rating) adapted ideas from PRECIS to build an even more versatile instrument [36]. The number of domains that are specifically related to the explanatory-pragmatic spectrum was reduced to six. These six domains and their respective definitions include eligibility criteria for trial participants, intervention flexibility, medical practice setting/practitioner expertise, follow-up intensity and duration, outcome(s), and participant adherence. Domains in PRECIS that were identified as redundant and focused on measures of study quality have been eliminated. In addition, detailed definitions of terms and more descriptive approaches for raters have been developed to facilitate consistency across raters.

10.3 CVOTs: From RCT to Pragmatic Trials

10.3.1 Regulatory Guidance

Diabetes mellitus is a metabolic disorder characterized by the presence of hyperglycemia (elevated blood glucose) due to defective insulin secretion, insulin action, or both. Over time, high blood sugar can damage blood vessels and the nerves which can raise the risk for CV disease. Type 2 diabetes mellitus (T2DM), previously known as adult-onset diabetes, is the most common type of diabetes. A number of anti-diabetic therapies were approved in Europe and in the US to ameliorate the symptoms of diabetes; however, the CV safety profile was less well understood. Concerns about this gap in assessment of CV safety grew after two highly controversial meta-analyses of CV risk were published [37, 38]. The U.S. Food and Drug Administration (FDA) convened an advisory committee meeting in July 2008 to discuss the role of CV risk assessments for the safety evaluation of drugs and biologics developed for the treatment of T2DM. Based, in part, on comments expressed at that meeting, the Agency issued a guidance for industry in December 2008 outlining recommendations on the evaluation of CV risk for new anti-diabetic therapies which led to dedicated CVOTs [39]. In particular, the U.S. FDA advised that pre-marketing phase 2 and phase 3 trials should rule out an 80% excess risk (upper bound of rate ratio of the 2-sided 95% confidence interval less than 1.8) for major adverse cardiovascular events (MACEs). If the upper bound of the 95% confidence interval is between 1.3 and 1.8, a dedicated post-marketing CVOT should be required to rule out a 30% excess risk after approval.

After the publishing of the 2008 U.S. FDA guidance, every novel anti-diabetic therapy approved has undergone a dedicated CVOT, typically involving 5,000–15,000 people with type 2 diabetes and high CV risk and planned to last 3–5 years [40]. Notably, none of the CVOTs completed and published to date have identified an increased risk of CV events; some of the CVOTs have instead demonstrated a reduced risk [41].

On October 24, 2018, the FDA held a 2-day Advisory Committee meeting to review the utility and impact of the FDA-required CVOT studies. The issues addressed by the FDA advisory board were: (1) the impact of the recommendations in the 2008 "Guidance for Industry" on the assessment of CV risk for drugs indicated to improve glycemic control in patients with T2DM; (2) the transferability of CV safety findings from members of a drug class that were studied to the entire class of drugs, and (3) whether an unacceptable increase in CV risk needs to be excluded for all new drugs to improve glycemic control in patients with T2DM, regardless of the presence or absence of a signal for CV risk in the development program [42]. The panel made several recommendations for future regulatory guidance and for CVOTs regarding anti-diabetic therapies, including requiring only a 1.3 noninferiority margin for regulatory approval, conducting trials for longer durations, considering head-to-head active comparator trials, increasing the diversity of patient populations, collection of safety data beyond CV events, and identifying ways to improve translation of trial results to general practice.

The U.S. FDA more recently revisited the recommendations and published new draft guidance in March 2020 to replace the December 2008 guidance [43]. The intention of this guidance is to provide clarity on the expectations for the development of drugs and biologics that improve glycemic control. The new draft guidance recommends that sponsors enroll a broader range of patients with comorbidities and diabetes-associated conditions, including patients with chronic kidney disease and older patients. However, very low-risk patients will still not be considered since an event-driven trial would take too long to conduct if patients with low-event rates are included. In an era of the increasing utility of RWD with sophisticated statistical analysis for regulatory decision-making, pragmatic trials can be a practical alternative for CVOT.

10.3.2 CVOT Pragmatic Trial

Interpretation of CVOT results is difficult. Ideally, similar blood glucose values (so-termed glycemic equipoise) should be achieved between comparator groups, to exclude the effect of glycemic control on CV events. However, glycemic equipoise has rarely been achieved in the CVOTs, and there have been many other differences between the active and control groups that could affect CV disease, including changes in body weight and the use of antihypertensive and lipid-lowering agents [44]. Additionally, comparison between trials or agents is complicated as MACEs accrued from different adverse CV events at different rates in different trials. This occurs mainly because of different inclusion and exclusion criteria, different combinations of CV risk factors, and/or different proportions of prior CV events, across trials [45]. Pragmatic trials conducted in larger and broader patient populations with active control comparators in usual care settings may provide better insights into CV events and better interpretation of results across different treatment options.

Both FDA advisory panels and the 2020 draft guidance are aligned with the pragmatic trial option, which allows broader patient populations, longer duration, focuses on safety events that are typical of patient experiences, etc. In fact, several completed CVOT trials can be viewed as pragmatic trials. Table 10.1 lists three such completed CVOTs that are considered as pragmatic trials in the literature. These studies incorporate several design features of

TABLE 10.1

Completed CVOT Pragmatic Trial

Study	Drug Name	Blinding	Sample Size	Median Duration	Comparator	Event Adjudication	Patient Population
TECOS [47]	Sitagliptin	Double-blind	N = 14,671	3.0 years	Placebo adding to usual care	Yes	Mean age: 65.5; Female: 29.3%; HbA1c: 6.5% to 9.0%; Prior CV disease 74.0%
EXSCEL [48]	Exenatide	Double-blind	N = 14,752	3.2 years	Placebo adding to usual care	Yes	Mean age: 62.0; Female: 38.0%; HbA1c: 7.3% to 8.9% Prior CV event: 73.1%
LEADER [49]	Liraglutide	Double-blind	N = 9,340	3.8 years	Placebo adding to standard care	Yes	Mean age: 64.3; Female: 35.7%; HbA1c: 8.7% +/- 1.6%; Established CVD (age >50): 81.3%

TABLE 10.2

Design Features of Classical CVOT and Current Pragmatic CVOT

Study Design Elements	Classical CVOT	Current Pragmatic CVOT
Comparator	Placebo (Double-blind)	Placebo, adding to usual care (Double-blind)
Enrollment	Extensive inclusion/exclusion criteria	Limited inclusion/exclusion criteria
Data Collection	Exclusively CRF	Focused CRF with EHR and/or claims
AEs	All	SAE and Prespecified Clinical Events of Interest
Monitoring	100% review	Less frequent review
Follow-Up	Extensive study site visit	Limited study site visit, Usual care visit, and Phone contact

a pragmatic trial. Table 10.2 compares these features for these three studies with classic CVOTs. It is noted that the trial designs are not fully pragmatic, partially due to the need to fulfill post-marketing CVOT requirements stipulated in the 2008 guidance. For example, a double-blind design and placebo control are selected for each of these three studies. To gain the greatest possible insight into whether a new therapy holds advantages over other specific diabetes drugs in terms of CV outcomes, it would be best to undertake head-to-head active-control trials. The only such study to date is the CAROLINA trial, which randomized patients with early T2DM and evidence of CV disease or high CV risk to Linagliptin versus Glimepiride, predominantly as add-on to metformin background therapy [46]. With the new regulatory guidance, it is now possible to switch to an open-label and usual care comparator in a future trial [15, 31, 43].

To illustrate the extent to which current CVOTs are pragmatic or explanatory, we briefly evaluated the TECOS trial using the PRECIS-2 tool which uses a five-point ordinal scale (ranging from very pragmatic to very explanatory) across nine domains of trial design, including eligibility, recruitment, setting, organization, intervention delivery, intervention adherence, follow-up, primary outcome, and analysis. The PRECIS scores across these nine trial design domains as well as the summary PRECIS wheel are presented in Table 10.3 and Figure 10.1. Note that the score is based on the authors' assessment. It is not representative of the TECOS sponsor's rating of this study.

For most of the criteria, the TECOS trial appears to be more pragmatic than a traditional RCT in which few restrictions or extra visits are required. The use of randomization and blinding does take it somewhat away from the practical experience but improves it by the presumed reduction with respect to potential biases. Current CVOT pragmatic trials are still using a focused eCRF to collect MACE events. In the future, data collected as part of routine

TABLE 10.3

Illustration of PRECIS Score in TECOS Trial (per authors' assessment)

Domain	Score	Rationale
Eligibility	4	Who is selected to participate in the trial?
		Not limited to patients with high risk of CV events. Patients with severe renal insufficiency are excluded.
Recruitment	5	How are participants recruited into the trial?
		Enrolled in the trial through sites which also provide their usual diabetes care.
Setting	4	Where is the trial being done?
		Multinational. Participants will be recruited and enrolled in the trial through sites which also provide their usual diabetes care. If not, additional visit to study site and communication between sites and usual care providers required.
Organization	3	What expertise and resources are needed to deliver the intervention?
		Training was provided to investigator and site staff in study procedures. There were no additional requirements for the study, other than usual care.
Flexibility (delivery)	3	How should the intervention be delivered?
		Double-blinded. Sitagliptin or matched placebo were used as part of a usual care regimen. To get follow up information, study nurse made the calls to patients and which was not the usual practice. Usual care of the patients does not involve treatment blinding.
Flexibility (adherence)	4	How should the intervention be delivered?
		Phone contact to review study therapy dosing instructions and reinforce adherence to medication. Usual care providers may adjust the AHA regimen.
Follow-up	3	How closely are participants followed up?
		Clinical events of interest would be recorded in the eCRF at the time of the study visit. Subject Follow-up: primary done at study site, all other laboratory values at usual care providers. Non-SAEs that were not part of prespecified outcomes were not collected.
Primary outcome	4	How relevant is it to participants?
		Primary outcome in the trial was CV outcome which was very relevant to patients. Outcome measurement had no relevance to the intervention. There was adjudication of CV event
Analysis	5	To what extent are all data included?
		Data analysis included intention-to-treat and per-protocol population. ITT analysis counts for 99.6% of those randomized.

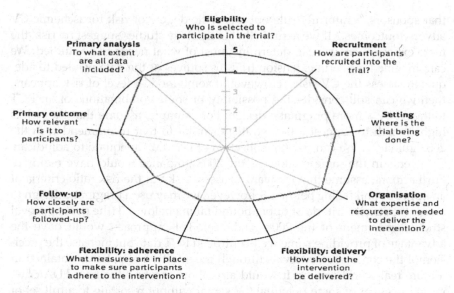

FIGURE 10.1
Illustration of the PRECIS wheel in the TECOS trial (per authors' assessment)

clinical practice could be used to detect CV outcomes in a pragmatic trial. Several real-world algorithms have been developed and can accurately identify cases with MACE [50, 51]. In comparing an EMR-based algorithm with physician review, a study (using statins as primary prevention) reported 90% to 97% positive predictive value in the identification of MACE cases [50]. In the Women's Health Initiative (WHI) study, acute myocardial infarction or coronary revascularization events derived from Medicare claims had good agreement with WHI outcomes when they were adjudicated by study physicians [51]. The adjudication process may still be needed to validate the CV outcome; however, conducting adjudication by itself may not affect study conduct so long as the committee remains separate from and not in communication with the sites regarding the adjudicated results. Other trials have published designs and have been initiated with a similar affinity for keeping the conduct close to actual practice with few extra data collection or design features, such as the Salford LUNG study [33].

The 2008 FDA guidance for a CVOT in diabetes had many features that under the PRECIS spectrum would render them not as pragmatic and of a high level of resource intensity. The updated FDA 2020 draft guidance allows for much more pragmatic choices in designing a study to address CV risks. In the draft guidance, the FDA recommends longer exposure and a broader range of patients with comorbidities and diabetes-associated conditions, including patients with chronic kidney disease and older patients. Unlike the agency's previous guidance, the new draft guidance does not recommend that sponsors "uniformly rule out a specific degree of risk for ischemic CV adverse outcomes." If we assume that early stage studies suggest no risk, the mere exposure criteria can determine most of what needs to be studied. We can imagine long-term extension trials as sources of the data needed to adequately assess the CV risks. However, if some modest level of risk appears, then we can still consider the possibility of some combinations of an RCT followed by a more pragmatic design. For example, assume that the risk is high enough to suggest a trial strategy similar to that recommended in the 2008 guidance. We can require collection of CV data adequate to adjudicate CV events in the same manner as the 2008 guidance would have required. Furthermore, assume that we want to assess risk but the risk ratio criteria of an upper bound falling below 1.8; however, we may use a pragmatic design to assess the stricter criteria of upper bound falling below 1.3 (the post-approval study requirement of the 2008 guidance). This approach would have the advantage of providing some well-controlled RCT data but increase the precision of the confidence intervals through the addition of patients obtained in a more real-world setting. It would also allow meeting the new FDA criteria in the event of some potential CV signal without resorting to a full set of two high-resource RCTs as required in the 2008 guidance. Variations in the strategy in the preceding example can be seen in terms of the PRECIS criteria, moving along the spectrum from a mostly real-world, quite pragmatic design approach to a nearly fully controlled RCT strategy with few pragmatic features.

10.4 Discussion

As traditional RCTs can be quite selective with respect to enrollment criteria, pragmatic trials reduce restrictive inclusion/exclusion criteria for trial participants and can better ensure that patient population under-represented in RCTs is enrolled to determine the efficacy and safety of drugs for these clinically relevant patients and their subgroups. Randomization in pragmatic trials increases the likelihood of unbiased comparisons and valid findings. This book chapter highlighted important considerations in trial design that can help to mitigate the risks to internal/external validity. We also provided insights from several pragmatic trials to examine CVOT for anti-diabetic therapy and the changing landscape for the treatment of diabetes and CV risk. The U.S. FDA 2008 guidance and newly published 2020 draft guidance for conducting a CVOT were also discussed as well as information regarding how to potentially fulfill regulatory requirements with a pragmatic trial design.

If the aim of current CVOTs is to simply rule out excessive CV risk, this can so be accomplished through traditional RCTs. Although pragmatic trials can achieve a large cost-saving per patient as compared with conventional RCTs, their total costs also include those related to heterogeneous populations in the real-world clinical setting which can require large sample sizes to ensure sufficient power. Considering the high investment in pragmatic trials, a cost advantage may only be achieved if the study design also enables the potential for additional benefits beyond demonstrating CV safety [52]. The CVOT can be coupled with other research questions to better inform decision-making. For example, the pragmatic trial can answer whether the new treatment has advantages over usual care with regard to adverse effects, costs, convenience, or the control of body weight while maintaining glycemic control if patients can achieve well-controlled T2DM using oral anti-diabetic therapies.

References

[1] M. S. Barnish, S. Turner, The value of pragmatic and observational studies in health care and public health, *Pragmatic and Observational Research* 8 (2017) 49.

[2] P. M. Rothwell, External validity of randomised controlled trials:"to whom do the results of this trial apply?", *The Lancet* 365 (9453) (2005) 82–93.

[3] N. Black, Why we need observational studies to evaluate the effectiveness of health care, *BMJ* 312 (7040) (1996) 1215–1218.

[4] C. Booth, I. Tannock, Randomised controlled trials and population-based observational research: Partners in the evolution of medical evidence, *British Journal of Cancer* 110 (3) (2014) 551–555.

[5] M. Calvert, J. Wood, N. Freemantle, Designing "real-world" trials to meet the needs of health policy makers at marketing authorization, *Journal of Clinical Epidemiology* 64 (7) (2011) 711–717.

[6] N. Freemantle, T. Strack, Real-world effectiveness of new medicines should be evaluated by appropriately designed clinical trials, *Journal of Clinical Epidemiology* 63 (10) (2010) 1053–1058.

[7] D. Schwartz, J. Lellouch, Explanatory and pragmatic attitudes in therapeutical trials, *Journal of Chronic Diseases* 20 (8) (1967) 637–648.

[8] M. G. Zuidgeest, I. Goetz, R. H. Groenwold, E. Irving, G. J. van Thiel, D. E. Grobbee, G. W. Package, Series: Pragmatic trials and real world evidence: Paper 1. introduction, *Journal of Clinical Epidemiology* 88 (2017) 7–13.

[9] H.-G. Eichler, B. Bloechl-Daum, E. Abadie, D. Barnett, F. König, S. Pearson, Relative efficacy of drugs: An emerging issue between regulatory agencies and third-party payers, *Nature Reviews Drug Discovery* 9 (4) (2010) 277–291.

[10] R. M. Califf, R. Platt, Embedding cardiovascular research into practice, *JAMA* 310 (19) (2013) 2037–2038.

[11] E. Irving, R. van den Bor, P. Welsing, V. Walsh, R. Alfonso-Cristancho, C. Harvey, N. Garman, D. E. Grobbee, G. W. Package, Series: Pragmatic trials and real world evidence: Paper 7. safety, quality and monitoring, *Journal of Clinical Epidemiology* 91 (2017) 6–12.

[12] T.-P. van Staa, L. Dyson, G. McCann, S. Padmanabhan, R. Belatri, B. Goldacre, J. Cassell, M. Pirmo-Hamed, D. Torgerson, S. Ronaldson, et al., The opportunities and challenges of pragmatic point-of-care randomised trials using routinely collected electronic records: Evaluations of two exemplar trials, *Health Technology Assessment* 18 (43) (2014) 1–146.

[13] FDA Guidance for Industry, Oversight of clinical investigations, a risk-based approach to monitoring, available at https://www.fda.gov/media/116754/download (2013).

[14] EMA, Reflection paper on risk based quality management in clinical trials, available at http://www.ema.europa.eu/docs/en_GB/document_library/Scientific_guideline/2013/11/WC500155491.pdf (2013).

[15] ICH, E19: Optimisation of safety data collection, available at https://database.ich.org/sites/default/files/E19_EWG_Draft_Guideline.pdf (2019).

[16] J. McCambridge, J. Witton, D. R. Elbourne, Systematic review of the hawthorne effect: New concepts are needed to study research participation effects, *Journal of Clinical Epidemiology* 67 (3) (2014) 267–277.

[17] J. B. Nieuwenhuis, E. Irving, K. O. Rengerink, E. Lloyd, I. Goetz, D. E. Grobbee, P. Stolk, R. H. Groenwold, M. G. Zuidgeest, et al., Pragmatic trial design elements showed a different impact on trial interpretation and feasibility than explanatory elements, *Journal of Clinical Epidemiology* 77 (2016) 95–100.

[18] J. J. Caro, K. J. Ishak, Optimizing the design of pragmatic trials: Key issues remain, *Journal of Comparative Effectiveness Research* 1 (4) (2012) 319–327.

[19] V. Gamerman, T. Cai, A. Elsäßer, Pragmatic randomized clinical trials: Best practices and statistical guidance, *Health Services and Outcomes Research Methodology* 19 (1) (2019) 23–35.

[20] A. J. Cook, E. Delong, D. M. Murray, W. M. Vollmer, P. J. Heagerty, Statistical lessons learned for designing cluster randomized pragmatic clinical trials from the nih health care systems collaboratory biostatistics and design core, *Clinical Trials* 13 (5) (2016) 504–512.

[21] J. J. Lee, X. Gu, S. Liu, Bayesian adaptive randomization designs for targeted agent development, *Clinical Trials* 7 (5) (2010) 584–596.

[22] P. Lapuerta, T. Simon, A. Smitten, J. Caro, C. S. Group, et al., Assessment of the association between blood pressure control and health care resource use, *Clinical Therapeutics* 23 (10) (2001) 1773–1782.

[23] J. H. Noseworthy, G. C. Ebers, M. K. Vandervoort, R. Farquhar, E. Yetisir, R. Roberts, The impact of blinding on the results of a randomized, placebo-controlled multiple sclerosis clinical trial, *Neurology* 44 (1) (1994) 16–16.

[24] J. B. Christian, E. S. Brouwer, C. J. Girman, D. Bennett, K. J. Davis, N. A. Dreyer, Masking in pragmatic trials: Who, what, and when to blind, *Therapeutic Innovation & Regulatory Science* 54 (2) (2020) 431–436.

[25] P. M. Welsing, K. O. Rengerink, S. Collier, L. Eckert, M. van Smeden, A. Ciaglia, G. Nachbaur, S. Trelle, A. J. Taylor, M. Egger, et al., Series: Pragmatic trials and real world evidence: Paper 6. outcome measures in the real world, *Journal of Clinical Epidemiology* 90 (2017) 99–107.

[26] B. Rubbo, N. K. Fitzpatrick, S. Denaxas, M. Daskalopoulou, N. Yu, R. S. Patel, U. B. Follow-Up, H. Hemingway, O. W. Group, et al., Use of electronic health records to ascertain, validate and phenotype acute myocardial infarction: A

systematic review and recommendations, *International Journal of Cardiology* 187 (2015) 705–711.

[27] A.-K. Meinecke, P. Welsing, G. Kafatos, D. Burke, S. Trelle, M. Kubin, G. Nachbaur, M. Egger, M. Zuidgeest, et al., Series: Pragmatic trials and real world evidence: Paper 8. data collection and management, *Journal of Clinical Epidemiology* 91 (2017) 13–22.

[28] G. E. Simon, S. M. Shortreed, R. C. Rossom, R. B. Penfold, J. A. M. Sperl-Hillen, P. O'Connor, Principles and procedures for data and safety monitoring in pragmatic clinical trials, *Trials* 20 (1) (2019) 690.

[29] ICH, Integrated addendum to ICH E6 (R1): Guideline for good clinical practice e6 (r2), available at https://database.ich.org/sites/default/files/E6_R2_Addendum.pdf (2016).

[30] M. Zwarenstein, S. Treweek, J. J. Gagnier, D. G. Altman, S. Tunis, B. Haynes, A. D. Oxman, D. Moher, Improving the reporting of pragmatic trials: An extension of the consort statement, *BMJ* 337 (2008) a2390.

[31] ICH, E2D (R1) EWG: Post-approval safety data management, available at https://database.ich.org/sites/default/files/E2D-R1_FinalBusinessPlan_2019_1118.pdf (2019).

[32] S. Meredith, M. Ward, G. Booth, A. Fisher, C. Gamble, H. House, M. Landray, Risk-adapted approaches to the management of clinical trials: Guidance from the department of health (DH)/medical research council (MRC)/medicines and healthcare products regulatory agency (MHRA) clinical trials working group, *Trials* 12 (1) (2011) 1–2.

[33] J. P. New, N. D. Bakerly, D. Leather, A. Woodcock, Obtaining real-world evidence: The salford lung study, *Thorax* 69 (12) (2014) 1152–1154.

[34] K. E. Thorpe, M. Zwarenstein, A. D. Oxman, S. Treweek, C. D. Furberg, D. G. Altman, S. Tunis, E. Bergel, I. Harvey, D. J. Magid, et al., A pragmatic–explanatory continuum indicator summary (PRECIS): A tool to help trial designers, *Journal of Clinical Epidemiology* 62 (5) (2009) 464–475.

[35] K. Loudon, S. Treweek, F. Sullivan, P. Donnan, K. E. Thorpe, M. Zwarenstein, The PRECIS-2 tool: Designing trials that are fit for purpose, *BMJ* 350 (2015) h2147.

[36] L. D. Alphs, C. A. Bossie, ASPECT-R—A tool to rate the pragmatic and explanatory characteristics of a clinical trial design, *Innovations in Clinical Neuroscience* 13 (1–2) (2016) 15.

[37] S. E. Nissen, K. Wolski, E. J. Topol, Effect of muraglitazar on death and major adverse cardiovascular events in patients with type 2 diabetes mellitus, *JAMA* 294 (20) (2005) 2581–2586.

[38] S. E. Nissen, K. Wolski, Effect of rosiglitazone on the risk of myocardial infarction and death from cardiovascular causes, *New England Journal of Medicine* 356 (24) (2007) 2457–2471.

[39] FDA Guidance for Industry, Diabetes mellitus—evaluating cardiovascular risk in new antidiabetic therapies to treat type 2 diabetes, withdrawed (2008).

[40] E. E. Regier, M. V. Venkat, K. L. Close, More than 7 years of hindsight: Revisiting the FDA's 2008 guidance on cardiovascular outcomes trials for type 2 diabetes medications, *Clinical Diabetes* 34 (4) (2016) 173–180.

[41] M. Singh, R. Sharma, A. Kumar, Safety of sglt2 inhibitors in patients with diabetes mellitus, *Current Drug Safety* 14 (2) (2019) 87–93.

[42] FDA Background Document, Endocrinologic and metabolic drugs advisory committee meeting, available at https://www.fda.gov/media/121272/download (2018).

[43] FDA Guidance for Industry, Type 2 diabetes mellitus: Evaluating the safety of new drugs for improving glycemic control, available at https://www.fda.gov/media/135936/download (2020).

[44] C. J. Bailey, C. Day, The future of new drugs for diabetes management, *Diabetes Research and Clinical Practice* 155 (2019) 107785.

[45] W. T. Cefalu, S. Kaul, H. C. Gerstein, R. R. Holman, B. Zinman, J. S. Skyler, J. B. Green, J. B. Buse, S. E. Inzucchi, L. A. Leiter, et al., Cardiovascular outcomes trials in type 2 diabetes: Where do we go from here? Reflections from a diabetes care editors' expert forum, *Diabetes Care* 41 (1) (2018) 14–31.

[46] J. Rosenstock, N. Marx, S. E. Kahn, B. Zinman, J. J. Kastelein, J. M. Lachin, E. Bluhmki, S. Patel, O.-E. Johansen, H.-J. Woerle, Cardiovascular outcome trials in type 2 diabetes and the sulphonylurea controversy: Rationale for the active-comparator carolina trial, *Diabetes and Vascular Disease Research* 10 (4) (2013) 289–301.

[47] J. B. Green, M. A. Bethel, P. W. Armstrong, J. B. Buse, S. S. Engel, J. Garg, R. Josse, K. D. Kaufman, J. Koglin, S. Korn, et al., Effect of sitagliptin on cardiovascular outcomes in type 2 diabetes, *New England Journal of Medicine* 373 (3) (2015) 232–242.

[48] R. R. Holman, M. A. Bethel, R. J. Mentz, V. P. Thompson, Y. Lokhnygina, J. B. Buse, J. C. Chan, J. Choi, S. M. Gustavson, N. Iqbal, et al., Effects of once-weekly exenatide on cardiovascular outcomes in type 2 diabetes, *New England Journal of Medicine* 377 (13) (2017) 1228–1239.

[49] S. P. Marso, G. H. Daniels, K. Brown-Frandsen, P. Kristensen, J. F. Mann, M. A. Nauck, S. E. Nissen, S. Pocock, N. R. Poulter, L. S. Ravn, et al., Liraglutide and cardiovascular outcomes in type 2 diabetes, *New England Journal of Medicine* 375 (4) (2016) 311–322.

[50] W.-Q. Wei, Q. Feng, P. Weeke, W. Bush, M. S. Waitara, O. F. Iwuchukwu, D. M. Roden, R. A. Wilke, C. M. Stein, J. C. Denny, Creation and validation of an emr-based algorithm for identifying major adverse cardiac events while on statins, *AMIA Summits on Translational Science Proceedings* 2014 (2014) 112.

[51] M. A. Hlatky, R. M. Ray, D. R. Burwen, K. L. Margolis, K. C. Johnson, A. Kucharska-Newton, J. E. Manson, J. G. Robinson, M. M. Safford, M. Allison, et al., Use of medicare data to identify coronary heart disease outcomes in the women's health initiative, *Circulation: Cardiovascular Quality and Outcomes* 7 (1) (2014) 157–162.

[52] O. Schnell, E. Standl, D. Catrinoiu, S. Genovese, N. Lalic, J. Skra, P. Valensi, A. Ceriello, Report from the 1st cardiovascular outcome trial (cvot) summit of the diabetes & cardiovascular disease (d&cvd) easd study group, *Cardiovascular Diabetology* 15 (1) (2016) 33.

11

Post-Market Safety Assessment Using Observational Studies and the Food and Drug Administration Sentinel System

Yong Ma

Center for Drug Evaluation and Research, US Federal Food and Drug Administration, Silver Spring, MD, USA

Rima Izem

Novartis, Basel, Switzerland

Ellen Snyder

Merck & Co, Inc, Kenilworth, NJ, USA

Wei Liu

Center for Drug Evaluation and Research, US Federal Food and Drug Administration, Silver Spring, MD, USA

CONTENTS

11.1 Introduction 234
11.2 History and Role of the Sentinel System 234
11.3 The Organizational Structure of Sentinel 236
11.4 The Sentinel Common Data Model, Distributed Database, and
     Distributed Regression Analysis 237
11.5 Drug Utilization Using the Sentinel System 239
11.6 Signal Identification in the Sentinel System 240
    11.6.1 Details of the TreeScan Method 242
    11.6.2 Signal Identification in a Single-Exposure Group 244
    11.6.3 Signal Identification in a Controlled Design 244
    11.6.4 Signal Identification in a Self-controlled Design 244
11.7 Signal Refinement in the Sentinel System 245
    11.7.1 The L1 Analysis 245
    11.7.2 The L2 Analysis 246
    11.7.3 The L3 Analysis 247
11.8 ARIA Sufficiency 248
11.9 Discussion: Future Outlook 249
References 250

Disclaimer: This manuscript reflects solely the views of the author and should not be construed to represent FDA's views or policies.

DOI: 10.1201/9780429488801-14

11.1 Introduction

The drug safety profile prior to approval is largely determined through the evaluation of safety data obtained from clinical trials. Although often considered not as rigorous as inference from clinical trials, causal inference based on observational studies is sometimes the only feasible or timely option post-marketing. Pharmaco-epidemiologists have developed general best practices guidelines for causal inference based on observational data, and in 2013, the US Food and Drug Administration (FDA) published a guidance entitled "Best Practices for Conducting and Reporting Pharmacoepidemiologic Safety Studies Using Electronic Healthcare Data Sets".[1] The evolving consensus is that when the best practices are followed in designing and analyzing a study, they can produce reliable and reproducible results to address a safety question.

The FDA guidance document recommends tailoring a study design and analysis to each causal inference question and to fit regulatory purpose. Thus, an ideal study will make judicious choices about the data source(s), comparator(s), and inclusion/exclusion criteria while maximizing the sample size and generalizability of the findings. The ideal study would also use validated outcomes and have adequate control for confounding by design or analysis.[2-5]

The above design considerations are relevant in many FDA evaluations of drug safety. First, they are crucial when the FDA determines whether a safety question can be assessed in the post-marketing setting or if it should be assessed pre-marketing. Second, they are considered when the FDA determines whether their own resources can answer any particular safety question or whether to issue a post-marketing requirement to an applicant to investigate a safety issue. Finally, these guidelines are at the forefront of any FDA evaluation of post-marketing studies including published studies and studies submitted by industry to fulfill a regulatory obligation. The rest of the chapter will introduce the Sentinel system, the in-house system for the FDA to investigate post-marketing drug safety questions with observational studies. We will discuss the unique statistical methodologies applied as well as how each of the design elements is implemented within the Sentinel system. While these methods and examples are specific to the FDA-conducted studies, the same principles and methods apply when the FDA requests studies from pharmaceutical industry or evaluates evidence from the literature.

11.2 History and Role of the Sentinel System

The Sentinel system is a large network of healthcare databases, built by the FDA for drug safety surveillance. The Sentinel system has existed for more than a decade. It started with the passing of the Food and Drug Administration

Amendments Act (FDAAA) by the United States Congress in 2007. The FDAAA calls for creation of an "active risk identification and analysis" system and its application to post-market monitoring of the safety of medical products using the pre-existing electronic health care data from multiple sources in the context of FDA regulatory decision making about product safety.

FDAAA required the system to have the data on at least 100 million people by 2012. In 2008, the FDA launched the Sentinel Initiative, and the following year, the FDA launched the Mini-Sentinel Pilot. In 2011, Mini-Sentinel reached the mandated 100-million-lives mark, and in the next year, Mini-Sentinel established a suite of reusable programming tools for routine queries. Finally, in 2016 the Sentinel system was launched. The Sentinel system was further expanded in 2019 adding two additional coordinating centers to the existing Sentinel Operations Center (SOC): The Innovation Center and the Community Building and Outreach Center to the existing Sentinel Operations Center. Figure 11.1 gives the timeline of the Sentinel system.

In 2016, the FDA officially integrated the Sentinel system into its routine drug safety operations. All FDA centers involved with medical products are now actively using the Sentinel system to monitor the safety of drugs, vaccines, biologics, and medical devices. Within Sentinel, there are several analysis systems monitoring different types of medical products:

1. The *Active Risk Identification and Analysis (ARIA)* system monitors the safety of drug products.

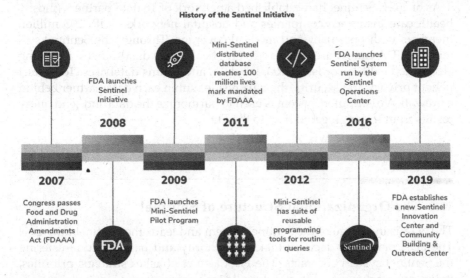

FIGURE 11.1
History of the Sentinel system

Source: https://www.sentinelinitiative.org/about

2. The *Post-market Rapid Immunization Safety Monitoring (PRISM)* system monitors vaccine safety.[6]

3. The *Blood Surveillance Continuous Active Surveillance Network (BloodSCAN)* supports a regulatory review of blood and blood products.[6]

4. The *CBER Biologics Effectiveness & Safety (BEST)* system was launched in October 2017 to assure the safety and effectiveness of biologic products including vaccines, blood and blood products, tissues, and advanced therapeutics.[6] Currently *BEST* has replaced *PRISM* and *BloodSCAN*.

FDA's Center for Devices & Radiological Health (CDRH) uses Sentinel as part of the post-marketing assessments to complement CDRH's other ongoing surveillance efforts, such as the FDA's manufacturer and user facility device experience, post-market mandated studies (for example, 522 post-market surveillance studies), and the National Evaluation System for Health Technology.[7]

The Sentinel Initiative also includes *FDA-Catalyst*, which supplements the Sentinel system. The *FDA-Catalyst* provides data coming from interventions or interactions with health plan members and/or providers. The *FDA-Catalyst* combines these data with data in the Sentinel infrastructure. An example would be the COVID MyStudies Mobile App for the electronic informed consent (eConsent) program, where the Sentinel Operations Center coordinates the branding and administration of the submission of secure and remote electronic consent.[8]

As of 2021, Sentinel had established a network of 16 data partners (mostly health care insurance companies and hospital networks) with 293 million members with pharmacy and medical coverage. Through the Sentinel system, the FDA can securely access diverse health care data sources, including electronic health records (EHRs), lab data, and claims databases. To protect patient privacy and security, the data reside within each data partner behind a firewall. A distributed system is used to harmonize the data and send query results from the aggregated data to the FDA.

11.3 The Organizational Structure of Sentinel

The FDA is in-charge of the Sentinel system and leads the Sentinel initiative. The FDA maintains decision-making authority and handles administering the Sentinel system contracts. The FDA also establishes strategic priorities, issues project requests, and reviews all work products.

The FDA collaborates closely with the SOC, led by the Harvard Pilgrim Health Care Institute. The SOC is the liaison that coordinates the efforts of conducting day-to-day operations of the Sentinel system.

There are two other centers within the Sentinel initiative, the Community Building & Outreach Center (CBOC) and the Sentinel Innovation Center (IC). Deloitte Consulting LLP established the CBOC and is currently operating it. The CBOC develops and engages the Sentinel community, expands the Sentinel system's user base, and strengthens Sentinel's capability to distribute knowledge based on real-world data (RWD). The Sentinel IC is a testbed to identify, develop, and evaluate innovative methods. The Sentinel IC's goal is to enhance Sentinel's capabilities with new analysis tools, novel data sources, and state-of-the-art approaches to automate key epidemiologic study operations.

11.4 The Sentinel Common Data Model, Distributed Database, and Distributed Regression Analysis

The data sources for observational studies usually come from one or more of the following: administrative claims data, electronic medical records, surveys, and registries. The choice of data source is determined by factors such as whether the data source captures the main design elements of interest, how reliable the data are, whether there is an opportunity for linkage with complementary data sources, and if the data source is sufficiently large.

FDAAA required the Sentinel system to have data on at least 100 million people; however, no single data source could satisfy such a requirement. Instead, data coming from multiple sources needed to be included without compromising patients' confidentiality. The Sentinel system uses the data coming from the 16 partners as of April 2021. Participating data partners hold data locally behind their firewalls; thus, Sentinel system is a distributed database. Data analysis is initiated by the SOC sending a query (i.e., analytic programming codes) to data partners, who subsequently return aggregated data. The aggregated data from each data partner are further aggregated into one final report for the FDA.

The founders of the Sentinel system recognized early on that the best way to leverage data from multiple sources in a timely way is to standardize the data sources. Consequently FDA/SOC builds, maintains, and updates the Sentinel Common Data Model (SCDM). While it takes considerable time for each new data partner to conform to this common terminology, the payoff is the ability to rapidly execute queries within the system.

The SCDM enables data partners to transform their data locally into the SCDM format. During this transformation, data partners remove or mask directly identifiable patient information. The standardized format enables them to execute routine querying tools, enabling the FDA to initiate its own medical product safety evaluation quickly, without writing new programs for each new study. The SOC coordinates the network of Sentinel data partners and leads to the development of the SCDM. Figure 11.2 describes the SCDM.

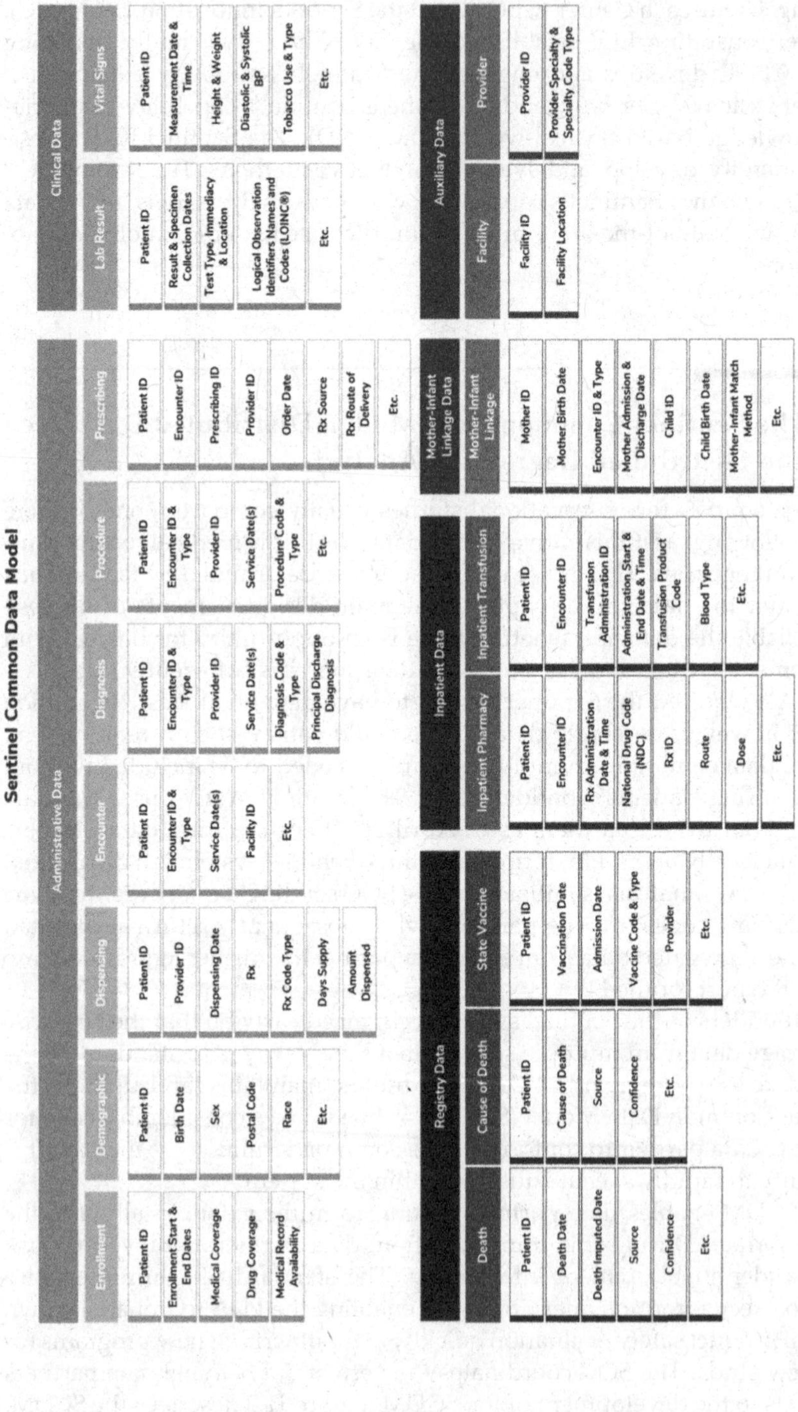

FIGURE 11.2

The Sentinel common data model

Source: https://www.sentinelinitiative.org/

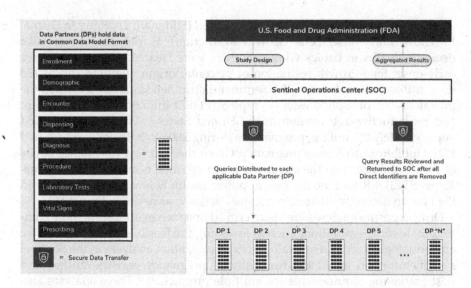

FIGURE 11.3
The Sentinel distributed databases

Source: https://www.sentinelinitiative.org/

The Sentinel system conducts regression analysis through Distributed Database, and Distributed Regression Analysis (DRA).[9] DRA is a set of statistical methodologies specifically developed to ensure valid regression estimates being generated without using individual-level data in a single location. Data analysis is performed in a "risk-set" format in DRA where aggregated data are transferred from data partners to SOC instead of individual level data. It has been demonstrated that the results using the risk-set approach can be mathematically equivalent to the data analysis using the pooled data set.[10] Figure 11.3 illustrates Sentinel's distributed data approach.

11.5 Drug Utilization Using the Sentinel System

Precise and timely drug utilization data are critical to public health and regulatory evaluations by the FDA. Thus, one of the first algorithms developed by FDA in the Sentinel system, the cohort identification and descriptive analysis (CIDA) program, identifies drug exposure and calculated exposure episodes. Counts of the number of patients exposed and the duration of exposure provide a real-world evidence (RWE) context for regulatory decisions because these numbers reflect those who will benefit or are at risk for a drug-related adverse event (AE) after the drug is approved in a broader population than the one recruited in pre-market clinical trials. For example, following the premature termination of the Dutch STRIDER (Sildenafil TheRapy In Dismal Prognosis Early-onset Fetal Growth Restriction)[11,12] trial due to excessive neonatal deaths in babies whose mothers were treated with sildenafil for early-onset fetal growth restriction of placental origin, the FDA conducted drug utilization analyses in the Sentinel distributed database to assess the prevalence of phosphodiesterase type-5 (PDE5) inhibitor use in pregnant and reproductive-age women in the United States.[13] The study identified approximately 3.3 million pregnancies during 2001 to 2018, 96 of which had PDE5 inhibitor use during pregnancy. Given the very low prevalence of use of PDE5 inhibitors and the inconsistency of elevated neonatal mortality rate across STRIDER centers, the risk to public health is low. These data helped the FDA to determine that no regulatory action was needed at present.

Drug exposure analyses are also a critical first step in informing more complex real-world evaluations such as assessing the feasibility of a prospective epidemiological study. For example, drug utilization analyses in the Sentinel system informed about the feasibility of pregnancy registry studies and post-marketing requirements for multiple products.[14,15] These analyses also provided potential contextual information for spontaneously reported AEs reported for these drugs.

These analyses can also be used to help predict the potential impact of different regulatory decisions on drug access and safety. For example, the Sentinel distributed database was used to characterize the use of higher versus lower dosage strength opioid analgesic products, including patient demographics, associated diagnoses, duration of use, concomitancy, and changing use patterns over time. These data were presented at an Advisory Committee meeting in June 2019 to inform about the discussion of the clinical need and risk mitigation for higher dosage opioid analgesic products.

In addition to CIDA algorithms for drug exposure, the FDA has developed multiple tools to investigate changes in patterns of drug utilization over calendar time and in different subgroups.[16,17] As in the examples presented earlier, drug utilization analyses are typically presented at most advisory committee meetings discussing post-marketing safety issues. Analyses typically stratify utilization by different characteristics impacting benefit or risk including formulation, dosage, indication, age, and pregnancy status. They also investigate the extent of exposure by calculating the duration of use, number of prescriptions per subject, and simultaneous exposure to other concomitant therapies.

11.6 Signal Identification in the Sentinel System

Signal identification is integral to post-marketing safety monitoring, including activities conducted using the Sentinel system. FDA uses the term signal to mean information that suggests a new potentially causal association, or a new aspect of a known association, between an intervention and an event or set of related events, that is judged to be of sufficient likelihood to justify further action to verify.[18]

In addition to safety signals arising from the pre-market development phase, most post-market safety alerts currently originate from another database, the FDA Adverse Event Reporting System (FAERS), a passive post-marketing surveillance system. However, the FDA is also building the analytical capabilities of the Sentinel system for signal detection, investigating multiple drug safety outcomes at once for the same exposure.

A safety alert is typically triggered for an unexpected safety event based on current knowledge or an unexpected magnitude of reports for an expected safety event. For example, non-arteritic anterior ischemic optic neuropathy (NAION) was an unexpected adverse event of PDE5 inhibitors found in the FAERS that led to the following drug safety communication.

"FDA ALERT [7/2005]: FDA has approved new labeling for Viagra, Levitra, and Cialis regarding post-marketing reports of vision loss related to NAION. Most, but not all, of these patients had underlying anatomic or vascular risk factors for development of NAION, including a low cup to disc ratio ("crowded disc"), age over 50, diabetes, hypertension, coronary artery disease, hyperlipidemia, and smoking. Given the small number of events, the large number of users of PDE-5 inhibitors and the fact that this event occurs in a similar population to those who do not take these medicines, it is not possible to determine whether these events are related directly to the use of PDE-5 inhibitors, to the patient's underlying vascular risk factors or anatomical defects, to a combination of these factors, or other factors. We cannot currently draw a conclusion of cause and effect. The FDA will continue to evaluate the issue. [...] As of May 18, 2005, a total of 43 cases of ischemic optic neuropathy among patients using the marketed PDE-5 inhibitors (sildenafil, tadalafil, and vardenafil) have been submitted to the FDA's Adverse Event Reporting System."

In contrast, bleeding events in patients treated with dabigatran reported in the FAERS were expected, but their magnitude exceeded the expectation based on pre-marketing clinical findings. These findings led to the following drug safety communication on December 7, 2011:

> The U.S. FDA is evaluating post-marketing reports of serious bleeding events in patients taking Pradaxa (dabigatran etexilate mesylate). Pradaxa is a blood-thinning (anticoagulant) medication used to reduce the risk of stroke in patients with non-valvular atrial fibrillation, the most common type of heart rhythm abnormality. [....] FDA is working to determine whether the reports of bleeding in patients taking Pradaxa are occurring more commonly than would be expected, based on observations in the large clinical trial that supported the approval of Pradaxa.

[…] FDA will communicate any new information on the risk of bleeding and Pradaxa when it becomes available.

The 2007 FDAAA mandates the use of data sources to "create a robust system to identify adverse events and potential drug safety signals". In Sentinel, signal identification, is often the result of screening numerous health outcomes that occur after exposure to a medical product. As false-positive signals can occur, the signal identification step is often followed by a clinical review and epidemiological safety studies to verify the signal.

The Sentinel system is less prone to reporting bias than FAERS. One of the first Sentinel investigations was of bleeding associated with dabigatran in the Sentinel system,[19] which led to a drug safety communication on November 2, 2012. "The U.S. FDA has evaluated new information about the risk of serious bleeding associated with the use of the anticoagulants (blood thinners) dabigatran (Pradaxa) and warfarin (Coumadin, Jantoven, and generics). […] This assessment was performed using insurance claims and administrative data from the FDA's Mini-Sentinel pilot of the Sentinel Initiative. The results of this Mini-Sentinel assessment indicate that bleeding rates associated with new use of Pradaxa do not appear to be higher than bleeding rates associated with new use of warfarin, which is consistent with observations from the large clinical trial used to approve Pradaxa (the RE-LY trial).[…]" Results from this investigation reassured the FDA and the public that the alert from FAERS were likely due to stimulated reporting rather than an excess bleeding risk of dabigatran compared to warfarin.

Figure 11.4 describes the overview of signal identification techniques available in the Sentinel system. Among them, the TreeScan method has been the most extensively evaluated. Three of the five methods described in Figure 11.4 are TreeScan-based. Two other methods, Information Component Temporal Pattern Discovery and Sequence Symmetry Analysis, are not TreeScan-based, and both use a self-controlled design.

11.6.1 Details of the TreeScan Method

TreeScan was developed in 2003 by Martin Kulldorff and colleagues.[20] It is a data mining tool simultaneously looking for excess risk in any of a large number of individual cells. It adjusts for multiple comparisons and uses a Poisson-based probability model or a binomial model. While the tree structure could be either in the exposure or in the outcome, we typically are trying to identify a specific AE associated with a drug of interest; therefore, we are examining an outcome tree.

Signal identification for a drug of interest could happen in a single-exposure group design, in a two-group design (i.e., with a comparator), or in a self-controlled design. Below we will describe each method in detail.

Overview of Signal Identification Techniques Utilized by the Sentinel System

Method	Study Design or Contrast	Test Statistic	Control for Multiple Testing	Adjustment for Trends in Healthcare Utilization
Information Component Temporal Pattern Discovery	Compares the rate of events in multiple pre-specified control and risk windows relative to the timing of a first dispensing using a self-controlled design, while adjusting for general dispensing patterns across the database	Ranks alerts based on the delta in Information Component between the risk and control windows	No, however, uses a shrinkage estimator to reduce false positives due to random variability or rare events	Yes
Propensity Score Based TreeScan	Compares the rate of events in a pre-specified risk window between persons newly exposed to a drug of interest who are matched by propensity score to a cohort of new users of a comparator drug	Ranks alerts based on the log-likelihood ration, a measure of observed versus expected counts, using a Bernoulli probability model	Yes, via Monte Carlo hypothesis testing	No
Self-Controlled TreeScan	Compares the rate of events in pre-specified control and risk windows within the same person			Optional
Sequence Symmetry Analysis	Compares whether an event occurs more frequently after exposure to a medication than before medication exposure using a self-controlled design	Ranks alerts based on magnitude of absolute difference in sequence orders and presented unadjusted p-values from chi-square tests	No	No
Tree-Temporal TreeScan	Compares the rate of events across multiple risk and control windows within the same person that do not require explicit pre-specification of the windows. Effectively combines the benefits of TreeScan with a temporal scan of many possible risk windows	Ranks alerts based on the log-likelihood ration, a measure of observed versus expected counts	Yes, via Monte Carlo hypothesis testing	Optional

FIGURE 11.4

Overview of signal identification techniques utilized by the Sentinel system

Source: https://www.sentinelinitiative.org/

11.6.2 Signal Identification in a Single-Exposure Group

In this case, we are interested in a newly marketed drug D. If no specific comparator is considered, all drugs used other than D in the database are grouped into the second group which serves as the comparator. The TreeScan method essentially compares an event with all other events to see if it is more likely to happen than expected. All AEs are first classified into a tree structure that follows a hierarchy such as the ICD-9 or ICD-10 coding system. We can make a hypothetical cut for each leaf of a tree and reduce the data into a 2×2 table. Here, we denote the total number of AEs in the exposed group as **C**, the total number of AEs in the database as **N**, and the number of the observed and expected number of AEs at a cut G as c_G and n_G. Then the expected number of AEs in each cell under the null hypothesis of no association can be estimated, conditioned on the total C number of AEs. Then we form the log-likelihood ratio statistic at each cut. The cut with the maximum likelihood constitutes the test statistic T as described in the formula below.

$$T = \left\{ c_G \log\left(\frac{c_G}{n_G}\right) + (C - c_G)\log\left(\frac{C - c_G}{N - n_G}\right) \right\} \times I\left(\frac{c_G}{n_G} > \frac{C - c_G}{N - n_G}\right)$$

The indicator I is used to capture only the AEs with increased risk compared to others.

There is no closed-form expression for T's distribution, and therefore, statistical significance is evaluated using Monte Carlo simulations.

11.6.3 Signal Identification in a Controlled Design

When a specific comparator is selected, the propensity score matching (PSM) method is often applied to control for confounding variables. For this situation, similar to signal identification in a single exposure group, we now use the total events in the two groups as N, and the above formula can still be used.

11.6.4 Signal Identification in a Self-controlled Design

In this case, a risk window and a control window are specified, and a patient is compared to himself/herself. The formula still applies except that now we are comparing two windows instead of two groups. Patient characteristics invariant in the two windows will be automatically controlled, but time-varying factors could still confound the study results. Confounding by time-varying factors can be adjusted by including the factors in the model if there is enough variation of the factors.[21]

11.7 Signal Refinement in the Sentinel System

The Sentinel system is also used to investigate an identified safety signal and, hence, serves as a "signal refinement" tool. Currently, data analysis for signal refinement is categorized into three levels, L1, L2, and L3.

11.7.1 The L1 Analysis

The L1 analysis can identify and extract cohorts of interest based on user-defined specifications including care setting, query period, exposures, outcomes, continuous enrollment requirements, washout period, incidence criteria, inclusion/exclusion criteria, relevant age groups, demographics, and other custom covariates. The L1 analysis usually calculates descriptive statistics for the cohort(s) of interest or unadjusted analyses such as a two-way contingency table of exposure relative to the outcome. These analyses can also minimally adjust for other characteristics such as the data partner, age group, sex, and year, by stratification. For example, the Sentinel investigation of bleeding and dabigatran[19] mentioned earlier was an L1 analysis. It showed the incidence of bleeding for dabigatran and of warfarin among all new users of either drug within the first few months of approval of dabigatran, stratified by age.

One way to adjust for confounding by indication is to consider an incident user cohort design. Drug exposure definition, ascertainment, and time of initiation are key in the L1 analysis. Before a L1 study can be initiated, the duration of exposure relevant for the outcome of interest needs to be determined. Also, the definition needs to be operationalized to address whether it is a point exposure or exposure over time, and whether the date of dispensing or the prescriptions' days-supplies should be used.

Inclusion and exclusion criteria are also operationalized to assess their impact on the generalizability of findings, control for biases, and impact on the sample size.

Safety outcomes identified in other observational studies could have come from a medical chart review, which is often considered the gold standard. However, the medical chart review in the Sentinel system is often not feasible or not available due to cost or timeliness. The outcome definition often comes from existing algorithms already validated or from a new or modified algorithm. If an existing algorithm is used, the transportability of the algorithm across settings is assessed. If a new or modified algorithm is used, the validity of the outcome definition in the study population should be determined before conducting a new study. Earlier in the mini-Sentinel phase, the acute myocardial infarction (AMI) validation project was conducted, and it helped establishing a process for validating the health outcome of interest in the Sentinel system.[22] ICD-9 codes including two decimal places were used to capture AMI and potential cases were evaluated using medical chart retrieval and review. Overall, the positive predicative value of the algorithm was 86% and ranged from 76% to 94% across four data partners.[23]

11.7.2 The L2 Analysis

The L2 analysis can identify cohorts of interest and perform more complex adjustments for confounding. It generates effect estimates and confidence intervals using regression methods such as linear regression, logistic regression, and time-to-event analysis (Cox's proportional hazard regression, etc.).

In an L2 analysis, details of the study design and data analysis, will be specified in a document called "Sentinel Concept Brief", similar to a usual study protocol specifying primary data analysis, secondary and sensitivity analyses, modeling approach and validations, sample size estimation, and how to handle missing data.

As with any observational study, biases introduced from confounders or effect modifiers can compromise the validity of a pharmacoepidemiologic safety study if not adequately addressed. Similarly, in the Sentinel system, all confounders and effect modifiers should have operational definitions and specific methods for their specified during the study design phase. For unmeasured confounders, a quantitative bias assessment could be used to evaluate their impact on the study results.[24]

To control for confounding by indication and channeling, a comparator drug with the same indication is often selected and used as a control in the study design. When feasible and relevant, multiple comparator groups can also be used. Other comparator choices include any drug for the same indication but is not associated with the outcome of interest (negative control drug) or self-control (patient as own control).

As for any study, a sufficient sample size and adequate study power are essential for L2 studies. The sample size and statistical power are often estimated before initiating an L2 study with the information extracted from a L1 analysis. The assumptions and the formulas used to calculate the sample size are often documented in the "Sentinel Concept Brief".

In the Sentinel system, propensity score (PS) methods are used to control for confounding. Commonly used methods include PSM, PS weighting, and PS stratification.

PSM was developed first and has been used for a long time. It is considered the most developed PS approach, and it is a popular choice in causal analysis for observational studies.[25,26] PSM provides a marginal estimate (the average effect in the matched population) and is intuitive in interpretation; it has less residual confounding than stratification and is less sensitive to extreme weights. However, PSM often suffers from sample loss, an issue with generalizability, and sometimes instability of model estimates.

At the time of this writing, PS weighting is also an option in the Sentinel system and provides an alternative tool. It also provides a marginal estimate of the average treatment effect in the full study population, hence minimizes sample size loss, and yields study results more generalizable compared to PSM. PS weighting suffers from being sensitive to extreme weights and is less intuitive than PSM, and PS weighting's performance is often poor if the study groups have little overlap.

PS stratification is also available in the Sentinel system. This method provides a conditional estimate. Its advantages include less sample loss and less sensitivity to extreme weights. When the stratification is 'fine', it performs just like PSM. However, it does suffer the drawback that when strata are few, there could be more residual confounding.

There are many L2 studies conducted using the Sentinel system. One example is the investigation of the association between dabigatran use and the risk of stroke, bleeding, and myocardial infarction.[27] This study used PSM to compare adults with atrial fibrillation initiating dabigatran to those initiating warfarin therapy. It was found that there were no significantly different rates of ischemic stroke or extracranial hemorrhage between the two drugs, but dabigatran showed less risk of intracranial bleeding and increased risk of myocardial infarction.

11.7.3 The L3 Analysis

The L3 analysis can identify cohorts of interest and perform complex adjustments for confounding as part of prospective sequential analysis. Sentinel is a collection of real-world dynamic distributed databases which get refreshed constantly. Sequential analysis in Sentinel is a prospective surveillance approach where pre-specified exposure-outcome pairs are evaluated periodically using frequently updated data. A safety signal is detected when there is a statistically significant excess risk, after adjusting for multiple sequential hypothesis testing. The key benefit of such a sequential analysis is its potential to detect a safety signal earlier than a one-time analysis. Sequential designs have been used extensively in randomized trials but are less frequently used in observational studies. The Sentinel system established a framework allowing a systematic, scalable, and transparent sequential design-planning process for medical product safety surveillance systems utilizing observational electronic health care databases.[28]

This sequential analysis was piloted with a comparison of rivaroxaban to warfarin among patients with atrial fibrillation using a group sequential method.[29,30] Each sequential look accrued information in two ways: through matching incident rivaroxaban users to incident warfarin users using variable ratio propensity score matching and through accruing follow-up for those already matched. To control for multiple looks and testing, this study used a Pocock boundary to split Type I error between looks. This pilot illustrated the complexity of pre-specifying a fixed strategy for control for confounding in sequential safety testing in the post-market setting when characteristics of the indicated population may change over time.

11.8 ARIA Sufficiency

Before the Sentinel system's existence, sponsors were asked to perform post-marketing safety analyses using observational data through post-marketing requirements (PMRs). The PMRs are usually issued at the time of approval but sometimes are issued after approval if the FDA becomes aware of new safety concerns. Under the FDAAA, the FDA may require the sponsor to conduct a PMR study if it is determined that neither the FAERS nor the ARIA system will be sufficient to address the safety concern.

To determine whether a medical product safety issue could be assessed in the Sentinel system, the FDA evaluates ARIA sufficiency by examining the level of evidence or intended purpose of the study (e.g., hypothesis generation, signal strengthening, or confirmation), whether ARIA is capable of meeting the requirement of the following study elements: proper study population, identification of the exposure, identification of the outcome(s), identification of covariates and available analytical tools.[31]

By the end of 2019, the FDA had undertaken analyses of more than 200 medical product safety issues (i.e., product-outcome pairs) identified between Fall 2015 and November 2019 to determine whether the capabilities of ARIA (i.e., the electronic data in the Sentinel Common Data Model and the existing Sentinel analytic tools) were sufficient to meet the specific study purpose for each safety issue.[32] The FDA determined ARIA to be sufficient to address approximately 40% of the product-outcome pairs.

The FDA found that the inability to identify the health outcome of interest was the most common reason for ARIA insufficiency. However, other factors could also be inadequate and lead to ARIA insufficiency.

Here we present two examples, one deemed ARIA insufficient and the other ARIA sufficient, to help illustrate how ARIA sufficiency is determined.

The FDA approved Xcopri (cenobamate) in November 2019 for the treatment of partial-onset seizures in adults. At the time of approval, data of pregnancy exposure during clinical trials were insufficient to inform the risk of maternal, fetal, and infant outcomes associated with the use of cenobamate. Female subjects who were pregnant or lactating were excluded from enrolling in the cenobamate clinical studies. In animal studies, administration of cenobamate during pregnancy or throughout pregnancy and lactation resulted in adverse effects on fetal development. It was therefore concluded that use in women of child-bearing age is a general concern.

After evaluating ARIA's capacity to address this concern, it was found that although the study population, drug exposures, safety outcomes, and relevant covariates might be correctly defined, ARIA lacked the analytical tools, i.e., the data mining methods for identifying birth defects and other pregnancy outcomes. ARIA was deemed insufficient for addressing this safety concern and a PMR was issued requiring a retrospective cohort study using claims or electronic medical record data with outcome validation, along with a pregnancy registry study.

Ibsrela (tenapanor hydrochloride) was approved in September 2019 by the FDA for the treatment of constipation-predominant irritable bowel syndrome (IBS). There is biologic plausibility for developing inflammatory bowel disease (IBD) among those exposed to tenapanor based on the drug's mechanism of action, preclinical data, and an imbalance observed with respect to some IBD symptoms in clinical trial data. The concern remains theoretical as confirmatory assessments with colonoscopies were not undertaken in the clinical trials but there remains the potential for a serious chronic risk. After evaluating all aspects of a planned study (the study population, drug exposures, safety outcomes, relevant covariates, and analytic tools), it was determined that the tools in ARIA are sufficient to assess an IBD risk among IBS patients treated with tenapanor. The FDA will monitor the accrual of tenapanor exposure to determine when adequate exposure has been captured in Sentinel to allow for the assessment and then perform an analysis to evaluate the risk.

11.9 Discussion: Future Outlook

At the time of this writing, the US is experiencing the unprecedented COVID-19 pandemic. The Sentinel system is being used to promote and protect public health by conducting key analyses such as near real-time monitoring of critical drugs used for the care of patients with COVID-19, assessment of the natural history of the disease, and validation of claims-based algorithms to identify in-hospital COVID-19 events within the FDA Sentinel system.[33] These and many other Sentinel activities can be found in the Sentinel's public website https://www.sentinelinitiative.org/.

Clinicians, pharmaco-epidemiologists and statisticians continue to work together on making full utilization of observational studies in safety decision making. The 21st Century Cures Act of 2016 mandates the FDA to evaluate the use of RWE to support regulatory decision-making, including approval of new indications for approved products and support post-approval study requirements. The implementation of the Cures Act has the potential to further broaden the use of RWE for regulatory decision-making. The Sentinel system will continue to evolve rapidly to support broadening the spectrum of using RWD to address clinical and regulatory questions. Sentinel published its five-year master plans in February 2021 which highlighted the areas of further development.[34] These activities include expanding data partners, data granularity, improving causal inference methods and signal detection capacities, and streamlining the workflow and operational efficiency. Sentinel is also planning to build on its capacities for causal inference using RWD for safety which should also help inform FDA's understanding of the use of RWD to measure effectiveness. Finally, the Sentinel System can significantly broaden its value by engaging and sharing with a broader user community.

Lately, interim results from the RCT DUPLICATE initiative, designed to emulate clinical trials using RWD informed the contexts when RWE may offer causal insights in situations where randomized controlled clinical trial data is either not available or cannot be quickly or feasibly generated.[35] Future research and development in this area will shed light on continued improvement in best practices.

References

1. FDA Guidance Document Best Practices for Conducting and Reporting Pharmacoepidemiologic Safety Studies Using Electronic Healthcare Data Sets. 2013 https://www.fda.gov/regulatory-information/search-fda-guidance-documents/best-practices-conducting-and-reporting-pharmacoepidemiologic-safety-studies-using-electronic
2. Chakravarty AG, Izem R, Keeton S, Kim CY, Levenson MS, Soukup M. The Role of Quantitative Safety Evaluation in Regulatory Decision Making of Drugs. *J Biopharm Stat*. 2016;26(1):17–29.
3. Marchenko O, Russek-Cohen E, Levenson M, Zink RC, Krukas-Hampel MR, Jiang Q. Sources of Safety Data and Statistical Strategies for Design and Analysis: Real World Insights. *Ther Innov Regul Sci*. 2018;52(2):170–186.
4. Ma H, Russek-Cohen E, Izem R, Marchenko OV, Jiang Q. Sources of Safety Data and Statistical Strategies for Design and Analysis: Transforming Data into Evidence. *Ther Innov Regul Sci*. 2018;52(2):187–198.
5. Izem R, Sanchez-Kam M, Ma H, Zink R, Zhao Y. Sources of Safety Data and Statistical Strategies for Design and Analysis: Postmarket Surveillance. *Ther Innov Regul Sci*. 2018;52:159–69.
6. https://www.sentinelinitiative.org/assessments/vaccines-blood-biologics
7. https://www.sentinelinitiative.org/assessments/devices-radiological-health
8. https://www.fda.gov/drugs/science-and-research-drugs/covid-mystudies-application-app

9. Her QL, Malenfant JM, Malek S, et al. A Query Workflow Design to Perform Automatable Distributed Regression Analysis in Large Distributed Data Networks. *EGEMS (Wash DC).* 2018;6(1):11.

10. Shu D, Yoshida K, Fireman BH, Toh S. Inverse Probability Weighted Cox Model in Multi-site Studies without Sharing Individual-level Data. *Stat Methods Med Res.* 2020;29(6):1668–1681.

11. Mohamoud MA FDA presentation, Advisory Committee on September 17th 2014. https://wayback.archive-it.org/7993/20170403223746; https://www.fda.gov/AdvisoryCommittees/CommitteesMeetingMaterials/Drugs/ReproductiveHealthDrugsAdvisoryCommittee/ucm404895.htm

12. Hawkes N. Trial of Viagra for Fetal Growth Restriction is Halted after Baby Deaths. *BMJ* 2018;362:k3247.

13. Liu W, Menzin TJ, Woods CM, et al. Phosphodiesterase Type 5 Inhibitor Use among Pregnant and Reproductive-age Women in the United States. *Pharmacoepidemiol Drug Saf* 2021; 30(2):126–134.

14. Bird ST, Gelperin K, Taylor L, Sahin L, Hammad H, Andrade SE, Mohamoud MA, Toh S, Hampp C. Enrollment and Retention in 34 United States Pregnancy Registries Contrasted with the Manufacturer's Capture of Spontaneous Reports for Exposed Pregnancies. *Drug Saf.* 2018;41(1):87–94.

15. Illoh OA, Toh S, Andrade SE, Hampp C, Sahin L, Gelperin K, Taylor L, Bird ST. Utilization of Drugs with Pregnancy Exposure Registries during Pregnancy. *Pharmacoepidemiol Drug Saf.* 2018;27(6):604–611.

16. Cocoros NM et al. A New Analytic Tool Developed to Assess Safe Use Recommendations. *Pharmacoepidemiol Drug Saf.* 2019;28(5):649–656.

17. Gagne JJ et al. Evaluation of Switching Patterns in FDA's Sentinel system: A New Tool to Assess Generic Drugs. *Drug Saf.* 2018;41(12):1313–1323.

18. FDA draft document Best Practices in Drug and Biological Product Postmarket Safety Surveillance for FDA Staff. 2019 https://www.fda.gov/media/130216/download

19. Southworth MR, Reichman ME, Unger EF. Dabigatran and Postmarketing Reports of Bleeding. *N Engl J Med.* 2013;368(14):1272–4.

20. Kulldorff M, Fang Z, Walsh S. A Tree-based Scan Statistic for Database Disease Surveillance. *Biometrics* 2003;59:323–331.

21. Whitaker HJ, Farrington CP, Spiessens B, Musonda P. Tutorial in Biostatistics: The Self-controlled Case Series Method. *Stat Med.* 2006;25(10):1768–1797.

22. Cutrona SL, Toh S, Iyer A, Foy S, Cavagnaro E, Forrow S, Racoosin JA, Goldberg R, Gurwitz JH. Design for Validation of Acute Myocardial Infarction Cases in Mini-Sentinel. *Pharmacoepidemiol Drug Saf.* 2012;21(Suppl 1):274–81.

23. Cutrona SL, Toh S, Iyer A, Foy S, Daniel GW, Nair VP, Ng D, Butler MG, Boudreau D, Forrow S, Goldberg R, Gore J, McManus D, Racoosin JA, Gurwitz JH. Validation of acute myocardial infarction in the Food and Drug Administration's Mini-Sentinel program. *Pharmacoepidemiol Drug Saf.* 2013;22(1):40–54.

24. McClure DL, Raebel MA, Yih WK, Shoaibi A, Mullersman JE, Anderson-Smits C, Glanz JM. Mini-Sentinel Methods: Framework for Assessment of Positive Results from Signal Refinement. *Pharmacoepidemiol Drug Saf.* 2014;23(1):3–8.

25. Pearl J. The foundations of Causal Inference. *Sociol. Methodol.* 2010;40:75–149.

26. Austin PC. An Introduction to Propensity Score Methods for Reducing the Effects of Confounding in Observational Studies. *Multivariate Behav Res.* 2011;46(3):399–424.

27. Go AS, Singer DE, Toh S, Cheetham TC, Reichman ME, Graham DJ, Southworth MR, Zhang R, Izem R, et al. (2017). Outcomes of Dabigatran and Warfarin for Atrial Fibrillation in Contemporary Practice A Retrospective Cohort Study. *Ann Intern Med.* 2017;167(12):845–854.

28. Nelson JC, Wellman R, Yu O, et al. A Synthesis of Current Surveillance Planning Methods for the Sequential Monitoring of Drug and Vaccine Adverse Effects Using Electronic Health Care Data. *EGEMS* (Wash DC). 2016;4(1):1219.

29. Martin D, et al. (2018). Sequential Surveillance for Drug Safety in a Regulatory Environment. *Pharmacoepidemiol Drug Saf. 2018;27*(7):707–712.

30. Chrischilles EA et al. (2018). Prospective Surveillance Pilot of Rivaroxaban Safety within the US Food and Drug Administration Sentinel System. *Pharmacoepidemiol Drug Saf. 2018;27*(3):263–271.

31. Guidance for Industry Postmarketing Studies and Clinical Trials — Implementation of Section 505(o)(3) of the Federal Food, Drug, and Cosmetic Act. 2011 https://www.fda.gov/media/131980/download

32. Nguyen MD, Maro JC, De Marco N, Money D, Pinheiro S, Kashoki M, Dal Pan G, Ball R. When is Real World Data Good Enough to Assess Drug Safety? Lessons Learned from the FDA's Sentinel System. In press.

33. https://www.sentinelinitiative.org/assessments/coronavirus-covid-19

34. FDA Sentinel system 5-year strategy 2019-2023 https://www.fda.gov/media/120333/download

35. Franklin JM, Patorno E, Desai RJ, Glynn RJ, Martin D, Quinto K, Pawar A, Bessette LG, Lee H, Garry EM, Gautam N, Schneeweiss S. Emulating Randomized Clinical Trials With Nonrandomized Real-World Evidence Studies: First Results From the RCT DUPLICATE Initiative. *Circulation* 2021;143(10):1002–1013.

12

Analysis Considerations for Real-World Evidence and Clinical Trials Related to Safety

Richard C. Zink
Lexitas Pharma Services, Inc, Durham, NC, USA

Melvin Munsaka
AbbVie, Inc., North Chicago, IL, USA

Birol Emir
Pzifer Inc., New York, NY, USA

Yong Ma
Center for Drug Evaluation and Research, US Federal Food and Drug Administration, Silver Spring, MD, USA

Judy X. Li
Bristol-Myers Squibb, San Diego, CA, USA

William Wang
Merck & Co, Inc, Kenilworth, NJ, USA

Dimitri Bennett
Takeda Inc, Tokyo, Japan; University of Pennsylvania, Philadelphia, PA, USA

CONTENTS

12.1 Introduction 254
12.2 Case Study on Chronic Hepatitis C 256
12.3 Statistical Methods to Account for Confounding for
   Observational Data 258
   12.3.1 Introduction 258
   12.3.2 What Are PSs? 258

DOI: 10.1201/9780429488801-15

12.3.3 PS Methods and Other Approaches for Dealing with
        Confounding...260
        12.3.3.1 PS Methods 260
        12.3.3.2 Additional Considerations in the Use of PSs..............260
        12.3.3.3 Other Methods for Confounding.................................264
12.4 Synthesis of Multi-Source Data..266
        12.4.1 Introduction 266
        12.4.2 Linking Data Sources 268
        12.4.3 External Control Arms 269
        12.4.4 Meta-Analysis of Safety 270
        12.4.5 Case Study on Safety Data Integration of RCTs and RWEs........271
12.5 Concluding Remarks..273
Acknowledgments...274
References...274

12.1 Introduction

Data from randomized controlled trials (RCTs) have traditionally played a primary role in regulatory decision-making. However, the cost of clinical development has skyrocketed in recent years. The Tufts Center for the Study of Drug Development estimates that the pre-tax capitalized cost per approved drug was $2.6 billion in 2013 dollars (DiMasi et al., 2016). Previous estimates in Year 2000 dollars put the cost at just over $800 million (DiMasi et al., 2003). If this trend continues, the costs of new therapies will place an extreme burden on consumers and health care systems, and it may make the development of new drugs in areas of unmet need, such as rare diseases, practically impossible to conduct.

The increasing cost of clinical development for sponsors coupled with challenges in studying hard-to-reach populations, rare diseases, and rare endpoints (such as for safety outcomes), along with a general lack of understanding as to how medical products are used in practice has motivated regulators and pharmaceutical industry sponsors to explore the utility of other sources of data. To empower the United States Food and Drug Administration (US FDA) to consider data outside of RCTs in its regulatory decisions, the United States Congress passed the 21st Century Cures Act in 2016 (US FDA, 2016). Two important features of the act specific to RWD include the ability to support the approval of new indications for already-approved drugs and the ability to support or satisfy post-approval study requirements. In late 2018, the US FDA published a framework for their real-world evidence (RWE) program (US FDA, 2018). In this document, the US FDA defines RWD and RWE as follows:

– RWD are "data relating to patient health status and/or the delivery of health care routinely collected from a variety of sources", and
– RWE is the "clinical evidence about the usage and potential benefits or risks of a medical product derived from RWD".

RWD sources can include structured and unstructured information derived from electronic health and medical records (EHR and EMR), administrative healthcare claims databases, hospital or death records, product or disease registries, pragmatic trials, and the many post-market surveillance databases of the US FDA, the European Medicines Agency (EMA), and the World Health Organization (WHO). Recently published evidence summarized recommendations to appropriately analyze, report, and interpret safety outcomes and the advantages and disadvantages of utilizing safety data from RCT and RWD sources (Zink et al., 2018; Izem et al., 2018; Marchenko et al., 2018; Ma et al., 2018). Recently, the FDA released a draft guidance document for submitting RWD to the agency (US FDA, 2019). Our focus in this chapter is to examine RWD in the context of safety outcomes. For example, Fralick and co-authors (2018) describe an example where RWD sources were used to select a population of patients similar to those from a clinical development program to compare telmisartan versus ramipril among numerous cardiovascular endpoints for a supplemental indication; efficacy endpoints were similar, though authors identified an increased risk of angioedema for patients on telmisartan.

There are numerous challenges for the study of safety outcomes, particularly in the context of RCTs (Zink et al., 2018). Here, we limit the challenges to those that can be addressed specifically through RWD sources. First and foremost, clinical trials to establish efficacy rarely have a sufficient sample size to appropriately power comparisons for uncommon safety outcomes. This problem is further exacerbated when looking for an increased risk of severe events within subgroups defined by patient and disease characteristics. Second, clinical trial populations often exclude patients with other major comorbidities, as well as those individuals taking one or more concomitant medications to address signs or symptoms of a disease, since both can potentially confound trial outcomes. Finally, RCTs for chronic indications are often too short in duration to adequately assess long-term safety outcomes. In contrast, post-market surveillance databases contain millions of reported adverse events, and the data partners involved with the FDA Sentinel initiative have over 300 million person-years of data available, including for patients who would normally not meet the inclusion criteria to participate in clinical trials (Ball et al. 2016; Marchenko et al. 2018).

However, there are challenges to using RWE generated by RWD in practice. One of the key challenges is the lack of randomization of patients to treatment interventions (though randomization is present in pragmatic trials). In these observational settings, comparisons between treatments require appropriate design and rigorous statistical methodologies to limit the effects of selection bias to assess safety and effectiveness. Another challenge of observational data involves the quality and validity of the evidence (Cameron et al., 2015). The Strengthening the Reporting of Observational studies in Epidemiology (STROBE) guidelines, Grading of Recommendations Assessment, Development, and Evaluation (GRADE) system, or Guidelines for Good Pharmacoepidemiology Practices (GPP) can be used to assess the quality and validity of observational data (Von Elm et al., 2007; Guyatt et al., 2011, 2011, 2011; Balshem et al., 2011; ISPE, 2015). Often these criteria help to determine the extent to which potential biases between treatments were addressed, and whether there is sufficient documentation describing statistical methodologies, data collection, and eligibility criteria. While the quality of evidence from RWD is important when analyzed on its own, sufficient details are required when analyzing these data in conjunction with data from randomized sources. For example, relevant demographics, medical history, and historical and concomitant medications are required to provide a meaningful estimand of the treatment effect.

This work is authored by the Safety Scientific Working Group (SWG) of the Biopharmaceutical Section of the American Statistical Association (ASA). In Section 12.2, we describe a case study of RWD/RWE in patients with chronic hepatitis C to motivate later discussions. In Section 12.3, we describe statistical methods to account for the biases and confounding inherent in observational studies and describe numerous examples related to safety outcomes. Decision-making using multi-source data is the focus of Section 12.4. We summarize our findings in Section 12.5 and discuss some ongoing challenges.

12.2 Case Study on Chronic Hepatitis C

HCV-TARGET is a collaboration between the Universities of North Carolina at Chapel Hill and Florida to study the real-world implications of the direct-acting antivirals (DAAs) for chronic hepatitis C viral (HCV) infection that began receiving regulatory approval in 2011 (Mishra et al. 2017). The study utilized patient medical records supplemented with patient-reported out-comes and data derived from biospecimens to inform the management of patients represented and under-represented in clinical trials (such as African American patients, or those with cirrhosis, decompensated cirrhosis, or those whom have received liver transplants), identify and remediate gaps in treatment guidelines, describe the management of adverse events, and iden-tify treatment-resistant variants of the virus. Data from HCV-TARGET were instrumental to support the initially off-label regimen of sofosbuvir and simeprevir (dropping peg-interferon and ribaviran), as well as a shortened treatment duration for the combination of sofosbuvir and ledipasvir (12 to 8 weeks) until data from clinical trials became available. HCV-TARGET provided evidence to answer treatment-specific safety questions surround-ing anemia and bradyarrhythmia, as well as the incidence of hepatocellular carcinoma (HCC) recurrence after DAA treatment. In 2015, the randomized pragmatic substudy PRIORITIZE (prioritizestudy.org) commenced in order to compare the safety and effectiveness of three commonly prescribed com-bination therapies: Harvoni (ledipasvir/sofosbuvir), Viekira PAK (ombitas-vir/paritaprevir/ritonavir/dasabuvr), and Zepatier (elbasvir/grazoprevir). PRIORITIZE enrollment closed in early 2018. Given the rapidly chang-ing environment for HCV treatments, HCV-TARGET was able to provide actionable evidence before the regulatory landscape for HCV treatments changed. In contrast, the results of some early post-marketing HCV clinical trial results were completed with the results already considered outdated (Mishra et al. 2017).

As described above, HCV-TARGET provided preliminary data of HCC recurrence following treatment with DAAs. These data were provided in response to a small observational study that showed an excessive relapse rate in DAA-treated patients (Reig et al., 2016). It was hypothesized that the rapid clearance of HCV perturbs the patient's immune response allowing cancer cells to proliferate. The EMA Pharmacovigilance Risk Assessment Committee (PRAC) recommended further study of this phenomenon in an observational setting (EMA PRAC, 2016). In 2017, a meta-analysis estimat-ing rate of HCC relapse was published (Cabibbo et al., 2017), and numer-ous papers providing evidence for and against the hypothesis of excess HCC recurrence after DAA therapy were published. Just recently, Singal et al. (2019) published an analysis of data from 31 health systems through-out the US and Canada that DAA therapy was not associated with HCC recurrence or early (within one year of complete response) HCC recurrence. Furthermore, the Direct Acting Antiviral Post Approval Safety Study (DAA-PASS, ClinicalTrials.gov Identifier: NCT03707080) is an ongoing real-world study to examine HCV+ DAA-naïve HCC patients with no evidence of tumors to compare recurrence rates between DAA-exposure and non-expo-sure in US and European patients.

12.3 Statistical Methods to Account for Confounding for Observational Data

12.3.1 Introduction

While RWE is often observational, randomization can be utilized in real-world studies. In fact, the US FDA framework emphasizes the importance of randomization for limiting bias, even in RWE studies (US FDA, 2018). The document summarizes ADAPTABLE, a randomized pragmatic trial using EHRs and claims data to study safety outcomes for two doses of aspirin, and VALIDATE-SWEDEHEART, a randomized registry study of two antiplatelet therapies to understand myocardial infarction and bleeding risks (Hernandez et al. 2015; Fröbert et al. 2013). Though the case study described in Section II was designed principally to answer questions surrounding efficacy, numerous safety endpoints will be summarized to understand the risks between the various treatments. This case study shares the same challenges for evaluating safety as it does for efficacy. The Salford Lung Study is a recent randomized pragmatic trial to compare a novel therapy against standard of care for patients at their general practitioner visits (Vestbo et al., 2016). To better understand serious adverse event (SAE) risks, study nurses at participating hospitals tagged safety issues in patient EMRs for further review by the principal investigator for determining event severity and causality. Finally, Zhao et al. (2016) described a randomized substudy within an observational HIV cohort that utilized matched patients from the observational cohort to create an informative prior in order to have a more efficient analysis of the randomized patients.

However, it is not possible in all situations to utilize randomization in studies of RWD sources. In the past, comparisons between treatment groups were made using traditional statistical models which adjusted for important covariates or confounding factors that may differ between groups due to the lack of randomization. Nowadays, analyses rely more on propensity scores (PSs) to help balance treatment groups by creating a "pseudo-randomization" of patients using their measured data. We discuss PSs in more details below. Finally, while the aforementioned examples utilized randomization to help limit bias, the treatments were provided in an open-label fashion which still allows for the potentially unobserved other forms of bias which PSs will not address. Greater understanding of how this lack of blinding may influence endpoints is needed.

12.3.2 What Are PSs?

An increasingly popular method compared to the traditional approach of covariate adjustment of treatment effects through statistical models, PSs are used to minimize the effects of bias on the estimate of the treatment effect

(Inacio et al., 2015). Propensities serve as "balancing scores" for measured confounders, and outcomes of interest are compared between groups while considering this balancing to reduce the potential bias of confounders. The basic idea behind PSs is to measure (or have information on) a number of characteristics for each patient for the time period before they are exposed and to use this information to create a model that predicts the probability of receiving the exposure of interest compared to a suitable control. More technically, a propensity can be defined as the probability of being exposed (treated, vaccinated, receiving surgery, etc.) based on patient (or hospital or regional) characteristics that are present before exposure occurs. The main advantage of the PS relative to traditional multivariable adjustment is the separation of the adjustment for confounding factors (which can be performed before endpoints become available) with the analysis of the treatment effect of the study outcome.

A PS is defined as the probability of a patient being assigned to a specific intervention, given a set of covariates. That is, it is the conditional probability of being assigned to a particular treatment given a vector of observed covariates (Rosenbaum & Rubin, 1983; Austin & Stuart, 2015). Suppose we have observed covariates $X = (X_1, X_2, ..., X_n)$ and a binary indicator of treatment group Z ($Z = 1$) (treated) and $Z = 0$ (if control) for each patient. The PS is defined for each patient as the probability that she or he received the treatment ($Z = 1$), given her or his observed covariates $PS = P(Z = 1|X)$. The probability each patient receives the specific treatment or control ranges from 0 to 1, that is, $0 < PS < 1$. The theory of PSs relies on two major assumptions, namely,

1. Every subject has a non-zero probability to receive either treatment.
2. There are no unmeasured confounders. In other words, all the covariates potentially related to treatment assignment are known.

As the PS summarizes all patient characteristics into a single covariate, it may reduce confounding (although it does not eliminate the potential for overfitting). PS methods aim to achieve some of the characteristics of RCTs by compensating for different patients having different probabilities of being assigned to the exposures under investigation. Thus, the aim of these methods is to attenuate problems of confounding of patient characteristics and assignment to an intervention typically found in observational studies. The PS allows one to design and analyze data from observational (non-randomized) studies so that it mimics some of the characteristics of a RCT. Specifically, conditional on the PS, the distribution of observed baseline covariates will mostly be similar between treated and untreated subjects.

In summary, PSs are useful in observational studies in emulating RCTs, reducing or eliminating confounding (selection bias), balancing treatment groups on covariates, and allowing for direct comparison of effects between treatments. Wang and Donnan (2001) discuss some considerations of PSs as they apply to drug safety studies.

12.3.3 PS Methods and Other Approaches for Dealing with Confounding

12.3.3.1 PS Methods

There are several PS methods that are available to address the confounding challenges in observation studies. Table 12.1 below summarizes these methods.

12.3.3.2 Additional Considerations in the Use of PSs

There are other considerations that need to be accounted for in practical applications of PSs including covariate selection and balance and accounting for unmeasured confounding. Additionally, there is increasing interest in the use of machine learning approaches in PSs. Below we describe some considerations for covariate selection and balance, unmeasured confounding, and machine learning in the context of PSs.

TABLE 12.1

Common PS Methods

PS Method	Description
Matching	In PS matching, each treated patient is allocated one control patient (in 1:1 matching), or multiple control patients (in 1:n matching), with the same PS or within a pre-specified caliber such that the scores differ only slightly. The treatment effect is then estimated in the matched population.
Stratification	Stratification on the PS involves grouping patients into mutually exclusive subsets based on their estimated PS. Ranked according to their estimated PS, subjects are then stratified into subsets using thresholds of the estimated PS. A common approach is to divide subjects into five equal-size groups using the quintiles of the estimated PS. Treatment effects are estimated for each stratum, and a stratified estimate is used to combine the information across strata.
Inverse Probability Weighting	Inverse probability of treatment weighting (IPTW) utilizes weights based on the PS to create a synthetic sample in which the distribution of measured baseline covariates is independent of treatment assignment. The use of IPTW is similar to the use of survey sampling weights that are used to weight survey samples so that they are representative of specific populations.
Covariate Adjustment	In covariate adjustment, the outcome variable is regressed on an indicator variable denoting treatment status with the estimated PS as a covariate. The choice of the regression model would depend on the nature of the outcome. For continuous outcomes, a linear model would be chosen; for dichotomous outcomes, a logistic regression model may be selected.

12.3.3.2.1 *Covariate Selection and Balance*

Identifying relevant variables (i.e., covariate selection and balance) that should be included in the PS model is one of the critical considerations when using PSs. Imbalance in some covariates is expected in the tails of the PS's distribution, which might include observations that are outside the range of common support. Here, balance refers to the ability of the PSs to make the distributions of covariates included in the model similar between the treatments, potentially within strata. Diagnostic plots (e.g., bar charts, box plots, and standardized mean differences plot) are often used to assess how successfully PSs create this balance. For example, it is recommended that standardized mean differences plot present values within ± 0.1. See Normand et al. 2001; Mamdani et al. 2005; Austin 2009; and Austin (2019) for more discussion. Some imbalances in the mean may indicate a need for the PS to be re-specified. In general, balance in key covariates is more crucial than balance in covariates that are less likely to impact the outcome. Of note, balance in mean values does not indicate balance in higher order moments. There is no agreed-upon best method of balancing the PS, and there is no rule regarding how much imbalance is acceptable. One can compute standardized differences which consider both means and variances. Another approach is to drop variables that are considered to be of less importance, recategorizing variables (e.g., making a continuous variable categorical or dichotomous), including interactions between variables or including higher order terms or splines of variables. Transformed variables may have different distributions across treatment and comparison groups that enable balance across groups. Other methods for identifying relevant variables include proxy adjustment and automated techniques for variable selection, such as the high-dimensional PS algorithm for variable selection.

After the PS is balanced within blocks across the treatment and comparison groups, some diagnostic checks should be performed to assess the level of balance of the individual covariates across treatment and comparison groups within blocks of the PS. Diagnostics checks for balance should be performed after the sample has been matched or weighted on the PS. In any event, it is most likely that the initial specification will most likely not be balanced. Diagnostic checks on covariate balance will ensure that the PS's distribution is similar across groups within each block and that the PS is properly specified. Additional discussions on covariate selection and balance can be seen in Arbogast & Ray (2011); Austin (2019); Benedetto et al. (2018); Li, Morgan, & Zaslavsky (2018); Elze et al. (2017); and Brookhart et al. (2006).

12.3.3.2.2 *Unmeasured Confounding*

In practical applications of PSs, there is often the assumption made that no unmeasured confounding exists in estimating treatment effects. In essence because of the limitation in the data, confounders included are not exhaustive.

Confounders that require detailed information on clinical parameters, life-style, or over the counter medications are often not measured in many data sources and can result in unmeasured confounding. Uncontrolled confounding can give rise to biased estimates in treatment effects. It is therefore important to examine the extent to which the resulting estimate is sensitive to unmeasured confounders. The use of directed acyclic graphs (DAGs) can be used to understand the relationships between the variables.

Sensitivity analysis techniques have been suggested for assessing the magnitude of these biases due to confounding. They provide a systematic approach for investigating the impact of residual confounding. The four basic approaches to sensitivity analysis include sensitivity analyses based on an array of informed assumptions, analyses to identify the strength of residual confounding that would be necessary to explain an observed drug-outcome association, external adjustment of a drug-outcome association given additional information on single binary confounders from survey data using algebraic solutions, and external adjustment considering the joint distribution of multiple confounders of any distribution from external sources of information using PS calibration. Sensitivity analyses and external adjustments can improve the understanding of treatment effects. Additional discussions of unmeasured confounding can be seen in the studies by Haneuse, VanderWeele, & Arterburn (2019); Zhang et al. (2018); Streeter et al. (2017); Faries et al. (2013); Byrd & Ho (2011); Groenwold et al. (2016); and Fewell et al. (2007).

12.3.3.2.3 *Machine Learning Approaches*

As noted above earlier, there are many questions and concerns and often heated debates on disadvantages in the use of PSs and methods in reducing selection bias and effective estimation of treatment effects. Recent advancements in statistical machine learning have offered alternatives to the use of PSs. Proponents of machine learning approaches argue that these methods are superior to existing parametric models. One method that has been used is classification and regression tree (CART) methods which encompasses a diverse number of classification and prediction algorithms (Breiman et al., 1984). In contrast to statistical approaches to modeling that assume a data model with parameters estimated from the data, CART tries to extract the relationship between an outcome and predictor(s) through a learning algorithm without a priori model assumptions. CART methods can help some of the challenges in PSs that require multiple assumptions. PS methods attempt to achieve covariate balance rather than estimating a true relationship.

Empirical studies have supported CART methods being more effective in achieving this goal when the true relationship between predictor(s) and PS is unknown or complicated (Lee, Lessler & Stuart, 2010). Some additional advantages of CART in estimating PSs include the capacity to handle categorical, ordinal, continuous, and missing data well, insensitivity to outliers and monotonic transformations of variables, and being able to model inter-actions and non-linearities. Additionally, and in practice, the PS is unknown and needs to be estimated and the incorrect PS estimation model produces biased treatment effect estimates, and there may not be a sufficient theoreti-cal or empirical basis to specify the PS estimation model. Random Forests as an automatic and non-parametric method have been suggested to deal with regression problems with many covariates and complex non-linear and interaction effects of the covariates. As machine learning approaches mature in the context of propensity, there is likely to be an increase in their use. For some additional discussions on the use of machine learning algorithms and covariate balance in PS matching and weighting, see the studies by Cannas & Arpino (2019); Goller et al. (2019); Sizemore & Alkurdi (2019); Tu (2019); de Vries, Smeden, & Groenwold (2018), Linden & Yarnold (2016); Cham (2013); and Lee, Lessler & Stuart (2010).

12.3.3.2.4 Use of PSs in Multiple Treatments

Most applications of PSs tend to focus on the settings when comparing esti-mates between two treatments. Some authors have attempted to extend the use of PSs to more than two treatments. Imbens (2000) extended past work to multiple treatments by defining a generalized PS as the conditional prob-ability of receiving a particular level of the treatment given the pre-treatment variables. Their methodology is based on the generalized PS for estimating treatment effects in the case of multiple treatments. The two methods they discussed include PS regression adjustment and PS weighting which they applied in a multiple treatment study.

McCaffrey et al. (2013) provided some guidance on the use of PSs in mul-tiple treatment settings including a step-by-step guidance implementing PS weighting for multiple treatments. They also proposed the use of general-ized boosted models (GBMs) for estimation of the necessary PS weights and defined the causal quantities that may be of interest to studies of multiple treatments and derive weighted estimators of those quantities. The authors also included a detailed plan for using the GBM to estimate PSs and using those scores to estimate weights and causal effects and provided tools for assessing balance and overlap of pretreatment variables among treatment groups in the context of multiple treatments.

For some additional discussions on the use of PSs in multiple treatments, see for example, the studies by Burgette, Griffin & McCaffrey (2020); Hu et al. (2020); Li & Li (2019); Zhang & Kim 2019; Lopez & Gutman (2017); Feng et al. (2012); McCaffrey et al. (2013); and Wyss et al. (2013).

12.3.3.2.5 Example Using PSs

Non-vitamin K oral anticoagulants (NOACs) are used for stroke prevention in patients with atrial fibrillation, but no clinical trials have compared them against each other for effectiveness or safety. The FDA leads an observational study of Medicare patients with non-valvular atrial fibrillation, comparing the three most commonly used NOACs, apixaban, dabigatran, and rivaroxaban among themselves and with warfarin users (Graham et al., 2019). The goal was to assess the risks of thromboembolic stroke, intracranial hemorrhage, major extracranial bleeding, and all-cause mortality of the three NOACs. The study used a two-step process to achieve covariate balances. At first, a 1:1 PS matching of warfarin to NOAC users with replacement was performed. This was followed by a second step where a multinomial logistic regression was used to estimate stabilized inverse probability of treatment weights for the four cohorts, using the three NOAC cohorts pooled together as the reference for weighting. The standardized mean differences were found to be ≤ 0.1, indicating negligible differences between groups. PS-adjusted Cox proportional hazard regression was used to estimate adjusted hazard ratios (HRs) and 95% confidence intervals (CIs) for the outcomes. Compared with warfarin, each NOAC was associated with reduced risks of thromboembolic stroke (20%-29% reduction; P = 0.002 [dabigatran], P < 0.001 [rivaroxaban, apixaban]), intracranial hemorrhage (35%-62% reduction; P < 0.001 [each NOAC]), and mortality (19–34% reduction; P < 0.001 [each NOAC]). The NOACs were similar for thromboembolic stroke, but rivaroxaban was associated with increased risks of intracranial hemorrhage (vs dabigatran: HR = 1.71; 95% CI, 1.35-2.17), major extracranial bleeding (vs dabigatran: HR = 1.32; 95% CI, 1.21–1.45; vs apixaban: HR = 2.70; 95% CI, 2.38–3.05), and death (vs dabigatran: HR = 1.12; 95% CI, 1.01–1.24; vs apixaban: HR = 1.23; 95% CI, 1.09–1.38). Dabigatran was associated with reduced risk of intracranial hemorrhage (HR = 0.70; 95% CI, 0.53–0.94) and increased risk of major extracranial bleeding (HR = 2.04; 95% CI, 1.78–2.32) compared with apixaban. The study concluded that, among patients treated with standard-dose NOAC for non-valvular atrial fibrillation and warfarin users with similar baseline characteristics, dabigatran, rivaroxaban, and apixaban were associated with a more favorable benefit–harm profile than warfarin. Among NOAC users, dabigatran and apixaban were associated with a more favorable benefit–harm profile than rivaroxaban.

See the studies by Vallarino et al. (2013) and Funch et al. (2019) for other safety examples that utilize PSs.

12.3.3.3 Other Methods for Confounding

Although there has been much progress in the use of PSs, there are still many questions and concerns and often heated debates on the disadvantages of these methods. For example, King and Nielsen (2019) have cautioned against the use of PS matching. Furthermore, the use of parametric models such as logistic regression to estimate PSs is based on the assumptions of variable selection (strong ignorability), the functional form and distributions of these variables, and the explicit specification of interactions. When these assumptions, which often cannot be empirically examined, are not met, covariate balance may not be achieved which will result in a biased treatment effect estimate. There are several alternative methods discussed in the literature, and some of these are summarized in Table 12.2.

TABLE 12.2

Other Methods for Confounding

Method	Description
Coarse Matching	Coarse matching refers to a situation in which the patient populations are sorted into a finite number of classes and then randomly matched within these classes. Often this involves categorizing continuous endpoints into a small number of groups with a goal of matching treatment and control patients for analysis.
Disease Risk Score	The disease risk score is another type of summary score which estimates disease probability conditional on non-exposure (Glynn, Gagne & Schneeweiss, 2012). Whereas a propensity score summarizes the way potential risk factors differ between users of alternative treatments to be compared, the disease risk score characterizes the relationship of risk factors with the study outcome. The disease risk score may be estimated in two ways. First, it can be calculated as a "full-cohort" disease risk score, constructed from a regression model relating the study outcome to the exposure of interest and the covariates for the entire study population. The fitted values for each study subject after setting the exposure status to "un-exposed" are then used to group subjects into strata in order to calculate a stratified estimate of the exposure effect. The disease risk score can also be estimated by fitting a model using the unexposed population only, with the fitted values then computed for the entire cohort. Summary measures of disease risk play a prominent role in guidelines for drug use and the evaluation of possible effect measure modification of new treatments or new indications for established therapies.
Instrumental Variables	Instrumental variable analysis is a method for controlling unmeasured confounding (Baiocchi, Cheng & Small, 2014). This type of analysis requires the identification of a valid instrumental variable, which is a variable that (i) is independent of the unmeasured confounding; (ii) correlated with the treatment; and (iii) affects the outcome only indirectly through its effect on the treatment. Instrumental variable methods have been used in studies to remove bias caused by possible unmeasured confounders (i.e., missing for all individuals) when estimating treatment effects. The validity of instrumental methods relies on including treatment effect confounders that are associated with the instrumental variables in the analysis. Ignoring missing values in such baseline variables may bias the result.
G-estimation	G-estimation is a method for estimating the joint effects of time-varying treatments using ideas from instrumental variable methods (Vansteelandt & Joffe, 2014). It allows for appropriate adjustment of the effect of a time-varying exposure in the presence of time-dependent confounders that are themselves influenced by the exposure. While standard regression requires no feedback between time-varying treatments and time-varying confounders, G-estimation does not.

(Continued)

TABLE 12.2

(Continued)

Method	Description
Targeted maximum likelihood estimation (TMLE)	TMLE is a semiparametric doubly robust method that improves the chances of correct model specification by allowing for flexible estimation using non-parametric machine learning methods (van der Laan MJ & Rubin, 2006; Luque-Fernandez et al., 2018). Doubly robust methods only require correct specification of either the outcome or the exposure model. Building on TMLE as a foundation, targeted learning is a statistical framework for handling causal inference on observational data (van der Laan & Rose, 2011).
Structural Nested Models	In assessing the effect of time-varying treatments, structural nested models (SNMs) are useful in dealing with confounding by variables affected by earlier treatments. These models often consider treatment allocation and repeated measures at the individual level (Almirall, Ten Have, Murphy, 2010; He, Stephens-Shields & Joffe, 2015).
Marginal Structural Models	Marginal structural models are an alternative to G-estimation. The marginal structural models are a class of causal models that allow for improved adjustment for confounding in situations of time-dependent confounding. They have two major advantages over G-estimation. The first is that it can be used for continuous, binary, count, and time-to-event outcomes, while G-estimation is limited to continuous outcomes and binary outcomes when the event rate is rare. The second major advantage is that marginal structural models resemble standard statistical models and common statistical software packages can be used, whereas G-estimation does not.

12.4 Synthesis of Multi-Source Data

12.4.1 Introduction

Data in a real-world setting can be collected through various sources including EHRs, claims and billing activities, disease registries, patient-generated data through the completion of surveys, or wearable or mobile devices. In contrast to well-controlled RCT data, the representativeness of these datasets often critically depends on the technology for data collection. The reliability (data accrual and data quality) and relevance of the underlying RWD sources, as well as the methodologies used to generate the evidence, are important criteria to evaluate the use of RWD together with RCTs in the context of regulatory decision-making (US FDA, 2018).

Evidence from RWD in conjunction with data from RCTs has multiple applications to safety. One of the applications is randomized clinical trials that use RWD to capture clinical outcomes or safety data, including pragmatic and large simple trials. Another common application is to bridge post-marketing with pre-marketing clinical trials by using RWD/RWE to fulfill the post-marketing requirement to further evaluate some certain safety concerns raised from the pre-marketing RCT studies as well as to gather key safety outcomes on important under-represented patient populations not studied during the development program, such as cirrhotic patients with HCV (US FDA, 2019).

Analyses conducted based on multi-source data can be challenging, especially with regard to safety. Variability in results from multi-source studies, the varying time frame over which data were collected, the quality of documentation for how selection bias was addressed for summary-level data, or the level of detail of underlying patient characteristics could allow for potential bias in the analyses. For example, AEs in clinical trials are coded using the MedDRA dictionary, while events from other sources may be coded using ICD-9 or ICD-10 codes. An overall coding step would be necessary to ensure that similar events are captured consistently for analysis. Understanding the different data source environments and identifying the potential sources of bias for comparing outcomes are also important. Due to the lack of rigorousness of the study design, certain key data may not be routinely collected, collected with less regularity, or collected in a manner that may not be particularly informative. On the other hand, as more data sources become available, it also increases the potential of duplicated information from overlapping sources.

For example, due to the inherent safety risk of performing a liver biopsy, information on the presence (or not) of liver cirrhosis may not be available. New technologies such as vibration-controlled transient elastography may be used to determine the likelihood of cirrhosis, though thresholds are often dependent on the underlying disease of the patient, and depending on the data source the specific measurements may not be available (though a code indicating cirrhosis is present may be). Furthermore, patients may have other signs or symptoms that are highly suggestive of cirrhosis such as presence of ascites, portal hypertension, splenomegaly, varices, or low platelet counts. In other words, identifying certain outcomes may be complex; modifications to traditional definitions may need to be made based on the granularity of the available data. Further assumptions may need to be made about the absence of certain outcomes. For example, suppose a physician remarks on various aspects of disease (or lack thereof) from diagnostic imaging through CT scans or MRIs. The physician will likely note the presence of abnormalities (e.g., "scan indicates features of cirrhosis"), and depending on patient health, they may explicitly note that certain abnormalities are not present (e.g., "patient has no evidence of cirrhosis"). However, if no mention of cirrhosis is made, particularly in diseases where cirrhosis is likely, is one to assume that the patient does not have cirrhosis? This seems to be a reasonable assumption (one that should be pre-specified prior to analysis), though sensitivity analyses of explicit presence or absence of certain clinical features could potentially include the "not mentioned" as "absent".

12.4.2 Linking Data Sources

There are an increasing number of available data sources, but not all sources contain all of the data that may be required to sufficiently answer a research hypothesis. For example, a healthcare claims database or a hospital charge-master data set may contain some patient demographic information, as well as very detailed records on procedures that were performed or diagnostic or event codes that were attributed to the patient, though the underlying data that contributed to a particular determination may not be available. For example, a patient may have an ICD-10 code indicating that they are hypertensive, or the patient may be taking anti-hypertensive medications (which would indicate hypertension), but these data sources will typically not contain specific values for a patient's systolic or diastolic blood pressures. Linking is a great way of connecting data between multiple sources to allow for a more detailed analysis.

Linking is a particular challenge in the US, since there is no single identification number that is used to identify a patient across healthcare systems (US FDA, 2018). Complicating things further, since many people tend to be insured through their employers, an individual may not have long-term follow-up in any single data source or may have gaps that are reflective of changes in insurers over time (Girman et al., 2019). However, it is possible to use combinations of available protected health information (PHI) in the data sources to link data and conclude with high certainty that they belong to the same individual. Linking may be performed deterministically, where all values for a predetermined set of identifiers must match in order for a match to be declared. This may be problematic as some data points may be missing or transcription errors may be present that prevent exact matches. However, unique identifiers (tokens) can be created which will allow for straightforward matching of patients between data sources using various sets of available identifiers. Alternatively, probabilistic methods consider that some identifiers have more discriminatory power than others and assign scores that indicate the likelihood of a match. Scores beyond a predefined threshold are determined to be correct matches.

Whatever linkage approach is ultimately employed, some assessment of the performance of the linkage methodology should be undertaken. Note that the success of linkage is dependent on the standardization of data across various sources including formats, prefixes, and punctuation. Patient nicknames can further complicate the analysis. For a thorough overview of linkage methodologies, see the study by Dusetzina et al. (2014). The study by Curtis et al. (2014) contains an example of deterministic linkage that includes physician visit dates in the algorithm for a rheumatoid arthritis registry. Finally, Tromp et al. (2011) performed a simulation study comparing deterministic and probabilistic methods.

12.4.3 External Control Arms

External or synthetic control arms are often utilized in studies of rare diseases where randomizing patients to a control arm would assign precious limited available patients to a non-experimental arm. This would result in less precise estimates of the treatment response as well as a smaller population in which to observe important safety endpoints for the novel therapy. Furthermore, external controls are often utilized in severe diseases, such as oncology, where there is no suitable control available and/or it may be deemed unethical to randomize patients to a control arm. Historical data available from past clinical trials or the medical literature may be used as a reference to understand and contextualize the observed benefit and risk of a single-arm trial. Alternatively, available patient-level data could be used to match patients from external studies to the patients in a single-arm trial, and this matched cohort could be analyzed similar to a typical trial; see the study by Chen et al. (2021) for a thorough summary of matching techniques. As an alternative to replacing the entire control arm, external data may be used to supplement a smaller concurrent control arm (Stuart & Rubin, 2008; Yuan et al., 2019). Such an approach provides an opportunity to evaluate the similarity of control findings from concurrent and historical sources. Goring et al. (2019) performed a recent review of US and European approval and found that of 43 products utilizing non-randomized designs, 4 involved matching to external controls with patient-level data, 12 referred to external controls, while 27 made no explicit mention to control data. See the studies by Viele et al. (2014), Lim et al. (2018), Lewis et al. (2019), and Ghadessi et al. (2020) for recent reviews of methodologies for incorporating historical data, including the potential warning from Lewis and coauthors where use of historical data alone would have resulted in concluding significant treatment benefits when there was none. The above cited papers tend to focus on efficacy evaluation. For safety evaluation, particular attention should be paid to ensure that the definition of safety endpoints is consistent with the external control, including dictionary coding and the relevant safety event adjudication (if any).

The FDA RWE framework (2018) raises important points about external control arms, some of which were alluded to in the preceding paragraph. It is important to consider how relevant past data, particularly data from the distant past, may be to relevant to ongoing investigations. For example, there can be changes to how medical practice is conducted, including changes to technologies, endpoints and how those endpoints are obtained, or the need or duration of follow-up care. Furthermore, the disease itself could be different due to the lack of treatments available at the time, such as the presence of novel mutations in viral genomes due to first-generation treatments (Sherman et al., 2006) or mutations in hematologic cancers (La Rosée P & Deininger, 2010). As described in the introduction and elsewhere, every data source should be assessed for its suitability for a particular research question and analysis; observational sources can rely on STROBE, GRADE, or GPP criteria. RWD sources may allow for the selection of a more contemporaneous set of patients to serve as external controls to an ongoing investigation, supplementing a control arm or serving as a control arm in its entirety. See Burcu et al. (2020) and Thorland et al. (2020) for thorough reviews of non-randomized data for external control arms.

12.4.4 Meta-Analysis of Safety

Meta-analysis is particularly important in the analysis of safety endpoints. First, one of the principal challenges inherent to safety analysis is the rarity of many safety outcomes. This rarity requires large sample sizes or long follow-up in order to observe a sufficient number of events to appropriately power comparisons between the treatment arms. Clinical development programs are generally of insufficient size or duration to identify rare outcomes; this partially explains why safety issues are observed once a drug has been on the market for some time. Second, meta-analysis allows for more precise estimates of treatment effects within important subgroups. Understanding treatment effects within subgroups allows the study team to assess the consistency and robustness of results obtained for the entire study population, and whether the estimated overall effect is broadly applicable to patients across the proposed indication. Finally, meta-analysis provides the opportunity to assess the consistency of treatment responses across studies. See the study by Zink et al. (2018) for more discussion on meta-analysis in the context of safety endpoints, particularly among multiple randomized studies. It is important to note that traditional meta-analysis allows for comparisons of treatments that are observed within each study. Network meta-analysis makes it possible to compare treatment arms that may not be available within the same study, making it possible to indirectly compare safety and efficacy endpoints between treatments (Efthimiou et al., 2016).

Similar to discussions above for external control arms, evidence from RWD can be used in conjunction with data from RCTs to conduct meta-analyses. For example, Prevost, Abrams and Jones (2000) discuss an example that combines data from RCTs, case-control studies, and one observational geographic study to assess the effectiveness of breast cancer screening in reducing mortality. The study authors suggest several approaches for considering the differences between randomized and observational sources:

1. Naïve: analyze studies without consideration for whether evidence is from randomized or observational sources. This approach is generally not recommended.

2. Informative priors: use observational data sources to develop more informative priors to analyze data from RCTs to improve analytical efficiency.

3. Model study types: include effects to account for differences between types of studies (e.g., randomized vs. observational) to assess potential differences between sources of evidence.

Schmitz, Adams, and Walsh (2013) describe similar approaches for analysis when applied to network meta-analyses. See the studies by Cameron et al. (2015), Verde and Ohmann (2015), and Efthimiou et al. (2017) for discussions of network meta-analysis with randomized and observational studies. Note that a meta-analysis of data pooled across multiple studies may increase power, but missingness in the outcome or failed randomization can introduce bias. The doubly robust TMLE can be applied in a meta-analysis to reduce bias in causal effect estimates and either meet or exceed the performance of other estimators in the literature (Gruber & van der Laan, 2013).

In a recently published framework for the synthesis of observational studies and RCTs (Sarri, Patorno, Yuan et al., 2020), the authors proposed a seven-step framework to systematically identify evidence for observational studies, critically appraise, and appropriately synthesize it with the results from RCTs. Authors highlight that the effect estimates from all the randomized and observational studies should not directly be combined in a meta-analysis without any type of statistical adjustment and offer recommendations on the most appropriate statistical approaches to follow in combining evidence from these sources; this framework potentially will improve transparency and build confidence in the use of observational studies and help in a decision-making process among regulators and other healthcare decision-makers. See Sarri, Patorno, and Yuan et al. (2020) for a thorough review of the framework for the synthesis of observational studies and RCTs.

Finally, repeating an important point made elsewhere: data sources should be assessed for quality, completeness, and sufficient documentation. Of particular concern, particularly when working with summary-level data, is to ensure that observational sources appropriately account for the potential of selection bias in the estimated treatment effect. All documentations should be used to assess the suitability of combining various data sources for meta-analysis, and the heterogeneity of the included studies should be assessed.

12.4.5 Case Study on Safety Data Integration of RCTs and RWEs

Historically, RCTs are designed to focus on internal validity for the cause-and-effect relationship of treatment and outcome in order to obtain regulatory approval (ICH, 2000). At the same time, observational studies are relevant for their external validity and generalizability (Berger, Sox, Willke et al., 2017). In this context, comparative effectiveness is used for making treatment options more relevant for the broader patient population. There are methods to aggregate the findings of multiple studies using traditional and network meta-analyses (DerSimonian & Laird, 1986; Lumley, 2002).

Although there may be some inherent limitations, these analyses still provide very valuable information.

Meanwhile, clinicians must develop practice guidelines and make individual-level treatment assignments using these aggregated population data. Is it possible to apply these data to individual patient treatments? In other words, could we quantitatively link RCT data with available observational data to have the benefit of the external validity of the observational data and internal validity of the RCT? Such analyses would better inform the clinician to make optimal treatment choices. In this context, one could still perform meta-analyses (Kunutsor, Wylde, Whitehouse et al., 2019).

Here, we summarize an example that can be applied to many other diseases (Alexander, Edwards, Savoldelli et al., 2017; Alexander, Edwards, Brodsky et al., 2018; Alexander, Edwards, Manca et al., 2018; Alexander, Edwards, Brodsky et al., 2019; Alexander, Edwards, Manca et al., 2019). This case study is based on pregabalin (Lyrica) for the treatment of peripheral neuropathy. One of the challenges in healthcare is that we have a diverse patient population with many trajectories in their medical histories that could be relevant to how they respond to the treatment. How can this history be used to customize care for patients? Can we utilize RCT data to predict Lyrica responders in real-world settings? This is particularly challenging because patient characteristics like pain scores are not systematically collected in clinical practice. However, information on a number of common domains is collected across RCTs and routine practice. The Virtual Lab project explored whether these common domains (such as laboratory measurements) could predict the treatment response.

Alexander, Edwards, and Savoldelli et al. (2017) implemented hierarchical clustering analysis to identify patient clusters in a large observational study to which RCT patients could be matched using the coarsened exact matching technique (Iacus, King G & Porro, 2012). Next, autoregressive moving average models (ARMAXs, Choi, 1992) were developed to estimate weakly pain scores (prior pain is an important predictor of current pain) for treated patients in each cluster in the matched dataset using maximum likelihood. The other aspect was to create a platform (an expandable software that sits on a physician's desktop) that is more accessible to a wide range of individuals, the Virtual Lab platform. This enabled users to run their own subgroups on virtual (simulated) patients and get descriptions on likely patient outcomes.

In the study by Alexander, Edwards, Brodsky et al. (2018), data were added from six RCTs to reduce covariate bias with the observational study and improve accuracy and to increase the variety of patients for pain response prediction. Novel 'virtual' patient pain scores were predicted over time using simulations to assign novel patients to a cluster and then applying cluster-specific regressions to predict pain response trajectories.

There are a few strengths and limitations of the approach taken. First, microsimulations allowed physicians to take patient-variability into account in making treatment assignments. Second, a platform makes it straightforward to add additional sources of data. Despite the benefits of the system in predicting pain from patient laboratory measurements, it was never used in the marketplace as a clinical support mechanism as the project utilized only study completers and Lyrica patients.

12.5 Concluding Remarks

This chapter was authored by the Safety Working Group of the Biopharmaceutical Section of the ASA. Our goals were to describe the ways in which evidence from RWD could be analyzed alone or in conjunction with data from RCTs, summarize methodologies that can be used to address selection bias of data from RWD sources or to analyze multi-source data, and provide relevant examples and case studies to further educate statisticians with real-world applications.

While progress has been made, there are still numerous challenges. We have described examples above where real-world cohorts take advantage of randomization to limit bias (Zhao et al. 2016; Hernandez et al. 2015; Fröbert et al. 2013; Vestbo et al., 2016). However, randomization was not blinded in these examples, and implementing blinded randomization in practice may add a number of logistical challenges and added cost to analyzing RWD. Cluster randomization could be utilized so that entire practices would be randomized to prescribing a single treatment, though it could make it more challenging to account for site-to-site differences in estimating treatment effects.

Due to space limitations, there were a few topics that were not addressed. First, the case study above in TARGET-HCV discusses supplementing EHR data with data from patient-reported outcomes, and new technologies make the collection of patient-level data through smartphones or other wearable devices easier than ever before. Izmailova, Wagner, and Perakslis (2018) provide an overview of the challenges and opportunities of wearable devices. On the surface, these devices can capture data easily, potentially in a more consistent manner than data from other observational sources, as well as in real-time allowing for the potential for intervention in the case of safety issues. One example for diabetes is the availability of continuous glucose monitors (CGMs) that make it possible to identify hypo- and hyperglycemic events. However, as illustrated by Vigers et al. (2019), the summary values computed for various CGMs on the market are not computed in a standardized fashion, though raw data obtained from the devices can allow for consistent computations. Therefore, studies that utilize multiple devices that measure similar data will need to consider how these data can be standardized. Patient or wearable devices make it possible to introduce randomized substudies into observational studies, allowing for a greater opportunity to manage bias in comparisons between groups (Li et al., 2020).

The second topic that was not addressed in detail involves analyses involving text-mining or natural language processing. These methodologies make it possible to understand and analyze free-text that may be more frequently a part of observational data sources. For example, EHRs may contain a copious amount of unstructured physician notes that could describe the occurrence and severity of various patient safety outcomes and data that may not be captured in a structured format more suitable for analysis. These unstructured data need to be analyzed and standardized into a format that allows for more straightforward analysis. Hernandez-Boussard et al. (2019) discussed the importance of this unstructured text in identifying cardiovascular and metabolic endpoints. Developing reproducible algorithms is critical for consistently identifying safety outcomes for a given population or even for the selection of appropriate patient populations for comparisons across multiple data sources. Such analyses have been conducted for numerous diseases (Liao et al., 2010; Ananthakrishnan et al., 2013; Corey et al., 2016).

Evidence from RWD has been used to answer questions surrounding safety for some time. Access to additional sources of data and rigorous statistical methodologies will allow us to explore safety concerns in greater detail. However, there is a cost to analyzing RWD and data from RCTs in conjunction. Cameron et al. (2015) state that "incorporation of both RCTs and non-randomized studies into [network meta-analysis] typically requires considerably more time, effort, and costs compared to including only RCTs". Certainly, having the appropriate in-house statistical expertise to synthesize data from multiple sources will require training beyond what the typical statisticians in the medical product industry receive.

Acknowledgments

The authors wish to thank the reviewers for their insightful comments which improved the content of this chapter.

References

Alexander J, Edwards RA, Savoldelli A, Manca L, Grugni R, Emir B, Whalen E, Watt S, Brodsky M & Parsons B. (2017). Integrating data from randomized controlled trials and observational studies to predict the response to pregabalin in patients with painful diabetic peripheral neuropathy. *BMC Medical Research Methodology* 17(1): 113.

Alexander J, Edwards RA, Brodsky M, Manca L, Grugni R, Savoldelli A, Bonfanti G, Emir B, Whalen E, Watt S & Parsons B. (2018). Using time series analysis approaches for improved prediction of pain outcomes in subgroups of patients with painful diabetic peripheral neuropathy. *PLoS One* 13(12): e0207120.

Alexander J, Edwards RA, Brodsky M, Savoldelli A, Manca L, Grugni R, Emir B, Whalen E, Watt S & Parsons B. (2019). Assessing the value of time series real-world and clinical trial data vs. baseline-only data in predicting responses to pregabalin therapy for patients with painful diabetic peripheral neuropathy. *Clinical Drug Investigation* 39: 775–786.

Alexander J, Edwards RA, Manca L, Grugni R, Bonfanti G, Emir B, Whalen E, Watt S & Parsons B. (2018). Dose titration of pregabalin in patients with painful diabetic peripheral neuropathy: simulation based on observational study patients enriched with data from randomized studies. *Advances in Therapy* 35: 382–394.

Alexander J, Edwards RA, Manca L, Grugni R, Bonfanti G, Emir B, Whalen E, Watt S, Brodsky M & Parsons B. (2019). Integrating machine learning with microsimulation to classify hypothetical, novel patients for predicting pregabalin treatment response based on observational and randomized data in patients with painful diabetic peripheral neuropathy. *Pragmatic and Observational Research* 10: 67–76.

Almirall D, Ten Have T & Murphy SA. (2010). Structural nested mean models for assessing time-varying effect moderation. *Biometrics* 66: 131–139.

Ananthakrishnan AN, Cai T, Savova G, Cheng SC, Perez RG, Gainer VS, Murphy SN, Szolovits P, Xia Z, Shaw S, Churchill S, Karlson EW, Kohane I, Plenge RM & Liao KP. (2013). Improving case definition of Crohn's disease and ulcerative colitis in electronic medical records using natural language processing: a novel informatics approach. *Inflammatory Bowel Disease* 19: 1411–1420.

Arbogast PG & Ray WA. (2011). Performance of disease risk scores, propensity scores, and traditional multivariable outcome regression in the presence of multiple confounders. *American Journal of Epidemiology* 174: 613–620.

Austin PC. (2009). Balance diagnostics for comparing the distribution of baseline covariates between treatment groups in propensity-score matched samples. *Statistics in Medicine* 28: 3083–3107.

Austin PC. (2019). Assessing covariate balance when using the generalized propensity score with quantitative or continuous exposures. *Statistical Methods in Medical Research* 28: 1365–1377.

Austin P & Stuart E. (2015). Moving towards best practice when using inverse probability of treatment weighting (IPTW) using the propensity score to estimate causal treatment effects in observational studies. *Statistics in Medicine* 34: 3661–3679.

Baiocchi M, Cheng J & Small DS. (2014). Instrumental variable methods for causal inference. *Statistics in Medicine* 33: 2297–340.

Ball R, Robb M, Anderson SA & Dal Pan G. (2016). The FDA's sentinel initiative—a comprehensive approach to medical product surveillance. *Clinical Pharmacology and Therapeutics* 99: 265–268.

Balshem H, Helfand M, Schünemann HJ, Oxman AD, Kunz R, Brozek J, Vist GE, Falck-Ytter Y, Meerpohl J, Norris S & Guyatt GH (2011). GRADE guidelines: 3. Rating the quality of evidence. *Journal of Clinical Epidemiology* 64: 401–406.

Benedetto U, Head SJ, Angelini GD & Blackstone EH. (2018). Statistical primer: Propensity score matching and its alternatives. *European Journal of Cardio-Thoracic Surgery* 53: 1112–1117.

Berger ML, Sox H, Willke RJ, Brixner DL, Eichler HG, Goettsch W, Madigan D, Makady A, Schneeweiss S, Tarricone R, Wang SV, Watkins J & Mullins CD. (2017). Good practices for real-world data studies of treatment and/or comparative effectiveness: Recommendations from the joint ISPOR-ISPE Special Task Force on real-world evidence in health care decision making. *Pharmacoepidemiology and Drug Safety* 26: 1033–1039.

Breiman L, Friedman J, Stone CJ & Olshen RA. (1984). *Classification and Regression Trees*. Boca Raton: Chapman & Hall/CRC.

Brookhart MA, Schneeweiss S, Rothman KJ, Glynn RJ, Avorn J & Sturmer T. (2006). Variable selection for propensity score models. *American Journal of Epidemiology* 163: 1149–1156.

Burcu M, Dreyer NA, Franklin JM, Blum Michael D, Critchlow CW, Perfetto EM & Zhou W. (2020). Real-world evidence to support regulatory decision-making for medicines: Considerations for external control arms. *Pharmacoepidemiology and Drug Safety* 1–8. DOI:10.1002/pds.4975.

Burgette L, Griffin BA & McCaffrey D. (2020). Propensity scores for multiple treatments: A tutorial for the MNPS function in the Twang package. Available at: https://cran.r-project.org/web/packages/twang/vignettes/mnps.pdf.

Byrd JB & Ho PM. (2011). The possibility of unmeasured confounding variables in observational studies: A forgotten fact? *Heart* 97: 1815–1816.

Cabibbo G, Petta S, Barbàra M, Missale G, Virdone R, Caturelli E, Piscaglia F, Morisco F, Colecchia A, Farinati F, Giannini E, Trevisani F, Craxì A, Colombo M & Cammà C. (2017). A meta-analysis of single HCV-untreated arm of studies evaluating outcomes after curative treatments of HCV-related hepatocellular carcinoma. *Liver International* 37: 1157–1166.

Cameron C, Fireman B, Hutton B, Clifford T, Coyle D, Wells G, Dormuth CR, Platt R & Toh S. (2015). Network meta-analysis incorporating randomized controlled trials and nonrandomized comparative cohort studies for assessing the safety and effectiveness of medical treatments: challenges and opportunities. *Systematic Reviews* 4: 147.

Cannas M & Arpino B. (2019). A comparison of machine learning algorithms and covariate balance measures for propensity score matching and weighting. *Biometrical Journal* 61: 1049–1072.

Cham HN. (2013). Propensity score estimation with random forests. Available at: https://repository.asu.edu/attachments/110663/content/Cham_asu_0010E_13074.pdf.

Chen J, Ho M, Lee K, Song Y, Fang Y, Goldstein BA, He W, Irony T, Jiang Q, van der Laan M, Lee H, Lin X, Meng Z, Mishra-Kalyani P, Rockhold F, Wang H & White R. (2021). The current landscape in biostatistics of real-world data and evidence: Clinical study design and analysis. *Statistics in Biopharmaceutical Research*. https://www.tandfonline.com/doi/full/10.1080/19466315.2021.1883474.

Choi BS. (1992). *ARMA Model Identification*. New York: Springer-Verlag.

Corey KE, Kartoun U, Zheng H & Shaw SY. (2016). Development and validation of an algorithm to identify nonalcoholic fatty liver disease in the electronic medical record. *Digestive Disease Science* 61: 913–919.

Curtis JR, Chen L, Bharat A, Delzell E, Greenberg JD, Harrold L, Kremer J, Setoguchi S, Solomon DH & Xie F. (2014). Linkage of a de-identified United States rheumatoid arthritis registry with administrative data to facilitate comparative effectiveness research. *Arthritis Care & Research* 66: 1790–1798.

DerSimonian R & Laird N. (1986). Meta-analysis in clinical trials. *Controlled Clinical Trials* 7: 177–188.

de Vries BLP, van Smeden M & Groenwold RHH. (2018). Propensity score estimation using classification and regression trees in the presence of missing covariate data. *Epidemiologic Methods* 7: 1–18.

DiMasi JA, Grabowski HG & RW Hansen. (2016). Innovation in the pharmaceutical industry: New estimates of R&D costs. *Journal of Health Economics* 47: 20–33.

DiMasi JA, Hansen RW & Grabowski HG. (2003). The price of innovation: new estimates of drug development costs. *Journal of Health Economics* 22: 151–185.

Dusetzina SB, Tyree S, Meyer AM, Meyer A, Green L & Carpenter WR. (2014). Linking data for health services research: a framework and instructional guide. (Prepared by the University of North Carolina at Chapel Hill under Contract No. 290-2010-000141.) AHRQ Publication No. 14-EHC033-EF. Rockville, MD: Agency for Healthcare Research and Quality. Available at: www.effectivehealthcare.ahrq.gov/reports/final.cfm.

Elze MC, Gregson J, Baber U, Williamson E, Sartori S, Mehran R, Nichols M, Stone GW & Pocock SJ. (2017). Comparison of propensity score methods and covariate adjustment: Evaluation in 4 cardiovascular studies. *Journal of the American College of Cardiology* 69: 345–57.

Efthimiou O, Debray TPA, van Valkenhoef G, Trelle S, Panayidou K, Moons KGM, Reitsma JB, Shang A & Salanti G. (2016). GetReal in network meta-analysis: A review of the methodology. *Research Synthesis Methods* 7: 236–263.

Efthimiou O, Mavridis D, Debray TPA, Samara M, Belger M, Siontis GCM, Leucht S & Salanti G. (2017). Combining randomized and nonrandomized evidence in network meta-analysis. *Statistics in Medicine* 36: 1210–1226.

European Medicines Agency Pharmacovigilance Risk Assessment Committee. (2016). Direct-acting antivirals for hepatitis C: EMA confirms recommendation to screen for hepatitis B. Available at: https://www.ema.europa.eu/en/documents/press-release/direct-acting-antivirals-hepatitis-c-ema-confirms-recommendation-screen-hepatitis-b_en.pdf.

Faries D, Peng X, Pawaskar M, Price K, Stamey JD & Seaman JW. (2013). Evaluating the impact of unmeasured confounding with internal validation data: An example cost evaluation in type 2 diabetes. *Value in Health* 16: 259–260.

Feng P, Zhou HX, Zou QM, Fan MY & Li XS. (2012). Generalized propensity score for estimating the average treatment effect of multiple treatments. *Statistics in Medicine* 31: 681–697.

Fewell Z, Smith GD & Sterne JAC. (2007). The impact of residual and unmeasured confounding in epidemiologic studies: A simulation study. *American Journal of Epidemiology* 166: 646–655.

Fralick M, Kesselheim AS, Avorn J, & Schneeweiss S. (2018). Use of health care databases to support supplemental indications of approved medications. *JAMA Internal Medicine* 178, 55–63.

Fröbert O, Lagerqvist B, Olivecrona G, Omerovic E, Gudnason T, Maeng M, Aasa M, Angerås O, Calais F, Danielewicz M, Erlinge D, Hellsten L, Jensen U, Johansson

AC, Kåregren A, Nilsson J, Robertson L, Sandhall L, Sjögren I, Östlund O, Harnek J, & James SK (2013). Thrombus aspiration during ST segment elevation myocardial infarction. *New England Journal of Medicine* 369: 1587–1597.

Funch D, Mortimer K, Ziyadeh NJ, Seeger JD, Li L, Norman H, Major-Pedersen A, Bosch-Traberg TH, Gydesen H & Dore DD. (2019). Propensity score matched approaches to estimate the association between liraglutide use and thyroid cancer. *Diabetes, Obesity and Metabolism* 21: 1837–1848.

Ghadessi M, Tang R, Zhou J, Liu R, Wang Chenkun, Toyoizumi K, Mei C, Zhang L, Deng CQ & Beckman RA. (2020). A roadmap to using historical controls in clinical trials – by Drug Information Association Adaptive Design Scientific Working Group (DIA-ADSWG). *Orphanet Journal of Rare Diseases* 15:69 1–19.

Girman CJ, Ritchey ME, Zhou W & Dreyer NA. (2019). Considerations in characterizing real-world data relevance and quality for regulatory purposes: A commentary. *Pharmacoepidemiology and Drug Safety* 28: 439–442.

Glynn RJ, Gagne JJ & Schneeweiss S. (2012). Role of disease risk scores in comparative effectiveness research with emerging therapies. *Pharmacoepidemiology and Drug Safety* 21: 138–47.

Goller D, Lechner M, Moczall A & Wolff J. (2019). Does the estimation of the propensity score by machine learning improve matching estimation? The case of Germany's programmes for long term unemployed. *IZA Discussion Papers* 12526, Institute of Labor Economics (IZA). Available at: http://hdl.handle.net/10419/207352.

Goring S, Taylor A, Müller K, Li TJJ, Korol EE, Levy AR & Freemantle N. (2019). Characteristics of non-randomised studies using comparisons with external controls submitted for regulatory approval in the USA and Europe: a systematic review. *BMJ Open* 9: e024895. https://doi.org/10.1136/bmjopen-2018-024895.

Graham DJ, Baro E, Zhang R, Liao J, Wernecke M, Reichman ME, Hu M, Illoh O, Wei Y, Goulding MR, Chillarige Y, Southworth MR, MaCurdy TE & Kelman JA. (2019). Comparative stroke, bleeding, and mortality risks in older Medicare patients treated with oral anticoagulants for nonvalvular atrial fibrillation. *American Journal of Medicine* 132: 596–604.

Groenwold RHH, de Groot MCH, Ramamoorthy D, Souverein PC & Klungel OH. (2016). Unmeasured confounding in pharmacoepidemiology. *Annals of Epidemiology* 26: 85–86.

Gruber S & van der Laan MJ. (2013). An application of targeted maximum likelihood estimation to the meta-analysis of safety data. *Statistics in Medicine* 69: 254–262.

Guyatt G, Oxman AD, Akl EA, Kunz R, Vist GE, Brozek J, Norris S, Falck-Ytter Y, Glasziou P, DeBeer H, Jaeschke R, Rind D, Meerpohl J, Dahm P & Schünemann HJ. (2011). GRADE guidelines: 1. Introduction-GRADE evidence profiles and summary of findings tables. *Journal of Clinical Epidemiology* 64: 383–94.

Guyatt GH, Oxman AD, Kunz R, Atkins D, Brozek J, Vist G, Alderson P, Glasziou P, Falck-Ytter Y & Schünemann HJ. (2011). GRADE guidelines: 2. Framing the question and deciding on important outcomes. *Journal of Clinical Epidemiology* 64: 395–400.

Guyatt GH, Oxman AD, Vist G, Kunz R, Brozek J, Alonso-Coello P, Montori V, Akl EA, Djulbegovic B, Falck-Ytter Y, Norris SL, Williams JW, Atkins D, Meerpohl D & Schünemann HJ. (2011). GRADE guidelines: 4. Rating the quality of evidence – study limitations (risk of bias). *Journal of Clinical Epidemiology* 64: 407–15.

Haneuse S, VanderWeele TJ & Arterburn D. (2019). Using the E-value to assess the potential effect of unmeasured confounding in observational studies. *Journal of the American Medical Association* 321: 602–603.

He J, Stephens-Shields A & Joffe MM. (2015). Structural nested mean models to estimate the effects of time-varying treatments on clustered outcomes. *The International Journal of Biostatistics* 65: 203–222.

Hernandez AF, Fleurence RL & Rothman RL. (2015). The ADAPTABLE trial and PCORnet: Shining light on a new research paradigm. *Annals of Internal Medicine* 163: 635–636.

Hernandez-Boussard T, Monda KL, Cresp BC & Riskin D. (2019). Real world evidence in cardiovascular medicine: Ensuring data validity in electronic health record-based studies. *Journal of the American Medical Informatics Association* 26: 1189–1194.

Hu L, Gu C, Lopez M, Ji J & Wisnivesky J. (2020). Estimation of causal effects of multiple treatments in observational studies with a binary outcome. *Statistical Methods in Medical Research* 29: 3218–3234.

Inacio MC, Chen Y, Paxton EW, Namba RS, Kurtz SM & Cafri G. (2015). Statistics in brief: An introduction to the use of propensity scores. *Clinical Orthopedics and Related Research* 473: 2722–2726.

Iacus SM, King G, Porro G. (2012). Causal inference without balance checking: coarsened exact matching. *Political Analysis* 20: 1–24.

Imbens GW. (2000). The role of the propensity score in estimating dose-response functions. *Biometrika* 87: 706–710.

International Council for Harmonisation. (2000). Guideline E10: Choice of Control Group and Related Issues in Clinical Trials. Available at: https://database.ich.org/sites/default/files/E10_Guideline.pdf.

International Society for Pharmacoepidemiology (ISPE). (2015). Guidelines for Good Pharmacoepidemiology Practices (GPP), Revision 3. Available at: https://www.pharmacoepi.org/resources/policies/guidelines-08027/.

Izem R, Sanchez-Kam M, Ma H, Zink RC & Zhao Y. (2018). Sources of safety data and statistical strategies for design and analysis: Post-market surveillance. *Therapeutic Innovation & Regulatory Science* 52: 159–169.

Izmailova ES, Wagner JA & Perakslis ED. (2018). Wearable devices in clinical trials: Hype and hypothesis. *Clinical Pharmacology & Therapeutics* 104: 42–52.

King G & Nielsen R. (2019). Why propensity scores should not be used for matching. *Political Analysis* 27: 1–20.

Kunutsor SK, Wylde V, Whitehouse MR, Beswick AD, Lenguerrand E & Blom AW. (2019). Influence of fixation methods on prosthetic joint infection following primary total knee replacement: Meta-analysis of observational cohort and randomised intervention studies. *Journal of Clinical Medicine* 8: 828.

La Rosée P & Deininger MW. (2010). Resistance to imatinib: Mutations and beyond. *Seminars in Hematology* 47: 335–343.

Lee BK, Lessler J & Stuart EA. (2010). Improving propensity score weighting using machine learning. *Statistics in Medicine* 29: 337–346.

Lewis CJ, Sarkar S, Zhu J & Carlin BP. (2019). Borrowing from historical control data in cancer drug development: A cautionary tale and practical guidelines. *Statistics in Biopharmaceutical Research* 11: 67–78.

Li F & Li F. (2019). Propensity score weighting for causal inference with multiple treatments. *Annals of Applied Statistics* 13: 2389–2415.

Li F, Morgan KL & Zaslavsky AM. (2018). Balancing covariates via propensity score weighting. *Journal of the American Statistical Association* 113: 390–400.

Li S, Psihogios AM, McKelvey ER, Ahmed A, Rabbi M & Murphy S. (2020). Microrandomized trials for promoting engagement in mobile health data collection: Adolescent/young adult oral chemotherapy adherence as an example. *Current Opinion in Systems Biology.* https://doi.org/10.1016/j.coisb.2020.07.002.

Liao KP, Cai T, Gainer V, Goryachev S, Zeng-Treitler Q, Raychaudhuri S, Szolovits P, Churchill S, Murphy S, Kohane I, Karlson EW & Plenge RM. (2010). Electronic medical records for discovery research in rheumatoid arthritis. *Arthritis Care and Research* 62: 1120–1127.

Lim J, Walley R, Yuan J, Liu J, Dabral A, Best N, Grieve A, Hampson L, Wolfram J, Woodward P, Yong F, Zhang X & Bowen E. (2018). Minimizing patient burden through the use of historical subject-level data in innovative confirmatory clinical trials: Review of methods and opportunities. *Therapeutic Innovation & Regulatory Science* 52: 546–559.

Linden A & Yarnold PR. (2016). Combining machine learning and propensity score weighting to estimate causal effects in multivalued treatments. *Journal of Evaluation in Clinical Practice* 22: 875–885.

Lopez MJ & Gutman R. (2017). Estimation of causal effects with multiple treatments: A review and new ideas. *Statistical Science* 32: 432–454.

Lumley T. (2002). Network meta-analysis for indirect treatment comparisons. *Statistics in Medicine* 21: 2313–24.

Luque-Fernandez MA, Schomaker M, Rachet B & Schnitzer ME. (2018). Targeted maximum likelihood estimation for a binary treatment: A tutorial. *Statistics in Medicine* 37: 2530–2546.

Ma H, Russek-Cohen E, Izem R, Marchenko O & Jiang Q. (2018). Sources of safety data and statistical strategies for design and analysis: Transforming data into evidence. *Therapeutic Innovation & Regulatory Science* 52: 187–198.

Marchenko O, Russek-Cohen E, Levenson M, Zink RC, Krukas-Hampel M & Jiang Q. (2018). Sources of safety data and statistical strategies for design and analysis: Real world insights. *Therapeutic Innovation & Regulatory Science* 52: 170–186.

McCaffrey DF, Griffin BA, Almirall D, Slaughter ME, Ramchand R & Burgette LF. (2013). A tutorial on propensity score estimation for multiple treatments using generalized boosted models. *Statistics in Medicine* 32: 3388–3414.

Mamdani M, Sykora K, Li P, Normand SL, Streiner DL, Austin PC, Rochon PA & Anderson GM. (2005). Reader's guide to critical appraisal of cohort studies: 2. Assessing potential for confounding. *British Medical Journal* 330: 960–962.

Mishra P, Florian J, Peter J, Vainorius M, Fried MW, Nelson DR & Birnkrant D. (2017). Public-private partnership: Targeting real-world data for hepatitis C direct-acting antivirals. *Gastroenterology* 153: 626–631.

Normand SLT, Landrum MB, Guadagnoli E, Ayanian JZ, Ryan TJ, Cleary PD & McNeil BJ. (2001). Validating recommendations for coronary angiography following acute myocardial infarction in the elderly: A matched analysis using propensity scores. *Journal of Clinical Epidemiology* 54: 387–398.

Prevost TC, Abrams KR & Jones DR. (2000). Hierarchical models in generalized synthesis of evidence: an example based on studies of breast cancer screening. *Statistics in Medicine* 19: 3359–3376.

Reig M, Mariño Z, Perelló C, Iñarrairaegui M, Ribeiro A, Lens S, Díaz A, Vilana R, Darnell A, Varela M, Sangro B, Calleja JL, Forns X & Bruix J. (2016). Unexpected high rate of early tumor recurrence in patients with HCV-related HCC undergoing interferon-free therapy. *Journal of Hepatology* 65: 719–726.

Rosenbaum PR & Rubin DB. (1983). The central role of the propensity score in observational studies for causal effects. *Biometrika* 70: 41–55.

Sarri G, Patorno E, Yuan H, Guo J, Bennett D, Wen X, Zullo AR, Largent J, Panaccio M, Gokhale M, Moga DC, Ali MS & Debray TPA. (2020). Framework for the synthesis of non-randomised studies and randomised controlled trials: a guidance on conducting a systematic review and meta-analysis for healthcare decision making. *BMJ Evidence Based Medicine*. Published online: 2020-12-09. doi:10.1136/bmjebm-2020-111493.

Schmitz S, Adams R & Walsh C. (2013). Incorporating data from various trial designs into a mixed treatment comparison model. *Statistics in Medicine* 32: 2935–2949.

Sherman M, Yurdaydin C, Sollano J, Silva M, Liaw Y, Cianciara J, Boron-Kaczmarska A, Martin P, Goodman Z, Colonno R, Cross A, Denisky G, Kreter B & Hindes R. (2006). Entecavir for treatment of lamivudine-refractory, HBeAg-positive chronic hepatitis B. *Gastroenterology* 130: 2039–2049.

Singal AG, Rich NE, Mehta N, Branch A, Pillai A, Hoteit M, Volk M, Odewole M, Scaglione S, Guy J, Said A, Feld JJ, John BV, Frenette C, Mantry P, Rangnekar AS, Oloruntoba O, Leise M, Jou JH, Bhamidimarri KR, Kulik L, Tran T, Samant H, Dhanasekaran R, Duarte-Rojo A, Salgia R, Eswaran S, Jalal P, Flores A, Satapathy SK, Wong R, Huang A, Misra S, Schwartz M, Mitrani R, Nakka S, Noureddine W, Ho C, Konjeti VR, Dao A, Nelson K, Delarosa K, Rahim U, Mavuram M, Xie JJ, Murphy CC & Parikh ND. (2019). Direct-acting antiviral therapy not associated with recurrence of hepatocellular carcinoma in a multicenter North American cohort study. *Gastroenterology* 56: 1683–1692.

Sizemore S & Alkurdi R. (2019). Matching methods for causal inference: A machine learning update. Available at: https://humboldt-wi.github.io/blog/research/applied_predictive_modeling_19/matching_methods/.

Streeter AJ, Lin NX, Crathorne L, Haasova M, Hyde C, Melzer D & Henley WE. (2017). Adjusting for unmeasured confounding in nonrandomized longitudinal studies: A methodological review. *Journal of Clinical Epidemiology* 87: 23–34.

Stuart EA & Rubin DB. (2008). Matching with multiple control groups with adjustment for group differences. *Journal of Educational and Behavioral Statistics* 33: 279–306.

Thorland K, Dron L, Park JJH & Mills EJ. (2020). Synthetic and external controls in clinical trials: A primer for researchers. *Clinical Epidemiology* 12: 457–467.

Tromp M, Ravelli AC, Bonsel GJ, Hasman A & Reitsma JB. (2011). Results from simulated data sets: probabilistic record linkage outperforms deterministic record linkage. *Journal of Clinical Epidemiology* 64: 565–572.

Tu C. (2019). Comparison of various machine learning algorithms for estimating generalized propensity score. *Journal of Statistical Computation and Simulation* 89: 708–719.

US Food and Drug Administration. (2016). 21st Century Cures Act. Available at: https://www.fda.gov/regulatory-information/selected-amendments-fdc-act/21st-century-cures-act.

US Food and Drug Administration. (2018). Framework for FDA's Real World Evidence Program. Available at: https://www.fda.gov/media/120060/download.

US Food and Drug Administration. (2019). Submitting Documents Using Real-World Data and Real-World Evidence to FDA for Drugs and Biologics: Guidance for Industry (Draft). https://www.fda.gov/media/124795/download.

Vallarino C, Perez A, Fusco G, Liang H, Bron M, Manne S, Joseph G & Yu S. (2013). Comparing pioglitazone to insulin with respect to cancer, cardiovascular and bone fracture endpoints, using propensity score weights. *Clinical Drug Investigation* 33: 621–631.

van der Laan MJ & Rubin D. (2006). Targeted maximum likelihood learning. *The International Journal of Biostatistics* 2: 1–38.

van der Laan MJ & Rose S. (2011). *Targeted Learning: Causal Inference for Observational and Experimental Data.* New York: Springer.

Vansteelandt S & Joffe M. (2014). Structural nested models and G-estimation: The partially realized promise. *Statistical Science* 29: 707–731.

Verde PE & Ohmann C. (2015). Combining randomized and nonrandomized evidence in clinical research: a review of methods and applications. *Research Synthesis Methods* 6: 45–62.

Vestbo J, Leather D, Bakerly ND, New J, Gibson JM, McCorkindale S, Collier S, Crawford J, Frith L, Harvey C, Svedsater H & Woodcock A. (2016). Effectiveness of fluticasone furoate–vilanterol for COPD in clinical practice. *New England Journal of Medicine* 375: 1253–1260.

Viele K, Berry S, Neuenschwander B, Amzal B, Chen F, Enas N, Hobbs B, Ibrahim JG, Kinnersley N, Lindborg S, Micallef S, Roychoudhury S & Thompson L. (2014). Use of historical control data for assessing treatment effects in clinical trials. *Pharmaceutical Statistics* 13: 41–54.

Vigers T, Chan CL, Snell-Bergeon J, Bjornstad P, Zeitler PS, Forlenza G & Pyle L. (2019). cgmanalysis: An R package for descriptive analysis of continuous glucose monitor data. *PLoS ONE* 14 (10): e0216851. doi:10.1371/journal.pone.0216851

Von Elm E, Altman DG, Egger M, Pocock D, Pocock SJ, Gotzsche PC & Vandenbroucke JP. (2007). Strengthening the reporting of observational studies in epidemiology (STROBE) statement: Guidelines for reporting observational studies. *British Medical Journal* 335: 806.

Wang, J & Donnan P. (2001). Propensity score methods in drug safety studies: Practice, strengths, and limitations. *Pharmacoepidemiology and Drug Safety* 10: 341–344.

Wyss R, Girman CJ, LoCasale RJ, Alan BM & Sturmer T. (2013). Variable selection for propensity score models when estimating treatment effects on multiple outcomes: A simulation study. *Pharmacoepidemiology and Drug Safety* 22: 77–85.

Yuan J, Liu J, Zhu R, Lu Y & Palm U. (2019). Design of randomized controlled confirmatory trials using historical control data to augment sample size for concurrent controls. *Journal of Biopharmaceutical Statistics* 29: 558–573.

Zhang D & Kim J. (2019). Use of propensity score and disease risk score for multiple treatments with time-to-event outcome: A simulation study. *Journal of Biopharmaceutical Statistics* 29: 1103–1115.

Zhang Z, Uddin MJ, Cheng J & Huang T. (2018). Instrumental variable analysis in the presence of unmeasured confounding. *Annals of Translational Medicine* 6: 1–11.

Zhao H, Hobbs BP, Ma H, Jiang Q & Carlin BP. (2016). Combining non-randomized and randomized data in clinical trials using commensurate priors. *Health Services Outcomes Research Methodology* 16: 154–171.

Zink RC, Marchenko O, Sanchez-Kam M, Ma H & Jiang Q. (2018). Sources of safety data and statistical strategies for design and analysis: Clinical trials. *Therapeutic Innovation & Regulatory Science* 52: 141–158.

Part D

Safety/Benefit-Risk Evaluation and Visualization

13

Trends and Recent Progress in Benefit-Risk Assessment Planning for Medical Products and Devices

Lisa R. Rodriguez

Center for Drug Evaluation and Research, US Federal Food and Drug Administration, Silver Spring, MD, USA

Cheryl Renz

AbbVie, Inc., North Chicago, IL, USA

Brian Edwards

NDA Org, London, UK

William Wang

Merck & Co, Inc, Kenilworth, NJ, USA

CONTENTS

13.1 Introduction 286
    13.1.1 Overview 286
13.2 Evolution of Benefit-Risk Planning During the Medical
    Product Development Life Cycle..287
    13.2.1 The Origins of Benefit-Risk Assessments 287
    13.2.2 Current Approaches to sBRAs 289
    13.2.3 Recent and Future Developments Applicable to
         Benefit-Risk Assessment Planning 292
13.3 Review of Published Recommendations Applicable to
    Benefit-Risk Assessment Planning 295
    13.3.1 EMA Perspective on Benefit-Risk 295
    13.3.2 FDA Perspective on Benefit-Risk 297
    13.3.3 Planning for Use of RWD Suitable
         for Benefit-Risk Evaluation 299
    13.3.4 Planning for Use of Patient Experience Data Suitable for
         Benefit-Risk Evaluation 301
13.4 Benefit-Risk Assessment Planning in Practice – Survey
    of the Industry Perspective..304

DOI: 10.1201/9780429488801-17

13.4.1 Understanding Industry's Approach to Benefit-Risk
     Assessment and Planning..304
13.4.2 Overview of Industry Survey Design and Participants...........304
13.4.3 Industry Survey Results: sBRA Implementation Measures.....305
     13.4.3.1 Coordination...305
     13.4.3.2 Data/Information Sources.............................306
     13.4.3.3 Tools 306
13.4.4 Industry Survey Results: sBRA Utilization
     for Decision-Making...306
13.4.5 Industry Survey Results: Challenges with sBRAs 308
13.4.6 Industry Survey Results: Future Opportunities for sBRAs 308
13.5 Future Work and Discussion..309
Acknowledgments 311
References..311

13.1 Introduction

A structured approach to benefit-risk evaluation is an essential tool for decision-making during pharmaceutical product development. A **structured benefit-risk assessment (sBRA) is a transparent, decision-making process for determining whether benefits outweigh the risks**. Early planning in the medical product development life cycle for benefit-risk assessments is particularly important so that medical product developers can identify and plan for all of the key decision-points. The focus of this chapter is considerations for incorporating such benefit-risk planning into the medical product development process.

13.1.1 Overview

There are many considerations for evaluating benefit-risk for decision-making in medical product development, and acknowledgement should be given to early efforts to provide a more structured process. **The evolution of benefit-risk assessments and implications for planning are described throughout Section 13.2**.

Currently, there is a gap between published recommendations and development needs because concerns for global development regulations and for internal planning are not aligned, particularly with respect to early benefit-risk considerations that may greatly impact the initial development of a product. **By assessing the regulatory landscape in Section 13.3, we have synthesized current recommendations and regulation** so that industry colleagues may utilize this assessment for benefit-risk planning purposes applicable across different regulatory regions. We note that to date there is no guidance by regulatory agencies for submission of a formal benefit-risk plan;

however, as we illustrate in this chapter, such a plan is useful for internal development purposes and for pre-emptive communication with regulatory authorities in order to improve the quality of decision-making during the development process.

Additionally, **as described in Section 13.4, we have assessed the current benefit-risk practices throughout the life cycle of development**. The survey findings are informative to identify challenges and opportunities for benefit-risk planning processes within organizations.

All of the inputs presented in this chapter, and **as summarized in Section 13.5, will be utilized in the future to develop a benefit-risk assessment plan (BRAP) template to be used by key stakeholders that aligns with global benefit-risk concerns throughout the medical product life cycle**. After further development activities, a BRAP template will be proposed in a future paper and described in a future edition of this book. The purpose of a BRAP will be to document a unified view of sBRAs throughout the medical product life cycle with flexibility to adapt to different decision-making environments. As discussed in Chapter 3, an aggregate safety assessment plan (ASAP) was developed by the ASA Biopharmaceutical Section Safety Evaluation Working Group, which focuses on safety planning. The proposed BRAP will align with the ASAP throughout medical product development. Certainly, **readers of this chapter may utilize the information presented here to devise their own BRAP that aligns with their existing processes.**

13.2 Evolution of Benefit-Risk Planning During the Medical Product Development Life Cycle

13.2.1 The Origins of Benefit-Risk Assessments

No one can put a precise date as to when benefit-risk assessment arrived or became central to assessing whether a medicine should receive approval or remain on the market if a safety concern arises with a product. The first version of the Declaration of Helsinki in 1964 states that "Every clinical research project should be preceded by careful assessment of inherent risks in comparison to foreseeable benefits to the subject or to others." However, across the world, the extrapolation into the licensing of new medicines and indications was neither straightforward nor consistent with balancing benefits and risks as more commonly happens now. For the US Food and Drug Administration (FDA), to merit approval of an NDA an applicant must demonstrate that the medical product is safe and provides 'substantial evidence' that the medical product is effective for use under the conditions described in its labeling (*Demonstrating Substantial Evidence of*

Effectiveness U.S. FDA Guidance 2019a). In the past, regulatory agencies may have evaluated or described safety and efficacy independently for a medicine on its own without direct comparisons. License applications in either the US or EU were rejected on grounds of quality, safety, or efficacy. Although this was intermittently referred to as benefits and risks, there was no internationally agreed approach to what this meant and how this should be achieved. Hence, there was a call during the 1980s for a harmonized approach to the licensing of new medicines. Breckenridge (1999) and Walker (2009) also provide an overview of changes to benefit-risk evaluation over the past decades.

In the late 1980s, the global pharmaceutical market was concentrated in the US, Europe, and Japan with all three regions moving in different directions as regards licensing of medicines and the criteria to apply. This was the main driver for setting up The International Conference on Harmonization of Technical Requirements for registration of Pharmaceuticals for Human Use which started in 1991. This led to an International Council for Harmonisation (ICH) guideline on the Common Technical document in 2000 which defines the presentation and content of the marketing authorization dossier in the US, EU, and Japan. In particular, the guideline describes in detail which criteria should be discussed in the 'Overview of Efficacy', 'Overview of Safety' and 'Benefits and Risk Conclusions' of the Clinical Overview Module 2.5 (*M4E Common Technical Document: Clinical Overview* ICH Harmonized Guideline, EMA 2005).

After some high-profile products were withdrawn, there were concerns that the way regulators waited until the frequency and severity of ADRs reached a certain level causing product withdrawal was not at all clear and that a 'better way' must be possible. In addition, there were product withdrawals when failure of effectiveness was demonstrated resulting in excess mortality such as with flosequinan (Packer et al. 2017). In addition, there were concerns from HIV and oncology patients that regulators were too risk averse and that the rationale for not approving potentially life-saving medicines was not clear.

There was a need for practical approaches to evaluating benefits and risks. The Council for International Organizations of Medical Sciences (CIOMS) IV working group produced a seminal report entitled "Benefit-risk balance for marketed drugs: evaluating safety signals" in 1998 (CIOMS 1998). This report examined the theoretical and practical aspects of how to determine whether a potentially major, new safety signal signifies a shift, calling for action, in the established relationship between benefits and risks. It also provided guidance for deciding what options for action should be considered and on the process of decision-making if such action be required.

Thus, for EU regulators who were building the new EU regulatory system with the EU Pharmacovigilance Working Party reporting into CPMP (as both were called then), this was an opportunity to update the regulatory approach to safety. Work on the EU pharmacovigilance guidelines (which became Volume 9- (Pharmacovigilance) of the rules governing medicinal products in the European Union) had already started in the early 1990s and was well advanced by 1995 although it was not until January 1999 that they were adopted by CPMP (EMEA, Notice to Marketing Authorization Holders Pharmacovigilance Guidelines CPMP/PHVWP/108/99corr). This was the first regulatory document to explicitly describe benefit-risk assessment in detail (See Section 6 Ongoing pharmacovigilance evaluation during the post-authorization period) which is summarized as follows:

> Overall benefit-risk assessment should take into account and balance all the benefits and risks referred to below. Benefit-risk assessment should be conducted separately in the context of each indication which may have impact on the conclusions and actions.

Section 6.2 entitled "Improving the benefit to risk balance" laid out principles which apply to this day redefining and underpinning post-authorization pharmacovigilance.

> The marketing authorization holder should aim to achieve as high as possible benefit balance for an individual product and ensure that the adverse consequences of a medicinal product do not exceed the benefits within the population treated. The benefit-risk profile of a medicinal product cannot be considered in isolation but should be compared with those of other treatments for the same disease.

> The benefit to risk can be improved either by increasing the benefits or by reducing the risks by minimizing risk factors (e.g. by contraindicating the use in patients particularly at risk or lowering dosage, introducing pre-treatment test to identify populations at risk, or monitoring during treatment for early diagnosis of hazards that are reversible). When proposing measures to improve the benefit to risk of a product, for example, by restricting use to a patient group most likely to benefit or to where there is no alternative, their feasibility in normal conditions of use should be taken into account.

Pre-authorization benefit risk was described in Directive 2001/83/EC on the Community code relating to medicinal products for human use, it is stipulated that a marketing authorization shall be refused if:

- The risk-benefit balance is not considered to be favorable
- Its therapeutic efficacy is insufficiently substantiated by the applicant
- Its qualitative and quantitative composition is not as declared

13.2.2 Current Approaches to sBRAs

As opposed to previous practice, the trend in recent benefit and risk evaluations has been toward more transparency via structured approaches. A sBRA may utilize **qualitative assessments** for decision-making (with some quantitative information) or **quantitative assessments** that numerically evaluate tradeoffs for decision-making. Most benefit-risk assessments for regulatory evaluation allow for qualitative assessments to support clinical decision-making for regulatory approvals. Frameworks as developed by the European Medicines Agency (EMA) or FDA are a useful tool for such purposes. This section focuses on what to consider for a sBRA and what a sBRA may look like, particularly with respect to EMA and FDA regulatory submissions.

Before applying a specific sBRA framework, as described in the 1998 CIOMS IV Benefit-Risk Working Group report, there are several factors that may influence the benefit-risk assessment process that should be kept in mind as "caveats" to understand the decision-making environment. The following **factors may impact any comparative benefit-risk assessment in medical product development** and are still currently relevant to define early within any benefit-risk planning effort:

- **Stakeholders** may offer very different perspectives that may potentially be part of a benefit-risk assessment.
- The **nature of the problem**, such as serious concerns that demand rapid attention (e.g. a pandemic), will need to be weighed against the need for additional data that might provide more certainty or confidence.
- There may be multiple intended **indications and populations** all of which help to determine the lens through which benefits and risks are compared.
- There may be **constraints of time, data, and resources** that do not allow for a systematic, thorough evaluation of benefits and risks.
- There may be trade-offs based on different **economic issues** and outcome measures, such as affordability to patients.

The EMA conducted a benefit-risk methodology project (2009–2014) and proposed moving forward with the PrOACT-URL decision framework (see Section 13.3.1), effects tables, and several methodological approaches (Hammond et al. 1999, EMA 2012a B-R Methodology Project, CHMP Working Group Report on B-R Assessment Models and Methods). The **EMA effects table** (Table 13.1) is one example of a sBRA framework. Such a framework provides a **consolidated view of the available evidence for favorable effects (benefits) and unfavorable effects (risks)**, along with a description of the **strength of evidence and any major uncertainties or limitations for each effect**. For a product under EMA review, the effects table appears in the Day 80 Assessment Report and is kept updated throughout the medical product lifecycle.

In 2013, the FDA Center for Drug Evaluation and Research (CDER) and Center for Biologics Evaluation and Research (CBER) developed a framework to organize reviews of a product's benefits and risks and to clearly

TABLE 13.1

EMA Effects Table

Effect	Short Description	Unit	Treatment	Control	Uncertainties/Strength of evidence	References
Favorable Effects						
		/				
Unfavorable Effects						

communicate benefit-risk assessment decisions. This **FDA framework is a qualitative approach focused on identifying and communicating the high-level key issues, evidence, and uncertainties** for a product under consideration for FDA approval, along with a summary of the conclusions that reflect the decision regarding product approval. Starting in 2015, the FDA CDER and CBER began using this framework as part of new molecular entity reviews and have expanded its use to be included in review templates for NDAs and BLAs (current version as displayed in Table 13.2). Refer to the 21st Century Cures Act, FDA PDUFA Plan FY 2018–2022, Lackey et al. 2021, and a white-paper by the Structured Benefit-Risk Working Group of the Biotechnology Industry Organization for further discussion of the implementation of this framework. See Section 13.3.2 for more about benefit-risk considerations across FDA centers, including the Center for Devices and Radiological Health (CDRH). The concepts in Table 13.2 are expanded upon in a more detailed benefit-risk assessment worksheet found in CDRH guidance (see Table 13.4).

Although the EMA and FDA frameworks share some similar components, a definitive approach to evaluating benefit-risk is limited by differences in global applicability to different legal and healthcare systems, a variety of

TABLE 13.2

FDA's Benefit-Risk Framework

Dimension	Evidence and Uncertainties	Conclusions and Reasons
Analysis of Condition		
Current Treatment Options		
Benefit		
Risk and Risk Management		
Conclusions Regarding Benefit-Risk		

benefit-risk viewpoints, lack of specific benefit-risk guidance from regulators, and other practical medical product development concerns throughout early to late stages. Considerations for these differences during drug development could influence planning efforts in order to provide early alignment with regulators on key benefit-risk issues.

For a comprehensive collection of other benefit-risk approaches, we refer the reader to the European Union's Innovative Medicines Initiative (IMI) Pharmacoepidemiological Research on Outcomes of Therapeutics by a European Consortium (PROTECT) project (2009–2015), which catalogued methods and graphical displays, developed innovative tools and standards, and provided test cases on its website (http://www.imi-protect.eu/), and the analysis by Hallgreen et al. 2016.

13.2.3 Recent and Future Developments Applicable to Benefit-Risk Assessment Planning

One unanswered question is how to identify benefit-risk considerations in one plan that can be used throughout medical product development and in alignment with regulatory needs. **The purpose of a formal BRAP would be to document a unified view of sBRA throughout the medical product lifecycle.** Planning for collection of information in order to update a sBRA throughout development is an important function of a BRAP. A BRAP would be a companion document to existing safety plans (such as an ASAP). A BRAP could be as simple as highlighting key benefits and risks for a medical product application, or it can be as comprehensive as a detailed plan aligned with other internal planning documents (such as a "Target Product Profile and Clinical Development Plan"), with updates at each decision point. The concepts and value of this structured thinking in a "life cycle approach" are well described in the study by Doan et al. (2019). However, we are not aware of a BRAP-type document that has been proposed that could be used for internal planning as well as be in a format that can be submitted to regulators.

As part of the ASA Biopharmaceutical Section Safety Evaluation Working Group, the BRAP Taskforce was created at the beginning of 2019 to align with the ASAP efforts in order to propose a future BRAP template that will incorporate planning for benefit-risk assessments. Recent efforts by the BRAP Taskforce are described in this section, as well as the taskforce's research efforts in Section 13.3 and industry survey in Section 13.4. A BRAP template will be proposed in the future by this taskforce, but readers are encouraged to use the information presented in this chapter to develop their own benefit-risk plan (generally referred to as a BRAP in this chapter) that would facilitate their organization's benefit-risk planning efforts.

The taskforce has identified that planning considerations critical to incorporate in a BRAP include **how to make benefit-risk assessments informative for regulatory decisions**, which can include any applicable recommendations as described in Section 13.3. The BRAP Taskforce has identified activities throughout the development lifecycle that can be informed by benefit-risk planning. **Each item listed in Figure 13.1 is a component that can be identified in a comprehensive BRAP**, and development activities can be summarized to the desired level of details useful for internal planning purposes. Additionally, new guidance on benefit-risk assessments from the FDA is expected in 2021, and other projects are currently ongoing to further develop approaches to sBRAs that will be informative for planning. Most recently, the CIOMS Working Group XII on "Benefit-Risk Balance for Medical Products" was launched at the end of 2019 to build on previous CIOMS findings in order to provide future practical recommendations on quantitative and qualitative approaches to the evaluation of benefit-risk. Recommendations from this working group are not yet available but anticipated in the coming years.

The taskforce has also identified that planning considerations particularly useful to incorporate in a BRAP include **when to make benefit-risk assessments and when to update a BRAP**. One approach to a BRAP would be to highlight key benefits and risks in the FDA framework just prior to submitting an NDA or BLA; however, one benefit of creating a BRAP is to be able to plan for benefit-risk assessments earlier and throughout medical product development. The BRAP Taskforce has identified key points at which a BRAP may be updated to reflect available information at key decision-points. Refer to Figure 13.1 below, which depicts benefit-risk considerations during the development lifecycle. This is an updated and modified version of the timeline presented during the 2019 Duke-Margolis Public Meeting on "Characterizing FDA's Approach to Benefit-Risk Assessment throughout the Medical Product Life Cycle," in order to incorporate additional considerations, such as device development and global regulatory processes.

For readers wishing to create their own BRAP, the BRAP Taskforce has identified the following characteristics that **a comprehensive BRAP may incorporate**:

a. A document incorporating points-to-consider, checklists, and process flow charts for **high-level benefit-risk planning** purposes

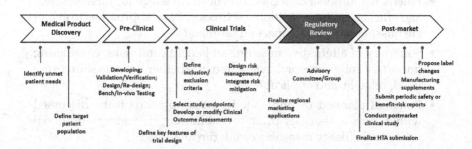

FIGURE 13.1
Benefit-risk considerations during the medical product lifecycle (updated by the BRAP Taskforce, based on the 2019 Duke-Margolis Public Meeting)

b. Facilitates **early planning** for development activities within an organization

c. Facilitates a **process of benefit-risk assessment planning throughout the life cycle** of a medical product, which is iterative as needed and **aligned with defined timepoints**

d. **Modularized** so that users can pick and choose parts of the BRAP template to utilize

e. Serves to **facilitate interactions** between regulatory agencies and industry sponsors at key points in the development and approval processes

f. Incorporates **global regulatory considerations** (See Section 13.3.1 and 13.3.2)

g. **Flexible** to incorporate special considerations on areas such as patient preference, patient experience of side effects/tolerability, clinical outcome assessment (COA)/patient reported outcome (PRO) data, REMS, real-world data (RWD), and other considerations as applicable (See Sections 13.3.3 and 13.3.4)

Another issue is **what to do when outside events impact planning**, such as an approval of a new product that changes the medical product development space, or other events, such as how the COVID-19 pandemic has affected ongoing studies. Meyer et al. (2020) pointed out several dimensions that should be considered in a development plan in the case of the COVID-19 pandemic, such as increased missed visits, loss to follow-up, delayed enrollment, changes in concomitant medications, lack of equivalence in methods of administration, and impacts on efficacy endpoints, estimands, and safety. These effects may also occur in a study in the case of other unexpected intercurrent events. Some **general recommendations that could be updated in a BRAP when such a major unexpected event occurs** include (Meyer et al. 2020):

- Rigorous evaluation of the probability of success of the current or future planned trials, which may impact future go/no-go decisions

- Potential additional data collection needs to adequately assess safety and efficacy, such as additional sources of data, approaches to evaluate missing data, and impact on estimands

- Summary of alternative measures of benefits and risks, considering important subgroups and evaluation of whether the data sources are valid, reliable, and applicable

- Potential planned or unplanned trial adaptations to be discussed with health authorities prior to unblinding, including additional safety and efficacy measures or interim analyses

- Considerations for changes to future study sites, study populations, or enrollment criteria in the development program

The remainder of this chapter will focus on the results of the BRAP Taskforce's efforts to identify current published recommendations applicable to benefit-risk assessment planning and efforts to assess how industry is currently utilizing sBRAs within their medical product development processes. These and future efforts will help to inform development of a BRAP template proposal.

13.3 Review of Published Recommendations Applicable to Benefit-Risk Assessment Planning

We conducted a literature review in order to identify published guidance from global regulatory authorities, published recommendations by global committees, and survey papers or books with implications for benefit-risk planning, process, and communication. We also considered publications related to special data types that would have a major impact on regulatory benefit-risk assessments, such as RWD and patient experience data, including as COAs, PROs, or patient preference information (PPI).

The global aspect of this literature search attempted to include perspectives from health authorities in the United States, Europe, Canada, Japan, Switzerland, Australia, and China, where available. Some general considerations are provided in an overview paper by McAuslane (2017) utilizing a Universal Methodology for Benefit-Risk Assessment framework applicable to Australia, Canada, Switzerland, and Singapore. However, there is little formal guidance about sBRAs published by various health authorities apart from those from the FDA and EMA; therefore, perspectives applicable to the United States and Europe are the focus of this section.

13.3.1 EMA Perspective on Benefit-Risk

The EMA's benefit-risk methodology project has proposed eight steps that should be considered when conducting a benefit-risk evaluation, as part of the PrOACT-URL framework (refer to Table 13.3 and https://protectbenefitrisk.eu/PrOACT-URL.html). Considering each of these steps would be important to include in a BRAP suitable for EMA submissions, this decision-making framework advocates the use of an effects table (see Table 13.1).

Another consideration for benefit-risk assessments that would impact planning is if the product is being developed in an area of high unmet medical need. Nicotera et al. 2019 provided an overview of a new adaptive pathway program that allows early dialogue on promising and innovative medical products in areas with a high unmet medical need. The EMA proposes

TABLE 13.3

EMA PrOACT-URL Framework

Problem	1. Determine the nature of the problem and its context
	2. Frame the problem
Objective	3. Establish objectives that indicate the overall purposes to be achieved
	4. Identify criteria for (a) favorable effects and (b) unfavorable effects
Alternatives	5. Identify the options to be evaluated against the criteria
Consequences	6. Describe how the alternatives perform for each of the criteria, that is, the magnitudes of all effects, and their desirability or severity, and the incidence of all effects
Trade-off	7. Assess the balance between favorable and unfavorable effects
Uncertainty	8. Report the uncertainty associated with the favorable and unfavorable effects
	9. Consider how the balance between favorable and unfavorable effects is affected by uncertainty
Risk tolerance	10. Judge the relative importance of the decision maker's risk attitude for this product
	11. Report how this affected the balance reported in step 9 above
Linked decisions	12. Consider the consistency of this decision with similar past decisions and assess whether taking this decision could impact future decisions

inclusion of more data sources beyond randomized controlled trials, including real-world evidence (RWE), pragmatic randomized controlled trials, and observational trials. Emphasis is also placed on the need to reduce the levels of uncertainty of risk and the expected benefits of new medical products. Planning for assessment of the benefit-risk balance of the medical product as information changes over the time is critical in this case.

Recently, the EMA has issued a strategic reflection document with a vision of goals from 2020 to 2025. Within this document, there are several key recommendations that will direct the future of regulatory goals related to benefit-risk and decision-making at the EMA and point to the need for medical product developers to invest in planning for benefit-risk assessments. In particular, the EMA will prioritize issues such as expanding benefit-risk assessment and communication, optimizing capabilities in modeling/simulation, and exploiting digital technology and artificial intelligence in decision-making. In addition, this will contribute to health technology assessment (HTA) preparedness and downstream decision-making for innovative medicines, reinforcing patient relevance in evidence generation and promoting the use of high-quality RWD in decision-making. Inclusion of some of these approaches in a medical product development plan will require increased interaction with the EMA to ensure appropriate implementation useful for regulatory decision-making. Use of RWD for benefit-risk assessment is discussed further in Section 13.3.3.

13.3.2 FDA Perspective on Benefit-Risk

In 2013, the commitment in PDUFA V was to implement FDA's Benefit-Risk Framework (Table 13.2) into the regulatory review process. A recent review paper by Lackey et al. (2021) summarized the role of the Benefit-Risk Framework for approvals of Drug and Biologics and how it has been used by FDA review teams to communicate approval decisions. The FDA was successful in implementing this high-level framework within the CBER and CDER primarily due to the flexibility of the format. The FDA CDRH has gone beyond this framework due to the device regulations that call for explicit benefit-risk evaluation (see section 513(a) of the Federal Food, Drug and Cosmetics Act).

In 2017, the FDA published guidance integrating the ICH guidelines for FDA submissions. In particular, the **FDA noted that a medical product sponsor should consider the following when completing Section 2.5.6 on Benefits and Risks Conclusions**:

- **Therapeutic Context**: Disease or condition to be treated, population to be treated, and benefits and risks of current therapies
- **Benefits**: Favorable effects, important characteristics of the product that may lead to improved patient compliance, population benefits, clinical importance and magnitude of the benefit, time course and variability of the benefit, and evidence strengths, limitations, and uncertainties
- **Risks**: Unfavorable effects, adverse events, drug interactions, risks to those other than the patient, risks based on class of drug, potential misuse/abuse, seriousness/severity, reversibility, tolerability, time course and variability, and evidence strengths, limitations, and uncertainties
- **Benefit-Risk Assessment**: Assess and weigh key benefits and risks, effect of uncertainties on interpretation of evidence or impact on the benefit risk assessment, impact of the therapeutic context, and approaches to risk management (e.g. labeling, registries, etc.)

Note that the above items can be assessed at different points in the medical product development life cycle and can be included in a BRAP document early in medical product development, in alignment with the FDA benefit-risk framework.

In particular, application of benefit-risk approaches at the FDA may be more prevalent within the context of device regulation at the CDRH. The CDRH released a draft guidance on benefit-risk assessment for medical devices in 2012 and two final guidances in 2019. The CDRH recommends several key factors to consider in order to identify uncertainty in the benefit-risk determination, considerations for breakthrough devices, and a very detailed benefit-risk assessment worksheet that may be used by CDRH staff to make determinations for a medical device under review. **Device sponsors may utilize this worksheet in order to adequately address benefit-risk concerns when planning studies and submissions to the CDRH.** Readers are referred to the appendices of the CDRH guidance for industry, "Factors to Consider When Making Benefit-Risk Determinations in Medical Device Premarket Approval and De Novo Classifications" for the worksheets for Benefits, for Risks, and for Benefit-Risk Assessments (excerpts also displayed in Table 13.4 below).

In 2018, the FDA released the PDUFA VI plan for "Benefit-Risk Assessment in Drug Regulatory Decision-Making," applicable through 2022. FDA's commitments to enhance benefit-risk assessment include refining templates and enhancing the clarity of the benefit-risk framework. For the time being, the benefit-risk considerations outlined in Table 13.4 provide a valuable resource for planning development activities. The FDA has also stated that it will continue to focus on including the patient's voice in medical product development and decision-making. Considerations for use of patient focused data for benefit-risk assessment are discussed further in Section 13.3.4.

TABLE 13.4

Benefit-Risk Assessment Questions Relevant to a Device (FDA CDRH)

Factors	Assessment Questions
Uncertainty	What information does the FDA have to assess benefit and risk?What is the quality of the information FDA is using (e.g. MDRs, literature, registry or clinical trial data, limited case studies, inspectional data, etc.)? Is the quality of information a reliable source for making an objective and unbiased benefit or risk decision? What is the uncertainty related to the current understanding of benefits and risks?
Mitigations	Could you identify ways to mitigate the risks such as using product labeling, establishing education programs, etc.?What is the type of mitigation proposed?Is the intervention related to design, labeling, or training?Has the manufacturer corrected the cause of the non-conformity?
Detectability	Can the user easily recognize the hazard to avoid the harm?Can the problem with the medical device be corrected before use by the user?
Failure Mode	Has the manufacturer identified the underlying cause?Has the firm submitted testing to the FDA?Has the FDA conducted testing? What were the results?
Scope of the device issue	Are the risks identified potentially inherent to similar medical devices of this type (i.e., industry-wide)?

TABLE 13.4

(Continued)

Factors	Assessment Questions
Patient impact	What are the risks to patients if the device is not available?Are patients better off if the device is available?What is the potential impact on patients related to the inspectional observation or regulatory non-compliance?Does the observation or violation directly relate to product quality?Does the observed regulatory non-compliance raise concerns regarding the firm's ability to produce safe and effective medical devices?
Preference for availability	Would patients and caregivers prefer to have access to the device?Are the benefits and risks adequately understood?
Nature of violations/ Non-conforming product	Was the violation systemic or non-systemic in nature?To what extent are the products non-conforming?
Firm compliance history	Has the same or a similar inspectional observation or regulatory violation been observed at the manufacturer in the past two years? In the past five years? In the past 10 years?Does the firm have a history of regulatory compliance and high quality device production?Has the firm demonstrated chronic and systematic regulatory non-compliance over time?Is the regulatory non-compliance significant enough that FDA would take regulatory action?Was the harm anticipated in the firm risk management documentation?Was the harm reported to the FDA by the firm quickly?Would providing notice to the firm assist in informing the firm of its legal responsibilities?

13.3.3 Planning for Use of RWD Suitable for Benefit-Risk Evaluation

A major concern with using RWD for health authorities is if it has been appropriately measured and appropriately characterizes the benefits and/ or risks. That is, there is a need to justify that RWD provide RWE of sufficient quality useful for decision-making. Planning for how to assess and utilize RWD within a BRAP will help to inform design of studies in a development program that will provide adequate justification to regulators for decision-making.

RWD may have utility in many contexts, although the use of RWD for healthcare practice and policy decisions is limited at present. RWD can be used to improve the efficiency of clinical trials (even if not used to generate RWE), provide evidence of current treatment options, or may provide supportive information for regulatory review of drugs or devices or to support a new indication of an already approved product (U.S. Food and Drug Administration 2018; and FDA CDRH and CBER 2017). The FDA guidance evaluates how RWD derived from different sources can be leveraged in support of scientific evidence for healthcare and regulatory decision-making. Appropriate incorporation of RWD can inform decisions involving the following, which can be indicated as objectives in a BRAP: (i) providing early indications of cost-effectiveness before large-scale trial data become available; (ii) directing future research based on the cost of reducing uncertainty; (iii) assessing subgroups, dosing schedules, and protocol deviations; (iv) informing strategic research and development along with pricing decisions; and (v) estimating the cost-effectiveness of complex pharmaceutical interventions (such as pharmacogenetics testing) (Swift et al. 2018).

Types of RWD used may include clinical health record data, lab test results, radiology images, provider notes, medication orders, medical claims, genomic data, biomarker data, mobile or wearable technology, health records, registries, social media data, health conditions related to the extended family, PROs, surveys, diaries, quality of life measures, and lifestyle factors. Planning for incorporation of such data in clinical trials is important and will be time-consuming if measures require validation by health authorities. One important aspect to include in a BRAP is the need for validation of digital technologies which may be a source of RWD especially if these are to be used as surrogate markers of benefit and risks (Ramamoorthy and Huang 2019).

The concept of "regulatory fitness" or "regulatory grade" is very important to consider when planning for inclusion of RWD. Ramamoorthy and Huang (2019) and Duke Margolis Center for Health Policy (2019) both recently published papers regarding the fitness of RWD for regulatory use. For FDA, EMA, and many other regulatory agencies, it is critical to determine when and how RWD analysis may lead to evidence for regulatory decision-making, and when it will not. Data and analysis approaches should be reliable, fit for purpose, and reproducible for regulatory-grade RWE. Breckenridge et al. (2019) have stated "regulatory grade" RWD require the following (which can be specified in a BRAP to inform study design decisions):

- Define the scientific question which should be meaningful and answerable with RWD
- Identify the appropriate study design
- Select the appropriate RWD to be used
- Define data standards/analytical methods and strategy
- Comply with regulatory standards, which may necessitate early meetings with regulators
- Audit trails available for regulators to identify sources and methods (as per CDER/CBER/CDRH guidance on *Use of Electronic Health Record Data in Clinical Investigations*)

There are other considerations for how sponsors may plan their studies when incorporating RWD. Such data should be collected in a systematic and standardized way to produce high-quality data that will support regulatory review. There are tradeoffs to be considered when making choices for design, data source, and analysis choices that can be included in a BRAP (Schneeweiss et al. 2016). To work with RWD across multiple sources, data may need to be put into a common format, sometimes referred to as a common data model, with common representation (terminologies, vocabularies, and coding schemes). If using data from different countries, the BRAP may specify differences in medical practice, health care delivery, and data reliability that could impact the relevance of RWD for decision-making. Ongoing collaboration between the FDA CDRH and the medical device innovation consortium (MDIC) has resulted in the development of the external evidence method (EEM) Framework for leveraging external data for pre-market regulatory medical device submissions. New innovative methods, novel statistical methodologies, and cataloging existing methods for leveraging data from different data sources illustrate how external data can potentially be utilized in regulatory decision-making.

13.3.4 Planning for Use of Patient Experience Data Suitable for Benefit-Risk Evaluation

During medical product development, it is important to enable patient-centered decision-making to better meet patient needs, but there has not been written guidance on how exactly to use patient preference or patient experience information in benefit-risk decision-making. A recent paper by Oncology practitioners from the FDA, MHRA, and Health Canada points out some of the differences in regulatory definitions for the PRO, health-related quality of life (HRQoL), PRO instrument, and treatment benefit from regulatory agencies in the US, Canada, and Europe (Kluetz et al. 2018). Having different definitions and regulations makes it difficult for sponsors to plan for trials with global considerations for labeling. There are several ongoing efforts to identify approaches to benefit-risk assessment of patient experience data. Some recommendations from recent publications are described in this section that are useful for planning purposes, and additional practical approaches from CIOMS Working Group XII are expected in the coming years.

There are many terms used to describe "patient experience" data. The EMA uses "PRO" as an umbrella term covering both single dimension and multi-dimension measures of symptoms, HRQL, health status, adherence to treatment, and satisfaction with treatment. The FDA CDER considers that patient-focused drug development (PFDD) may involve several different types of measures that can be used for benefit-risk evaluations, including COAs. A COA is an assessment of a clinical outcome and can be made through reports by a clinician, a patient, and a non-clinician observer or through a performance-based assessment, usually from interventional trials. There are differences to consider based on the sources of information, such as data coming from interventional trials or other sources. The range of patient experience data that would fit within the FDA definition of PFDD data includes patient registry data, natural history study data, patient focus group or meeting reports, patient survey data, COA data collected during clinical trials, and elicited patient preference data. The FDA CDRH also defines PPI as "qualitative or quantitative assessments of the relative desirability or acceptability to patients of specified alternatives or choices among outcomes or other attributes that differ among alternative health interventions." In particular, PPI may come from a variety of sources, including interviews with patients.

In 2005, the EMA published a broad reflection paper on HRQoL measures and later published Appendix 2 to the guideline on the evaluation of anticancer medical products in men: The use of PRO measures in oncology studies (2016) that identified important considerations related to PROs in clinical trials. The EMA outlined broad principles of scientific best practice and provided guidance on the value of PRO data in the development of medicinal products for the treatment of cancer, which addressed some question of how a sponsor in the oncology space may implement such data in planning for clinical trials to assess benefits and risks. This EMA guideline notes that **PRO endpoints should be incorporated into protocol development at the earliest stage and should be explicitly stated as a clinical trial objective or hypothesis in the study protocol and statistical analysis plan**. To this end, a BRAP may incorporate PRO or COA endpoints at a very early stage in order to plan for evidence gathering and hypothesis generation throughout development.

In 2012, the FDA CDRH published a guidance document on benefit-risk determinations, which made it clear that PPI is an important factor to consider when evaluating medical devices for market approval. A 2015 paper by Ho et al also outlines the experiences from CDRH in incorporating patient-preference evidence. This paper describes the first study submitted to CDRH which was designed to provide regulators with quantitative data on patient preferences in support of regulatory benefit-risk tradeoff determinations. The MDIC also conducted a patient-centered benefit-risk project and issued a 2015 report with a framework for incorporating information on patient preferences for benefit-risk. Most recently, a current project to look at Patient Preferences in Benefit-Risk Assessments during the Drug Life Cycle (PREFER) aims to make recommendations about how and when to include PPI. One recent review paper from the PREFER project by Janssens et al. 2019 identified several ways in which PPI may be used, including early development and go/no-go decision-making, defining subgroups with different benefit-risk trade-offs when designing trials, and informing expanded indications or populations. Whichello et al. (2020) (also from PREFER project) also recently published a paper on including PPI, identifying decision-points in the medical product life cycle where PPI can be informative. Within an organization, these **decision points that may be informed by PPI include**

prioritizing targets, leads, studies and assets; optimizing assets; informing regulatory submissions; and managing the medical product life cycle.

At the time of this publication, the FDA CDER/CBER was developing a series of four methodological PFDD guidance documents useful for product development and submissions for regulatory decision-making (see https://www.fda.gov/drugs/development-approval-process-drugs/fda-patient-focused-drug-development-guidance-series-enhancing-incorporation-patients-voice-medical). This effort will address topics related to incorporating COAs into endpoints for regulatory decision-making including COA-related endpoint development, defining meaningful within-patient score changes, and collection, analysis, interpretation, and submission of data. A 2019 public workshop on developing Guidance 4 identified several estimand attributes that will be considered when using COA-based endpoints for regulatory decision-making. Based on the FDA's estimand framework, **a sponsor may consider the following approaches to better plan research objectives in order to address benefits and risks of patient experience data**:

- Clearly define the scientific research question
- Define the target population through inclusion/exclusion criteria to reflect the targeted patient population for approval
- Clearly define the endpoint of interest using well-defined measurement tools
- Address intercurrent events in the analysis plan, in alignment with the research question
- The population level summary should be based on the scientific research question

Summary of Regulatory Guidance Global Trends

Apart from a few meetings or abstracts found on the internet, there was no clear guidance from regulatory authorities in the rest of the world regarding benefit-risk assessments. Health Canada recently published a guideline (Health Canada 2019) to assist market authorization holders in developing a qualitative benefit-risk assessment for marketed medical products when requested. The document provides an outline of how to respond to a Health Canada request for a benefit-risk assessment, but it is not clear how often this assessment is requested in practice and it is intended only for post-marketing.

Primarily based on the available guidance from the FDA and EMA and publications from cross-industry global collaborative groups (such as ICH, CIOMS, and IMI-PROTECT), **the taskforce identified the following trends**:

- Globally, there is a trend in the pharmaceutical sector to use structured frameworks for qualitative evaluation of benefit-risk in regulatory decision-making for medical product approvals

- For some medical devices and diagnostics, the benefits and risks may be defined on an individual patient-level basis; some quantitative decision-making can also be used, but it is much less common
- Regulatory agency viewpoints are evolving, and trends show consideration of RWD and patient experience data as supplementary information in decision-making; and making these data fit for regulatory consideration requires early planning
- Regulated industry may integrate global regulatory agencies' benefit-risk framework considerations in their development lifecycle, but there are challenges to alignment, which current consortia are hoping to address with future recommendations

Further information regarding findings from the BRAP Taskforce literature review will be available in a future paper.

13.4 Benefit-Risk Assessment Planning in Practice – Survey of the Industry Perspective

13.4.1 Understanding Industry's Approach to Benefit-Risk Assessment and Planning

We will also summarize learnings from the industry survey conducted by the taskforce, highlighting future opportunities for benefit-risk assessment planning as identified by survey respondents. Optimization of benefit-risk decision-making is paramount to medical product development and the provision of medicines and devices as new treatment options for patients.

As highlighted in the literature review (Section 13.3), the structured approach to benefit-risk assessment (sBRA) fosters processes, discussion, analyses, decision-making, and then documentation which provides an integrated benefit-risk assessment focused on a product's key or most important benefits and risks, often based on a specific indication, and describes how the product addresses an unmet medical need or impacts the current treatment paradigm.

The BRAP Taskforce conducted a survey (during January–March 2020) of global pharmaceutical and medical device companies' representatives involved in benefit-risk efforts within their companies to gain an understanding about current practices and implementation.

13.4.2 Overview of Industry Survey Design and Participants

Both regulators and pharmaceutical and medical device companies (industry) are finding ways to plan for and then implement sBRAs. To understand how some pharmaceutical and device companies are utilizing sBRAs, a

TABLE 13.5

Pharmaceutical Companies Providing Information about sBRAs via an Online Survey[a]

Company Category (by estimated # of employees)	All Companies	Companies Utilizing sBRA[b]	Companies Not Utilizing sBRA[c]
Small (<10, 000)	4 (15%)	2	2
Medium (10,000–<50,000)	11 (42%)	9	2
Large (>50,000)	11(42%)	10	1

[a] The survey was sponsored by the American Statistical Association (ASA) – DIA Safety Working Group BRAP Taskforce (Conducted Q1 2020).
[b] For companies utilizing sBRAs, the most common therapeutic areas of development were CNS/neurology, immunology, oncology, and rare diseases.
[c] For companies not utilizing sBRAs, the most common therapeutic areas of development were immunology and oncology.

survey was conducted by the BRAP Taskforce, on a voluntary and anonymous basis, of various pharmaceutical and device companies (Table 13.5) about sBRAs. Of the 26 companies that participated in the survey, 21 (80%) reported the use of sBRAs.

The survey focused on five key areas of benefit-risk assessment efforts:

- Implementation measures
- Utilization for decision-making
- Visualization options
- Challenges
- Future opportunities

A future paper will cover the detailed results of this survey. The following sections are based on the survey results and highlight how some pharmaceutical and medical device companies (industry) are adapting and applying a structured approach to benefit-risk to aid their decisions and medical product development efforts.

13.4.3 Industry Survey Results: sBRA Implementation Measures

Respondent companies that have implemented sBRAs provided information about various measures that support implementation within their respective companies.

13.4.3.1 Coordination

Most companies that conduct sBRAs have a dedicated benefit-risk group or specialist who leads or coordinates the sBRA effort. The functional representative or department most commonly responsible for the sBRA efforts is safety/pharmacovigilance, whereas in a few companies, members of clinical development, statistics, or epidemiology are responsible for the sBRA effort. Additionally, in all companies, a sBRA is a cross-functional effort with specific functional involvement dependent on the specific activity (Table 13.6).

13.4.3.2 Data/Information Sources

A variety of data or information sources are being used to support sBRAs. As expected, all companies use information from clinical trials. The majority also use information from published literature, spontaneous adverse event reports, observational data (defined in the survey as electronic health records, prescription databases, and registries), and RWD.

13.4.3.3 Tools

Most companies (~80%) have document templates and/or sBRA training in place to assist with conduct of sBRAs. Some companies have standard operating procedures or best practice manuals to guide sBRA implementation. Only a few companies prepare and use statistical analysis plans that are aligned or integrated for sBRA and safety evaluations. Many companies use benefit-risk frameworks (e.g. FDA Benefit-Risk Framework [see Table 13.2] EMA PROACT-URL [see Table 13.3]) to support their sBRA documentation.

Most companies are using Forest plots for key benefits and key risk displays. Many are using value trees (Keeney and Raiffa 1993) to display key benefits and key risks and some use effects tables (see Table 13.1). A value tree is an explicit visual map of the attributes or criteria of decisions, as described in Chapters 5 and 15.

13.4.4 Industry Survey Results: sBRA Utilization for Decision-Making

Companies are using sBRAs to inform various decisions during medical product development. sBRAs are most commonly used to inform decisions about submission and subsequently to support preparation of the Clinical Overview Section 2.5.6. In the post-market period, sBRAs are most commonly used to support preparation of periodic safety reports (PSURs/PBRERS, PADRs). In addition, sBRAs support preparation for regulatory meetings/advisory committee meetings.

Most companies are using sBRAs for decision-making with internal safety/benefit-risk committees or governance groups. Very few companies are using sBRAs for decision-making about preclinical activities, early phase development, and independent data monitoring committees or payers/reimbursement/HTA activities.

TABLE 13.6

Pharmaceutical Departments or Functions Involved in sBRA Activities

ACTIVITY	FUNCTION	Clin Dev	PV/Safety	Reg	Statistics	Med Affairs	Epi	HEOR
Selects Key Benefits and Risks by Indication	Most							
	Some							
Prepares / Documents Qualitative sBRA	Most							
	Some							
Decides / Conducts Quantitative sBRA	Most							
	Some							
Conducts / Oversees BR Patient Preference Studies	Most							
	Some							
Decides / Implements Risk Minimization	Most							
	Some							

Abbreviations: Clin Dev = Clinical Development; Reg = Regulatory; Med Affairs = Medical Affairs; Epi = Epidemiology; HEOR = Health Economics and Outcomes Research

Shading for "Most" means that >70% reported the department/function was involved in the specific sBRA-related activity. Shading for "Some" means that 30–70% companies reported the department/function was involved in the specific sBRA-related activity.

Some companies are using quantitative methods (defined in the survey as "e.g. multiple-criteria decision analysis [MCDA]") to aid benefit-risk decision-making, noting use is in a limited or on an as needed basis.

13.4.5 Industry Survey Results: Challenges with sBRAs

Companies participating in the survey noted that one of the biggest challenges to sBRAs is limited understanding and awareness about the need and added value of sBRAs across their Research and Development (R&D) organization. They highlighted that their colleagues are not aware of regulators' current expectations and that they tend to underestimate the importance of sBRA efforts. They emphasized the importance that sBRA efforts be endorsed by senior leaders across their R&D organization for it to gain traction and become the "default approach" for submission preparations with support from cross-functional team experts. They also emphasized the need for sufficient staff with expertise to conduct sBRAs as well as the need for training and procedures to be in place to support sBRAs.

Companies also highlighted particular challenges related to quantitative benefit-risk methods. Here, companies mentioned that they face a lack of acceptance of quantitative methods within their companies. They cited as key hurdles to quantitative efforts a lack of easy to use software tools/analytical tools to conduct these assessments, the complexity to elicit weights for certain methods, limited experience on how to interpret quantitative results, and limited examples to demonstrate how quantitative assessments aid sBRAs or benefit-risk decision-making.

13.4.6 Industry Survey Results: Future Opportunities for sBRAs

The respondent companies offered a variety of opportunities to advance the use and implementation of sBRAs at their companies. **Key opportunities for future sBRAs** as reported by the companies include:

- Practical use cases and scenarios for products in different stages of development
- Application of sBRAs for HTA
- Guidance on how to use sBRAs for post-approval benefit-risk decisions
- How and when to implement patient preference evaluations
- How to use PPI to support benefit-risk decision-making
- Establishment of an expectation that regulators and industry should discuss the planned approach to sBRA at the pre-submission stage
- Feedback from regulators on whether a company's sBRA within a submission was useful

13.5 Future Work and Discussion

The regulatory landscape on benefit-risk evaluation is evolving. This is best reflected in the ICH guidance M4E (R2) and ICH guidance IE2C(R2). These regulatory guidances serve to provide harmonized approaches for the benefit-risk assessment of the medical product for its intended use. These guidances focus on the "What" should constitute the benefit-risk assessments, what are the key benefits, and what are the key risks, as well as the benefit-risk context and discussion. However, there is much work to be done on "How" to achieve these objectives. The survey by the ASA safety working group BRAP Taskforce on the benefit-risk industry practice was able to show good adoption of regulatory guidance across industry. Companies are using various frameworks, where the FDA framework is most popular. We also noticed that the planning of the formal benefit-risk assessment is not a common practice.

To better inform benefit-risk evaluation in medical product development, we suggest that the future lies with aligning perspectives and early benefit-risk planning within a formal BRAP (referred to as a BRAP in this chapter). However, collaboration with regulators is needed in order to make a BRAP applicable to regulatory concerns and aligned with the regulatory process. Additionally, incorporation of two distinct areas of research – patient perspectives and RWD – into the benefit-risk determination of a product has been an important topic in recent years. It will remain a future priority to find more ways to engage with patient advocates and include patient perspectives into B-R evaluations and to plan for regulatory-grade RWD (if applicable to the development program). One approach the BRAP taskforce will take is to include recent recommendations on evaluating RWD, COAs, and PPI into the BRAP template. However, more thought and discussion are needed regarding how to incorporate the patient perspective on benefit-risk trade-offs that impact decision-making. We hope to see more public workshops focused on this issue.

In addition to the literature review and industry survey, the BRAP Task force is currently interviewing a group of well-known and global key opinion leaders (KOLs) in benefit-risk assessment. The main purpose of these interviews is two-fold: (1) seeking the KOL's opinion on how benefit-risk assessment will develop in the future, with regard to regulatory requirements, methodology research and uptake by industry globally, and (2) gathering the KOL's viewpoint on benefit-risk assessment planning in their respective organizations. The results of these interviews will be published in a future paper.

The work presented in this chapter will serve as the foundation for our working group to develop a BRAP template that facilitates the planning of benefit-risk assessments throughout product development life cycle. We intend that such a document or template: (1) defines a process of benefit-risk assessment planning throughout the life cycle of a medical product, which is iterative as needed; (2) incorporates points-to-consider, checklists, and process flow charts for high-level planning purposes; (3) provides a modularized approach so that users can pick and choose parts of the BRAP to utilize; and (4) offers flexibility to incorporate special considerations on areas such as patient preference, patient experience of side effects/tolerability, COA/PRO data, REMS, and RWD. Such a template will connect safety and benefit-risk planning activities together. It will also hopefully serve to facilitate interactions between regulatory agencies and industry sponsors at key points in the global medical product development and approval process.

For readers developing their own BRAP, below are the key take-aways from this chapter that could be considered when creating such a planning document:

- Define the decision-making environment, such as stakeholders, objectives, constraints, caveats, and different trade-offs for stakeholders to consider. Implementation of the EMA PrOACT-URL framework may be useful for this purpose.
- Consider the applicable regulatory paradigm for the product under development, such as the EMA or FDA structured benefit-risk frameworks that should be informed by the BRAP at key points in development – both require identification of key benefits, key risks, available evidence, and uncertainties that can be summarized in a BRAP. Benefit-risk considerations specific to the FDA CDRH have already been identified in guidance and could be generally applicable to many medical products under development (see Table 13.4).
- Details regarding how, when, and what information will be collected in the development program and/or clinical studies that will be useful for informing a sBRA at key time points in the development process can be included in a BRAP (see Figure 13.1 for suggested milestones).
  - o The level of detail can be dependent on the role the BRAP will play for internal planning purposes and for interaction with health authorities. As indicated by the survey results, alignment with existing internal procedures is important to be able to implement a BRAP and add value, but a BRAP can be useful to inform a variety of decisions during development.
  - o Note that RWD can potentially support development and regulatory decision-making, but the appropriateness is determined by the relevance to the question at hand and the reliability and reproducibility of the data.

o Note that the role of patient experience data (e.g. COA/PRO/PPI) varies by the regulatory region and the scientific research question. Early incorporation of planning activities for these data in the development lifecycle is important.

- Provide a plan for when and how the BRAP will be updated and how it may be used for interactions with regulatory authorities and other stakeholders. Consideration of who is part of the core benefit-risk management team that is responsible for updating the BRAP is also essential, but this depends on the structure of the organization.

Acknowledgments

This chapter reflects the views of the authors and should not be construed to represent FDA's views or policies. The authors acknowledge that this chapter has been informed by the work of the BRAP Taskforce. The work of the BRAP Taskforce focuses on connecting safety and benefit-risk planning activities together for high-level considerations for regulatory interactions and internal planning. The objectives of the BRAP Taskforce include assessment of the global regulatory landscape, assessment of the current industry perspective, and external alignment with other groups and researchers, which has informed the authors' work toward a BRAP template. The current results from the research efforts of the BRAP Taskforce are presented in this chapter, along with future directions of benefit-risk planning efforts.

BRAP Taskforce Members: Brian Edwards (co-chair), Lisa R. Rodriguez (co-chair), Andreas Brueckner, Susan Colilla, Michael W. Colopy, Martin Gebel, Jürgen Kübler, Leila Lackey, Kerry Jo Lee, Xuefeng Li, Oksana Mokliatchouk, Cheryl Renz, Jeff Roberts, Arianna Simonetti, Susan Talbot, Lothar T. Tremmel, William Wang, Max G. Waschbusch, Hong Yang, Sammy Yuan

References

Almenoff JS, Pattishall EN, Gibbs TG, et al. (2007). "Novel statistical tools for monitoring the safety of marketed drugs." *Clinical Pharmacology & Therapeutics*, 82, 157–166.

Biotechnology Industry Organization. "A Lifecycle Approach to FDA's Structured Benefit-Risk Assessment Framework." https://www.bio.org/FDAwhitepaper

Breckenridge A. (1999). "Clinical pharmacology and drug regulation." *British Journal of Clinical Pharmacology*, 47, 11–12.

Breckenridge AM, Breckenridge RA, & Peck CC. (2019). "Report on the current status of the use of real-world data (RWD) and real-world evidence (RWE) in drug development and regulation." *British Journal of Clinical Pharmacology*, 85(9), 1874–1877.

CHMP Working Group Report on Benefit-Risk Assessment Models and Methods (2007). EMEA/CHMP/15404/2007.

CIOMS (1998). "Benefit-Risk Balance for Marketed Drugs: Evaluating Safety Signals," Report of CIOMS Working Group IV, Geneva.

CIOMS (2019). Benefit-Risk Balance for Medicinal Products, CIOMS Working Group XII.

Doan T, Renz C, Bhattacharya M, Lievano F, & Scarazini L. (2019). *Pharmacovigilance: A Practical Approach* (Missouri: Elsevier).

Duke-Margolis Public Meeting. (2019). *Characterizing FDA's Approach to Benefit-Risk Assessment Throughout the Medical Product Lifecycle.* https://healthpolicy.duke.edu/events/public-meeting-characterizing-fdas-approach-benefit-risk-assessment-throughout-medical

European Medicines Agency (EMEA). (1999). "Notice to Marketing Authorization Holders Pharmacovigilance Guidelines," CPMP/PHVWP/108/99corr

European Medicines Agency. (2005). "Reflection paper on the Regulatory Guidance for the Use of Health-Related Quality of Life (HRQL) Measures in the Evaluation of Medicinal Products." July 2005. EMEA/CHMP/EWP/139391/2004.

European Medicines Agency. (2008). "Reflection Paper on Benefit-Risk Assessment Methods in the Context of the Evaluation of Marketing Authorization Applications of Medicinal Products for Human Use." EMEA/CHMP/15404/2007

European Medicines Agency (2012a). *"Benefit-risk methodology project work package 3 report: Field tests.* London: European Medicines Agency." EMA/718294/2011

European Medicines Agency (2012b). Benefit-risk methodology project. Work package 4 report: benefit-risk tools and processes. EMA/297405/2012.

European Medicines Agency (2016). "Appendix 2 to the guideline on the evaluation of anticancer medicinal products in man: The use of patient-reported outcome (PRO) measures in oncology studies." EMA CHMP.

Hallgreen CE, Mt-Isa S, Lieftucht A, et al. (2016). "Literature review of visual representation of the results of benefit–risk assessments of medicinal products." *Pharmacoepidemiology and Drug Safety*, 25: 238–250.

Hammond JS, Keeney, RL & Raiffa H. (1999). *Smart choices: A Practical Guide to Making Better Decisions.* Crown Business.

Health Canada. (2019). "Format and content for post-market drug benefit-risk assessment in Canada - Guidance document".

Ho MP, Gonzalez, JM, Lerner, HP, Neuland CY, Whang JM, McMurry-Heath M, Hauber AB, Irony T. (2015). "Incorporating patient preference evidence into regulatory decision making." *Surgical Endoscopy* 29: 2984–2993.

ICH Harmonised Guideline. (2003). "M4E Common Technical Document for the Registration of Pharmaceuticals for Human Use: Clinical Overview and Clinical Summary of Module 2 and Module 5: Clinical Study Reports," *EMA*.

ICH M4E(R2) guideline on enhancing the format and structure of benefit-risk information in ICH efficacy update July 2017.

ICH guideline E2C (R2) on periodic benefit-risk evaluation report (PBRER). January 2013.

IMI PREFER (Patient preference in benefit risk assessment during the drug life cycle). https://www.imi.europa.eu/projects-results/project-factsheets/prefer.

IMI PROTECT (Pharmacoepidemiological Research on Outcomes of Therapeutics by a European Consortium). http://www.imi-protect.eu/

Janssens R, Huys I, van Overbeeke E, Whichello C, Harding S, Kübler J, & Veldwijk J. (2019). "Opportunities and challenges for the inclusion of patient preferences in the medical product life cycle: a systematic review." *BMC Medical Informatics and Decision Making*, 19(1), 189.

Keeney, RL, & Raiffa H (1993). *Decisions with Multiple Objectives: Preferences and Value Tradeoffs* (New York: Cambridge University Press).

Kluetz PG, O'Connor DJ, & Soltys K (2018). "Incorporating the patient experience into regulatory decision making in the USA, Europe, and Canada." *The Lancet Oncology*, 19(5), e267–e274.

Lackey L, Thompson G, & Eggers S (2021). "FDA's benefit-risk framework for human drugs and biologics: Role in benefit-risk assessment and analysis of use for drug approvals." *Therapeutic Innovation & Regulatory Science*, 55, 170–179.

McAuslane N, Leong J, Liberti L, & Walker S (2017). "The benefit-risk assessment of medicines: Experience of a consortium of medium-sized regulatory authorities." *Therapeutic Innovation & Regulatory Science*, 51, 635–644.

Medical Device Innovation Consortium (MDIC). (2015). "Medical device innovation consortium (MDIC) patient centered benefit-risk project report: A framework for incorporating information on patient preferences regarding benefit and risk into regulatory assessments of new medical technology."

Medical Device Innovation Consortium (MDIC). (2021). "External evidence methods (EEM) framework."

Meyer RD, et al. (2020). "Statistical issues and recommendations for clinical trials conducted during the COVID-19 pandemic." *Statistics in Biopharmaceutical Research*, 12(4), 399–411.

Nicotera G, Sferrazza G, Serafino A, Pierimarchi P (2019). "The iterative development of medicines through the European medicine agency's adaptive pathway approach." *Frontiers in Medicine*, 6(148). doi:10.3389/fmed.2019.00148.

Packer M, Pitt B, Rouleau J, Swedberg K, DeMets DL, & Fisher L (2017). "Long-term effects of Flosequinan on the morbidity and mortality of patients with severe chronic heart failure: Primary results of the PROFILE trial after 24 years." *JACC Heart Fail*, 5(6), 399–407.

Ramamoorthy, A., & Huang, S.-M. (2019). "What Does It Take to Transform Real-World Data Into Real-World Evidence?" *Clinical Pharmacology & Therapeutics*, 106(1), 10–18.

Santoro, A., Genov, G., Spooner, A., Raine, J., & Arlett, P. (2017). "Promoting and Protecting Public Health: How the European Union Pharmacovigilance System Works." *Drug Safety*, 40(10), 855–869.

Schneeweiss, S., et al. (2016). "Real world data in adaptive biomedical innovation: A framework for generating evidence fit for decision-making." *Clinical Pharmacology & Therapeutics*, 100(6), 633–646.

Soekhai V, Whichello C, Levitan B, Veldwijk J, Pinto CA, Donkers B, Huys I, van Overbeeke E, Juhaeri J, & de Bekker-Grob EW (2019). "Methods for exploring and eliciting patient preferences in the medical product lifecycle: A literature review." *Drug Discovery Today*, 24(7), 1324–1331.

Swift, B., Jain, L., White, C., Chandrasekaran, V., Bhandari, A., Hughes, D. A., & Jadhav, P. R. (2018). "Innovation at the intersection of clinical trials and real-world data science to advance patient care." *Clinical and Translational Science*, 11(5), 450–460.

U.S. Food and Drug Administration. 21st Century Cures Act, TITLE III—DEVELOPMENT, Subtitle A--Patient-Focused Drug Development, SEC. 3001. PATIENT EXPERIENCE DATA. https://www.govinfo.gov/content/pkg/PLAW-114publ255/html/PLAW-114publ255.htm

U.S. Food and Drug Administration CDRH Guidance. (2016). Patient Preference Information – Voluntary Submission, Review in Premarket Approval Applications, Humanitarian Device Exemption Applications, and De Novo Requests, and Inclusion in Decision Summaries and Device Labeling.

U.S. Food and Drug Administration CDRH (2016). Factors to Consider Regarding Benefit Risk in Medical Device Product Availability, *Compliance, and Enforcement Decisions.*

U.S. Food and Drug Administration CDER/CBER (2020). Guidance on Patient-Focused Drug Development: Collecting Comprehensive and Representative Input https://www.fda.gov/media/139088/download

U.S. Food and Drug Administration (2013) Structured Approach to Benefit-Risk Assessment in Drug Regulatory Decision-Making. Draft PDUFA V Implementation Plan. Fiscal Years 2013–2017. Available online at http://www.fda.gov/downloads/ForIndustry/UserFees/PrescriptionDrugUserFee/UCM329758.pdf.

U.S. Food and Drug Administration. Benefit-risk assessment in drug regulatory decision-making: PDUFA VI Implementation Plan (FY 2018-2022). https://www.fda.gov/media/112570/download

U.S. Food and Drug Administration E2C(R2) Periodic Benefit Risk Evaluation Report (PBRER) Guidance for Industry. July 2016.

U.S. Food and Drug Administration CDER/CBER/CDRH Guidance for Industry. (2018). "Use of Electronic Health Record Data in Clinical Investigations".

U.S. Food and Drug Administration (2019a). CDER and CBER Guidance for Industry, "Demonstrating Substantial Evidence of Effectiveness for Human Drug and Biological Products."

U.S. Food and Drug Administration (2019b). CDRH Guidance for Industry, "Factors to Consider When Making Benefit Risk Determinations in Medical Device Premarket Approval and De Novo Classifications".

U.S. Food and Drug Administration Patient-Focused Drug Development Guidance Series for Enhancing the Incorporation of the Patient's Voice in Medical Product Development and Regulatory Decision Making https://www.fda.gov/drugs/development-approval-process-drugs/fda-patient-focused-drug-development-guidance-series-enhancing-incorporation-patients-voice-medical

U.S. Food and Drug Administration Public Workshop December 2019 https://www.fda.gov/drugs/development-approval-process-drugs/public-workshop-patient-focused-drug-development-guidance-4-incorporating-clinical-outcome)

U.S. Food and Drug Administration (2019d). CDRH Guidance for Industry, "Consideration of Uncertainty in Making Benefit-Risk Determinations in Medical Device Premarket Approvals, De Novo Classifications and Humanitarian Device Exemptions".

U.S. Food and Drug Administration (2018). Framework for FDA's Real-World Evidence Program.

U.S. Food and Drug Administration, CDRH and CBER (2017). Use of Real-World Evidence to Support Regulatory Decision-Making for Medical Devices.

Mussen F., Salek S., & Walker S. (2019). Chapter 1: Concept and scope of benefit-risk evaluation of medicines in Benefit-Risk Appraisal of Medicines, *Benefit-Risk Appraisal of Medicines: A Systematic Approach to Decision-making* (Chichester, West Sussex, UK; Hoboken, NJ: Wiley-Blackwell).

Whichello C., Bywall K.S., Mauer J., Watt S., Pinto C.A., van Overbeeke E., Huys I., de Bekker-Grob E.W., Hermann R., & Veldwijk J. (2020). "An overview of critical decision-points in the medical product lifecycle: Where to include patient preference information in the decision-making process?" *Health Policy* 124(2020): 1325–1332.

14

Estimands in Safety and Benefit-Risk Evaluation

Shuai S. Yuan
GlaxoSmithKline, Upper Province, PA, USA

Yodit Seifu
Amgen Inc, Thousand Oaks, CA, USA

William Wang
Merck & Co, Inc, Kenilworth, NJ, USA

Michael Colopy
UCB Bioscience Inc, Morrisville, NC, USA

CONTENTS

14.1 Introduction 318
    14.1.1 Overview of the Estimand Framework 318
14.2 Safety Estimands and Challenges 319
    14.2.1 Question for Safety 319
    14.2.2 Estimand for Safety Assessment in Contrast with
           Efficacy Estimand 320
           14.2.2.1 Treatment...321
           14.2.2.2 Population...321
           14.2.2.3 Variable (outcome)...322
           14.2.2.4 Intercurrent Events 323
           14.2.2.5 Summary Measure..328
14.3 BR Estimands and Challenges 332
    14.3.1 Estimands for Population-Level BRA: Demonstrated in
           the Context of CTD and PBRER..334
           14.3.1.1 Attributes of Estimands for Efficacy 334
           14.3.1.2 Attributes of Estimands for Safety 335
           14.3.1.3 Attributes of Estimand for BRA.................................338
           14.3.1.4 Other Considerations for Population-Level BRA.......339

DOI: 10.1201/9780429488801-18

14.3.2 Estimand for Patient-Level BRA: Demonstrated
in the Context of Patient's Perspective on BRA..........................340
      14.3.2.1 Attributes of Estimand for Individual BRA.................342
14.4   Summary and Discussion...344
References..346

14.1 Introduction

14.1.1 Overview of the Estimand Framework

In order to have a successful clinical development, it is crucial to define clearly and accurately the scientific questions of interest that the clinical program is aiming to answer. Defining the right question of interest prior to the start of study will help the trialists to have a clear and consistent development plan, trial design, and interpretation of the clinical trials. An unclear or improperly defined question will lead to unsound trial design or even failed clinical program. The process of defining the right questions of interest to be answered by the clinical trials could be a time-consuming and evolving journey and requires a cross-disciplinary team working together.

Toward the goal of defining the treatment effect with clarity and accuracy, ICH E9(R1) (European Medicines Agency, 2020) outlines the framework of estimand, the attributes of an estimand, how estimand will impact the clinical trials, and the usages of estimands. Estimand is a precise description of the treatment effect reflecting the clinical question posed by the trial objective. In principle, the estimand should be aligned with the trial objective and determine the main estimator with one or more sensitivity analyses to support its robustness. Attributes of estimand include treatment, population, endpoint, intercurrent event, and summary statistics of the treatment effect. Each element needs to be determined precisely before the trial starts. The treatment is the intervention under investigation by the sponsors of the clinical trials. The population is the targeted patients for the treatment reflecting the question of clinical interest, which is usually determined by exclusion and inclusion criteria of the study protocol. Of note, E9(R1) also mentions population defined by principal stratum according to the potential occurrence (or non-occurrence, depending on context) of a specific intercurrent event post-randomization. Endpoint (variable) is the parameter used to measure the treatment effect and is determined by the clinical questions of interest, therapeutic context, the nature of the disease, and treatment under investigation. Intercurrent events, as an important focus of the ICH E9(R1) guidance, are the post-randomization events that potentially affect the existence and interpretation of the treatment effect and the inference on the treatment effect of the treatment of interest. One should distinguish between missing data and intercurrent events. The former is due to unwanted events in clinical trials that should be avoided, for example, missing a disease assessment due to withdrawal from the study; the latter, as a comparison, is not always a bad thing that should be prevented in clinical trials as they could be just part of clinical practice, such as treatment discontinuation. ICH E9(R1) provides five strategies for handling intercurrent events, namely, treatment policy, hypothetical, composite variable, while-on-treatment, and principal stratum strategies. Finally, the summary statistics will be used to infer the existence and measure the magnitude of the treatment effect at the population level. The estimand framework will impact all steps of clinical trials from planning to interpreting the final data and provides a basis for communication within sponsor's organization and between sponsors and regulators.

The current E9 (R1), however, is mainly written from the perspective of assessment of efficacy in randomized clinical trials (RCTs). So far in the literature, the estimand framework has rarely been considered for safety assessment of the treatment in clinical trials, which is the other side of the coin that plays an equally important role as the efficacy when the benefit and risk of a treatment are evaluated. Although some attributes of estimand of efficacy may be common between efficacy and safety such as treatment, there are more noticeable differences between the efficacy and safety when the estimand framework is applied to safety. This chapter will try to expand the estimand framework to safety and combine them together to tackle the estimand of benefit-risk assessment (BRA) of the treatment.

14.2 Safety Estimands and Challenges

14.2.1 Question for Safety

Understanding the safety profile of a drug is critical for any clinical development program. In this chapter, by the term "drug", we refer to drugs, biologics, and gene/cellular therapy. Assessment of adverse events (AEs) is a core part of evaluating the drug safety profile. An AE is referred to "Any untoward medical occurrence in a patient or clinical investigation subject administered a pharmaceutical product" by ICH E6 (R2), including unfavorable and unintended sign, abnormal laboratory findings, symptoms, or disease associated with the exposure to a medicinal product regardless of the causal relationship (Guidance for Industry, 2018).

As in the efficacy setting, defining the right question for safety should be the first step toward the right safety estimand. At different stages, the safety questions could be different though. For drugs intended for human use, at the pre-clinical stage, a common question for safety is "what are the potential safety concerns for the investigational product if the product is used in humans". The safety data at this stage are mainly derived from non-clinical trials (e.g., in vivo study) and historical data of other drugs in the same class. Those sources of data can infer the important potential risks (e.g., toxicity) of the investigational medical product (IMP), which can be documented in the investigational brochure for further assessment. Once the IMP proceeds into the clinical stage, clinical trials will be conducted to assess the safety and efficacy profile of the IMP. Safety and toxicity are usually the primary objectives of Phase 1 trials. For example, in many Phase 1 oncology trials, a main goal will be to find maximum tolerated dose (MTD) in terms of the dose limiting toxicity criteria and identify a recommended phase 2 dose (RP2D) thereafter. A good estimation of the MTD will give the upper limit of the dose level for late-phase studies in order to protect the patients' safety. At the end of Phase 2, one or more dose levels will be selected based on the benefit and risk profile of the IMP. Safety will be considered together with efficacy in the context of BRA. A typical question will be, what dose level(s) will have the most benefit and outweigh the risks so that the best dose(s) can be carried forward to Phase 3 for confirmatory assessment. Once into Phase 3, the relevant questions include whether the safety profile of the drug is acceptable and if the drug has a favorable benefit-risk (BR) profile when comparing with the control arm. During Phase 2 and 3 stages, as there are more safety data collected, new safety signals could be found; pre-defined risks from historical data and early phase studies could be validated or refuted. After entering into the post-marketing stage, safety signaling and validation will both continue to monitor if there are any previously unknown risks emerging or elevation of the known risks, either from long-term safety follow-up, new studies conducted for additional indications, or more importantly the massive real-world data such as database from the vaccine adverse event reporting system (VAERS) and FDA adverse event reporting system (FAERS).

In order to answer and track the questions of interest for safety at different stages, safety monitoring, assessment, and management should be conducted from the preclinical stage to all phases of clinical trials and continue through the post-marketing stage (Guidance for Industry: Premarketing Risk Assessment 2005c, Development and Use of Risk Minimization Action Plans 2005a, Good Pharmacovigilance Practices and Pharmacoepidemiologic Assessment 2005b). The Estimand framework can be used to guide the data collection, analysis, and interpretation of safety assessment across the lifecycle of the drug development.

14.2.2 Estimand for Safety Assessment in Contrast with Efficacy Estimand

This section will discuss the challenges and differences of defining each component of safety estimands in contrast with the efficacy estimand.

Suggestions and considerations of safety estimands are provided based on the current practice and publications from the pharmaceutical society with guidance from E9(R1).

14.2.2.1 Treatment

The treatment for safety estimand is the same as the treatment associated with the efficacy estimand. As a common practice of clinical trials, the treatment start date (same or slightly different than the randomization date in an RCT) determines the start of the safety data collection, safety monitoring, and signal detection in a trial. When collecting AEs, a common approach is to only consider treatment emergent adverse events (TEAEs) which are defined as the AEs occurring after the administration of the IMP. The starting point of administering a single drug is easy to define. However, it could be tricky if the investigational product is composed of multiple drugs or procedures. For example, the combination of drug of chemotherapy (e.g., pemetrexed and platinum) and immuno-oncology drug (e.g., Keytruda) is approved to treat advanced non-small cell lung cancer (https://www.keytrudahcp.com). Usually, the two drugs will not be administered at the same time. In such cases, recording of TEAEs starts with initiation of any component of the study drug. The type and mechanism of treatment will determine the safety event collection timeframe in a clinical trial. For example, the duration for collecting safety events for drugs is usually up to five times the half-life of the IMP (Guidance for Industry, 2014); for vaccine, the World Health Organization (WHO) recommends at least six months follow-up post-vaccination (WHO 2010). The long-term safety follow-up duration also needs to consider the treatment type. For example, the Food and Drug Administration (FDA) recommends a year or more of follow-up in early-phase trials for CAR-T therapy, with a long-term safety follow-up lasting for 15 years (Guidance for Industry, Considerations for the Design of Early-Phase Clinical Trials of Cellular and Gene Therapy Products 2015a).

14.2.2.2 Population

In principle, for a given protocol, the target population for safety assessment and efficacy assessment should be the same, as it is generally determined by inclusion and exclusion criteria of the protocol. However, in the statistical analysis plan of a protocol, there could be a subtle difference on what subjects will be counted to the denominator (i.e., the analysis set) for assessing safety and efficacy. The analysis set of safety assessment is usually defined on so-called "as treated population" which could be different from the analysis set that is used to assess the efficacy. As a typical example in the statistical analysis plan, "ITT population" is used to assess the efficacy but may not be for safety assessment.

The ITT principle implies that all the randomized subjects should be included for the analysis, and all randomized subjects will be followed to the pre-planned trial end time, regardless of what happened after randomization so that the follow-up times for both the treatment and control are equal. However, it is not always interpretable to analyze safety based on the ITT principle. For example, a subject randomized to the treatment arm might be wrongly given the control drug. If the subject subsequently had an AE, according to the ITT principle, this case of AE should be counted towards the treatment arm. However, it is hard for physicians to interpret the result from such an analysis. The "as-treated population", as a comparison, will correct those assignment errors and count the subject towards the arm they actually received. Such mismatch between the safety analysis set and efficacy analysis set, however, would not impact the interpretation of the respective estimates regarding the targeting population. Either the "as treated population" or "ITT population" is actually only a representative subset of the target population. Therefore, the interpretation should be for the target population, but not for those analysis subsets. And correcting the assignment error seems not to change the definition of the target population. Admitting that such correction could introduce imbalance between the treatment arm and control arm, whether the analysis result of safety and efficacy assessment on the selected subsets are truly representing those for the target population is a different issue which could potentially be answered by the five estimand strategies as discussed below. This is also discussed in the context of population-level BRA (Section 14.3.1).

14.2.2.3 Variable (outcome)

Variable should be obtained for each patient in the population and defined to address the clinical question of interest. In study protocols, in comparison to the efficacy endpoints, the safety endpoints are often not clearly defined. Many attributes of safety can contribute to such phenomena as discussed by O'Neill (2008).

In a typical clinical trial, there are hundreds of safety parameters including AEs, laboratory test results, vital signs, physical exams, and patient reported outcomes (PROs). The assessment of AEs cannot be solely based on frequency, instead it should be multi-faceted in order to fully characterize the risks of a drug. We need to consider the severity, duration, temporal relation of the AE with respect to treatment, and reversibility of the AEs. Very importantly, the association detected between the occurrence of AEs and the treatment does not mean that there is a causal relationship since 1) there are hundreds of safety events occurring in a typical clinical trial which will lead to false positive findings due to the multiplicity issue and 2) there may be other confounding factors contributing to the associations. Therefore, the causality assessment of the safety variable with regard to the treatment(s) is critical and requires clinical and biological judgment. AEs are often retro-spectively defined, except for those important identified AEs which may be prespecified in the protocol. Many safety outcomes are not well understood before the Phase 3 studies. For rare events, this phenomenon is even more evi-dent as the clinical trial populations are always limited, unlikely to reveal all the rare events. Some rare and serious events can only be observed after the drug is followed up in the real world for long enough time. Follow-up time associated with AEs has to be considered when computing the likelihood of developing an AE. The longer the follow-up time is, the more chance one may observe an event. In many trials, the follow-up time for safety outcomes is often too short or unequal between treatment arms, which leads to biased estimation of the risk difference or ratio when comparing risks between the two arms if not adjusted for exposure time. Another interesting point is about adjudication of AEs of special interest. In the cases where independent adju-dication is conducted, the adjudication will still be limited to subjects that the investigator judged to have such events. Therefore, the total number of investigator-reported events will very likely be reduced after blind adjudica-tion which may underestimate the true risk of the safety outcome. Therefore, defining variables for safety estimand is a long and evolving process which needs collaboration of a multi-disciplinary team; and choices of safety vari-ables and approaches of assessing safety variables could be changed when more safety data are accumulating.

Of note, there is a very formal process for safety signal management used in pharmacovigilance setting. Using the safety signal management process of EU as an example, this process typically includes signal detection, signal val-idation, signal confirmation, signal analysis and prioritization, signal assess-ment, and recommendation (Potts et al., 2020). An AE could be detected but not confirmed at current time; an initially detected AE could be refuted later in the process; and safety signals detected also have different priorities at each different stage. So at different stages of the signal management, the defi-nition and strategy to handle the safety estimands may change even for the same safety signal(s).

14.2.2.4 *Intercurrent Events*

Intercurrent events and the strategies to handle the intercurrent events are discussed extensively in E9 (R1) with focus on efficacy in a RCT. Examples of intercurrent events for safety, similar to those for efficacy, include treat-ment discontinuation or switch, rescue medication or concomitant medica-tion, withdrawal from the study, death or other events that either prevent the event of interest from occurring, or change the variable value. In RCTs, the intercurrent event happening after randomization could result in imbal-ance between the treatment arm and control arm. The approach to account for intercurrent events when comparing treatment arms will be critical as

TABLE 14.1

Definition of Strategies to Handle Intercurrent Events

Treatment policy	The treatment policy strategy uses the value of variable regardless of the intercurrent events, adhering to the ITT principle.
While-on-treatment	The while-on-treatment strategy only considers the event or value of the variable "prior to the occurrence of the intercurrent event" and ignores the data after the intercurrent event. Any events after the treatment switch or rescue medication will not be counted. Alternatively, the event (e.g., ADR) occurs "while the patient is exposed to treatment" can be counted.
Hypothetical	The hypothetical strategy will assume a value that the variable would have taken in the hypothetical scenario where the intercurrent event had not happened. This would often require imputing the value to what would be expected without the intercurrent event.
Composite	The composite strategy combines the intercurrent event into the variable of interest. For example, a typical event-free survival would measure the time from randomization to progression, death, or subsequent cancer therapy whichever occurs first. Thus, it combines death and subsequent therapy as the event of interest.
Principal stratum	The principle stratum strategy will stratify the subjects according to the potential occurrence of the intercurrent event either under one of the treatment arms or both arms. The analysis would be conducted only on the principle stratum of interest. According to ICH E9 (R1), the principal stratum would not be defined based on the actual intercurrent event, but the counterfactual occurrence of the intercurrent event. The causal inference approach beyond the conventional treatment effect estimation in RCTs will be needed as the true mechanism of the intercurrent event is not known (Unkel et al. 2019).

different approaches will lead to different estimates of the treatment effect size. Sometimes the impact could be large and lead to a change in the conclusion on the BR balance of the drug. Five strategies are discussed in the guidance as shown in Table 14.1.

The intercurrent events for efficacy and for safety could be the same or very different. For example, death could be an intercurrent event for both efficacy and safety variables in a drug for type 2 diabetes mellitus. As a contrast, in cancer drug trials, death will be the outcome of interest for efficacy, and an intercurrent event for selected safety outcomes of interest. Since safety has many different dimensions and outcomes, it is challenging to specify a one-size-fit-all estimand strategy for both efficacy and safety outcomes. One may need to consider the objective, length and design of the trial, the mechanism of action of the treatment, the types of the events to be assessed, and the BRA strategy (see Section 14.3 for more discussion) in order to find good approaches to handle the intercurrent events for safety. For example, the estimand strategy for known AEs of special interest or major morbidity/mortality that were found from past trials can be defined with precaution and more consideration on the intercurrent events that may impact their occurrence, while the estimand strategy for unexpected or unknown AEs cannot be fully defined before a trial starts and hence a rough estimand with a crude rate (see Section 14.2.2.5) as the summary measure can be used. The choices of the estimand with respect to the intercurrent events should be informed by cross-disciplinary discussion regarding what the relevant clinical safety questions of interest are and what questions can be reliably answered with statistical analyses with the chosen estimand strategy. Sensitivity analysis using different approaches or different assumptions will be warranted to assess the robustness of the primary analysis of the safety endpoint as pointed out by E9(R1).

The focus of E9(R1) was on efficacy estimands. Only a few authors have discussed the estimand framework for safety (e.g., Akacha et al. 2016, Yang et al. 2019, Rockhold et al. 2019, Unkel et al. 2019). The two estimands "treatment policy" and "while-on-treatment" are probably the most well-known and popular approaches for safety endpoints. Akacha et al. (2016) discussed the estimand problem from the angle of adherence; specifically, they propose using separate estimands for patients who did not adhere to treatment due to safety and/or lack of efficacy from the estimand on adherers. They claim that it is more meaningful and interpretable to consider benefit and risk of treatment in patients that adhere to treatment. Their argument, although focusing on efficacy, touches upon the safety questions in RCTs and could be used for choosing the safety estimand. In contrast, Yang et al. (2019) consider the two estimand frameworks for safety assessment with respect to principles of the design of RCTs. They discuss the bias that could be introduced by naïve on-treatment analysis which ignores the events post-treatment. Through comparing the short-term risk estimate versus the long-term risk estimate in real examples, they show the risk of the drug could be underestimated or overestimated by on-treatment analysis. They suggest that according to the type of safety event, the duration of the study (short term vs long term), mechanism of the action of the drug, and the specific questions being asked, both on-treatment analysis (or on-treatment plus some days) and ITT analysis could be used for safety.

ITT has many merits, primarily driven by the need to keep the validity of randomization. Randomization in RCTs is critical to achieve balance between treatment arms in the distribution of known and unknown cofounding variables and minimize the bias that could be introduced as a result of imbalance and confounding, hence, rendering clear interpretation of the results. In E9(R1), the treatment policy estimand is the one complying with ITT principle, which will ignore any intercurrent events post-randomization. For safety, however, people have found it difficult to interpret the result from the ITT approach under certain intercurrent events. Randomization error is one reason, as it will result in mismatch between the treatment assigned and the treatment actually received by patients. If the ITT principle is blindly used which will ignore those randomization errors, the causal relationship between the AE and the treatment will not be able to be assessed correctly. Adherence is another example as discussed by Akacha et al. (2016). Since the incidence of safety events is correlated with drug exposure, the incidences of those treatment-related AEs among adherers and non-adherers are supposed to be different due to their different exposure. Simply mixing them together using a naïve percentage to summarize the safety profile of the adherers and non-adherers certainly masks some important information from the data. A similar argument can apply to treatment switching or rescue medication post-randomization. If AEs occur after treatment switching or rescue medication, without proper clinical judgment on the causal relationship between the AEs and drugs that are taken by the subjects, it is hard for clinicians to interpret such naive percentages of AEs calculated based on treatment policy estimand. An AE after treatment switching or rescue medication could be attributed to the experimental treatment or control, the rescue medication alone, or the combination of the prior treatment and newly added rescue medication. One possible solution for that is, as proposed by Akacha (2016), detangling the safety profile according to the adherence and summarizing risk together with the benefit of the subjects in a tripartite framework which considers the percentage of patients non-adherent due to safety, percentage of patients non-adherent due to lack of efficacy, and treatment effect in adherers; or as mentioned in the study by Yang et al. (2019), summarizing the incidence of the AEs while-on-treatment plus additional exposure time. As an illustrative example, the on-treatment period could be extended to cover the x-number of days after the last dose of the drug where 'x' is a relatively short period of time. Here, we propose to use an estimand which is a compromise between the treatment policy and while-on-treatment estimand for risk assessment. We name it "while-at-risk" as the term is more relevant for safety assessment. This was alluded already in the while-on-treatment strategy section of E9(R1). We extend the idea here and make it more specific for safety assessment. Such a while-at-risk strategy should meet the following three requirements:

- It keeps the principle of the while-on-treatment strategy for its ease of interpretation in the context of safety assessment. So the comparison between treatment arms will be based on the treatment they actually received, instead of treatment arms they were randomized to. It is true that this seems to violate the randomization principle. However, such assignment error is not the purpose of the pre-planned randomization code and usually due to human error. Correcting the assignment error by assigning the subject to the actual treatment received may introduce a little imbalance of the confounding factors between treatment arms at the first glance. However, such imbalance could be offset as long as the assignment error is not systematic and just occur in a small number of subjects.

- It takes into account the fact that the safety events will be affected by the total exposure time and the mechanism of the drug (e.g., symptom relieving vs disease modifying). So in RCTs with both the treatment and control, equal and long enough follow-up after discontinuation of treatment for both the treatment arm and control arm should be used for the while-at-risk period in order to capture the potential events after the rescue medication/treatment switching. For some symptom-relieving drugs with a short half-life the while-at-risk period is typically defined to be the last date of treatment plus five times half-life of the drug which could be enough to capture the events attributable to the treatment. For a disease-modifying drug that may cause changes in the body permanently or a prophylactic drug that could stay in the body for long time, such a while-at-risk period could be years depending on the type of the drug. For example, as mentioned earlier, for CAR-T or other cellular or gene therapy, the current FDA guidance for safety follow-up is one year short-term plus 15 years long-term follow-up (Guidance for Industry: Considerations for the Design of Early-Phase Clinical Trials of Cellular and Gene Therapy Products, 2015a). The long-term safety data may be obtained by setting up registries or by utilizing other long-term safety databases such as FAERS or VAERS.
- For AEs occurring after rescue medication or treatment switching, clinical judgment should be utilized to assess the causal relationship between those AEs and the treatment(s). A naïve percentage by putting all occurrence into numerator may over-estimate risk, especially for rare events. A commonly used approach may be to consider the treatment-related AE only.

14.2.2.4.1 *Missing Data for Safety*

Missing data pose a unique challenge for safety assessment. Although missing data should be distinguished from intercurrent events, missing data itself could be a consequence of intercurrent events. However, the mechanism of missingness for safety is not the same as efficacy. Very likely, some intercurrent events with impact on safety assessment may be due to the lack of efficacy, such as treatment discontinuation/switching or rescue medication. In these situations, the missingness of safety data is not at random anymore. And it is usually impossible/challenging to impute the missing value of safety variables including the occurrence of safety events. So the hypothetical strategy is difficult to be used for safety in this scenario. Composite estimand may be a choice if the intercurrent event can be viewed as the consequence of the AEs. It is also a good practice to continue collecting safety events after treatment discontinuation to mitigate the missing data problem, especially when the drug effect is of long term.

14.2.2.4.2 Intercurrent Event Due to Pandemics

A special topic of estimand is the intercurrent events associated with pandemic such as COVID-19 and the impact of the government, institutional, and community responses to the pandemic. Take the COVID-19 pandemic as an example herein, the intercurrent events related to COVID-19 pandemic include study treatment discontinuation/interruption due to supply interruptions or COVID infection, concomitant treatment/procedure or rescue medication for disease related to COVID-19, non-adherence to protocol-mandated visits and laboratory/diagnostic testing, change of sites, change of event reporting from physical visit to remote visit, AEs, or even death due to COVID-19. Many researchers discussed the potential solutions to deal with the impact to the planned estimand strategy (Meyer et al. 2020, Nilsson et al. 2020, Akacha et al. 2020). As the COVID-19 pandemic is an unexpected factor to impact clinical studies which may not be common in medical practice and real-world application, some strategies like treatment policy, composite, or principle stratum may not be appropriate for handling of the intercurrent events due to COVID-19. The while-on-treatment and hypothetical estimand are of more interest to the regulators or sponsors than other estimands. More discussions can be found in the study by Meyer et al. 2020.

14.2.2.5 Summary Measure

To summarize the risk level of AEs for a drug, different risk metrics are warranted. Table 14.2 summarizes some common statistical measures for absolute risk (within-arm) and relative risk (between-arm) of AEs. For other types of safety parameters such as laboratory test results, vital signs, physical exams, or PROs, the considerations of summary measures could be different.

The choices of the summary measure depend on the question of interest and the type of risks. For example, if one is interested in the probability that a subject experiences the AE in a given time period, the crude rate (i.e., AE percentage) can be used. However, the crude rate ignores the subject's exposure duration to the drug. If the exposure is different between the treatment and control arms in a RCT, the risk difference or ratio based on the crude rate may be biased estimators of the relative risk. Incidence proportion, as an alternative to the crude rate, will only count the number of events in a specific risk period in the numerator to match the exposure time in both arms.

When the question of interest is to consider whether the AE occurred at least once in a unit of exposure time (measured by person time), the exposure-adjusted incidence rate can be used. It is a simple approach taking into account the impact of exposure on AE occurrence and measures the average number of AE occurrences per person-year (or per other person-time units). When the risk (measured by hazard) is constant over time, the exposure-adjusted incidence rate is appropriate. Note that the numeric value of the exposure-adjusted incidence rate could be impacted by the choice of time

TABLE 14.2

Summary Metrics for Safety Estimand in an RCT

Absolute Risk Metric per Treatment Group	Calculation	Comments	Relative Risk Metric Between Treatment Groups	Comments
Crude rate	a/n a: number of patients expiring the event n: total number of subjects	• Easy to calculate, understand, and interpret • Measure the probability AE occurs in a given time period • Ignore the impact of exposure • Lack of temporal information	risk difference, risk ratio, odds ratio	• risk difference is always defined • risk ratio or odds ratio may not always be defined for rare events • risk ratio and odds ratio do not reflect background risk
Exposure-adjusted incidence rate	a/E a: see above E: total exposure (person time)	• Easy to calculate, understand, and interpret • Incorporate temporal and exposure information • Assuming constant risk over time	rate difference, rate ratio	• Similar to 1st row
Incidence proportion (cumulative Incidence)	$\dfrac{a_{0\le t\le t}}{n}$ a_u: total number of subjects having event(s) in a period, eg. [0, t] n: see above	• Measure the number of patients experiencing at least one AE in a specific time interval • Not properly adjusted for exposure	Proportion difference, proportion ratio	• Similar to 1st row
Kaplan-Meier approach	$\hat{S}(t) = \Pi_{t_i \le t}\left(1 - \dfrac{d_i}{n_i}\right)$ $S(t)$: survival function of T t_i: time point of event d_i: number of events at time t_i n_i: number at risk at time t_i	• Incorporate temporal and exposure information • Consider censoring due to intercurrent events and competing events • Focus on the first instance of AE • Valid even when risks varies over time • Implicitly assume that everyone will have AEs	Difference or ratio of median time to event(mTTE), Difference or ratio of 1-KM probability	• mTTE is of clinical interest, easy to understand and interpret • Median may not always be defined • 1-KM probability measures the probability that the first AE occurs up to specific time

(Continued)

TABLE 14.2

(Continued)

Absolute Risk Metric per Treatment Group	Calculation	Comments	Relative Risk Metric Between Treatment Groups	Comments
Restricted Mean Survival Time	$E[min(T, \tau)]$ T: time to event τ: cutoff time	• Equal to area under survival curve • Measure the average time to event up to specified cutoff • Incorporate temporal and exposure information • Well defined irrespective of risk pattern	RMST ratio or difference	• Easy to understand by non-statistician
Hazard rate	$\lambda(t) = \dfrac{f(t)}{S(t)}$ $f(t)$: the density of T $S(t)$: the survival function of T	• Not easy to interpret to non-statistician	Hazard ratio	• Often under proportional hazard assumption

unit, hence becoming arbitrarily large or small. The incidence proportion and incidence rate are widely used in epidemiology.

If the question of interest is in time to the first instance of AE, one may use the time to event approach, including the Kaplan Meier approach, restrictive mean survival, or hazard rate. This category of methods explicitly considers the follow-up time and censoring during the course of the clinical trials. It is a recommended approach by health technology assessment (HTA) agencies of Germany (i.e., IQWiG) where treatment policy estimand is preferred. Unkel et al. (2019) had a very good discussion about the time-to-event approach in the presence of varying follow-up time and competing events.

In E9 (Guidance for Industry 1998), the AE proportion, incidence rate, and survival analysis are all discussed. The popular practice in reporting AE to regulatory agencies is using the AE proportion and incidence rate. When the drug is a long-term treatment and a substantial proportion of intercurrent events such as treatment withdrawal or death is expected, "survival analysis methods should be considered and cumulative AE rates calculated in order to avoid the risk of underestimation".

When it is of primary interest to compare the risk profile of two active treatments, the comparisons typically are not only restricted to the relative risk difference or relative risk elevation, but also include how these risk difference/elevation could be balanced by relative gain/loss in efficacy. Therefore, such comparisons should occur in the context of BRA. Both the risk ratio and risk difference are legitimate approaches to use. The risk ratio is often more sensitive to risk elevation than the risk difference, so it could be a better tool for signal detection. Regulatory agencies are interested in evaluating the BR profile of the new treatment in a holistic way and across the lifecycle of the drug. The absolute risk measures including frequency or percentages and relative comparison between the new treatment and placebo are widely acceptable by regulators for such purposes. In comparison, HTA bodies are interested in the relative comparisons of the BR profile of the new treatment in comparison with the standard of care or other specific treatments for the purpose of pricing, reimbursement, and market accesses. The relative risk measures based on different risk metrics could be good fits for HTA usage.

A special consideration can be given for risk estimation in the presence of pandemic like COVID-19. COVID 19 could increase the AE rate for certain events. Typical examples include cough or fever. When comparing the incidence of the AEs between treatment arms, the risk difference point estimate could be the same if the COVID-19 impacts both arms similarly. However, the variance of such point estimate may be changed due to the increased noise level. The risk ratio, odds ratio, or log odds ratio based on the crude rate may regress toward the null (value=1) if a similar number of events were added to the numerator and denominator of the ratios. If the time to event approach is used for risk assessment, the median time to AEs will be shorter. The hazard ratio will change (the direction of change may not be clear), and the magnitude of the changes, however, depends on the true hazards of both arms. So one should be cautious when interpreting the ratios of certain AEs if the trial is impacted by the COVID-19 pandemic.

14.3 BR Estimands and Challenges

The safety of the drug is assessed not only for safety signal detection and evaluation but also for BRA throughout the drug development life cycle. Most noticeably, these assessments happen in four settings, namely, when the company is making a go or no go assessment of the drug to late-stage development, when submitting a dossier for a drug or device for regulatory approval, when performing periodic assessment of safety of the product via the periodic benefit-risk evaluation report (PBRER), and finally when submitting a dossier to the HTA agency.

In the context of BRA, the clinical question of interest may or may not be addressed by a single clinical trial or by using only clinical trial data. Furthermore, the BR questions of interest may differ depending on the stakeholder. Table 14.3 presents some clinical questions associated with BRA in the four cases where BRA is made during the development lifecycle. The last set of clinical questions in Table 14.3 present a patient's perspective. There is an overlap in these BR questions. For example, the three BR questions of interest to patients also fall under the more general question "how the medicinal product addresses a medical need?"

The FDA Guidance document "M4E: The CTD-Efficacy" (2017) outlines how both qualitative and quantitative methods can be used for BRA. Generally, the efficacy results observed for each indication are compared to safety findings that use the while-on-treatment strategy for handling intercurrent events. Of note, the FDA Guidance document on the CTD (Guidance for Industry 2017) states that a descriptive approach that explicitly communicates the interpretation of the BRA will be adequate. Furthermore, subgroup analysis results may be discussed to assess the benefit and risk in a special population such as elderly patients or patients who have a special health condition.

In addition, in 2019 the FDA's Center for Devices and Radiologic Health (CDRH) has issued a detailed guidance document for BR determination (Guidance for Industry 2019). This guidance document explains the principal factors the FDA considers when making BR determination in the pre-market review of certain medical devices. The factors outlined include benefit and risk factors as well as additional factors such as patient perspectives and alternate treatment/ diagnostic options and include hypothetical worksheet examples.

TABLE 14.3

Examples of Clinical Questions of Interest for BRA

Stakeholder	Clinical Question Associated with BRA
BRA in CTD Section 2.5.6	"How the severity of disease and expected benefit influence the acceptability of the risk of the therapy"[a] "How the medicinal product addresses a medical need"[a]
PBRER	"…a comprehensive, concise, and critical analysis of new or emerging information on the risks of the medicinal product, and on its benefit in approved indications, to enable an appraisal of the product's overall BR profile"[b]
HTA agency[c]	Does the new product have additional benefits against a comparator?
Patients[d]	(1) Can the patient tolerate the drug? (2) If the patient tolerates the drug and takes the drug what is the probability the drug meets the efficacy objective? (3) If the patient takes the drug long enough to see the efficacy effect, what is the safety risk?

[a] Section 2.5.6.4, FDA Guidance for Industry: M4E: The CTD-Efficacy (2017)
[b] Section B, FDA Guidance for Industry: E2C (R2) Periodic Benefit-Risk Evaluation Report (PBRER) (2016)
[c] This refers to Gemeinsamer Bundesausschuss assessment of the new drug as presented in the study by Leverkus and Chuang-Stein (2016)
[d] Akacha et al. (2016)
CTD: Common Technical Document.

There are two general types of BRA. The first type tries to assess BR by obtaining a separate estimate of the efficacy and safety estimands for the population of interest and finally assessing BR by contrasting these "population" estimates qualitatively and quantitatively. Hence, the BR estimate or assessment depends on other parameter estimates. In this chapter, we will call such BRAs population-level BRA. For the second type of BRA, the estimand of interest relies on assessing the BR at the subject level. In this case, what we are trying to estimate is a BR parameter. Thus, we will call such BRAs patient-level BRA.

The purpose of the BR conclusion section of the CTD (Guidance for Industry 2017) is to integrate and explain the BRA of the drug for its intended use, and it is recommended to be based on weighing key benefits and key risks of the product. Similar instructions are also provided in the FDA guidance document "E2C (R2) Periodic Benefit-Risk Evaluation Report (PBRER)" (2016). Hence, the FDA guidance documents of the CTD and PBRER appear to recommend utilizing population-level BR measures for assessing BR in these documents. This does not, however, preclude the use of individual-level BRA (Guidance for Industry 2016). In fact, patient-level BRAs are better at detecting associations between benefit and risk parameters and can help in the overall BRA.

14.3.1 Estimands for Population-Level BRA: Demonstrated in the Context of CTD and PBRER

In this section, the population-level BR estimands will be discussed. Risk (safety) and benefit (efficacy) estimands are derived separately, and these estimands are then qualitatively and quantitatively compared (e.g., graphically, numerically, or mathematically) to assess BR balance. The context of this BR estimand is BRA as outlined in the CTD and PBRER.

The IMI-PROTECT Consortium group reviewed 47 different BR approaches that were extracted from identified reviews and papers (Mt-Isa et al. 2013). The approaches they reviewed were grouped into frameworks, metrics, estimation techniques, and utility. Descriptive methods can be largely incorporated under the framework group. Examples of such methods are value trees and effects tables.

The consideration that needs to take place when evaluating the appropriate estimands for a population-level BRA will be demonstrated via an example and an effects table. Consider the drug Noctiva, a desmopressin product, approved in 2017 for the treatment of nocturia. The International Continence Society defines nocturia as one or more nighttime voids preceded and followed by sleep (Van Kerrebroeck et al. 2010). The approval of this drug was based on the results of two double-blind pivotal studies DB3 and DB4 (FDA Briefing Document 2016). In both trials, Noctiva-treated patients were compared with placebo-treated patients. These two studies included a two-week placebo lead-in period. The purpose of this lead-in period was to identify placebo non-responders. At the end of these two weeks, both placebo lead-in responders and non-responders were randomized to Noctiva arms or a placebo arm (FDA Briefing Document 2016).

Efficacy (benefit) in these studies was measured by a reduction in nocturic events. The most important safety concern for desmopressin products is hyponatremia, which can result in seizures, coma, and death (FDA Briefing Document 2016). Hence, in this example, the risk that will be assessed for BR will be hyponatremia. The effects table (Table 14.4) presents the benefit and risks for the two pivotal studies, DB3 and BD4. The comparison is between the Noctivia 1.5 mcg arms and the Placebo arms.

Since population-level BRA is based on a separate assessment of benefit (efficacy) and risk (safety), the efficacy and safety estimands will be first considered separately (see Section 14.2.2 for the components of estimand).

14.3.1.1 Attributes of Estimands for Efficacy

The population targeted by the efficacy question of interest, for the nocturia indication, is adults with nocturia who are 50 years of age or older. In contrast, the efficacy analysis population was adults with nocturia who are 50 years of age or older and have a minimum number of post-randomization assessment of nocturic events, as recorded in patients' diary (defined to be intent-to-treat population in DB3 and DB4 protocols). Hence, it included both the lead-in placebo-responder and non-responder patients. The co-primary variables of interest were changed from the baseline in the mean number of nocturic episodes per night and the binary variable, whether or not a subject has achieved a 50% reduction from the baseline in the number of nocturic episodes (FDA Briefing Document 2016). The population-level summaries were the adjusted mean difference for the change from the baseline variable and the proportion difference between the treatment and placebo arm for the binary co-primary endpoints. The potential intercurrent events were patient treatment discontinuation or disruption. Additional missing data can also arise when patients who were on treatment fail to complete the voiding events diary for a particular day or visit. In this study, which was submitted to regulatory authority prior to the ICH E9(R1) addendum on estimands and sensitivity analysis came into effect, there was no distinction made between voiding information that cannot be collected on treatment and missing data. For the primary analysis that was employed in DB4, any missing voiding diary data were imputed using the multiple imputation method. This multiple imputation application follows the hypothetical strategy for handling intercurrent events. This strategy did not differentiate between on-treatment voiding information that could not be collected due to intercurrent events and on treatment missing voiding data. In the estimand framework, intercurrent events that lead to data being unable to be collected may be treated differently from missing voiding dairy data. If the reason for treatment discontinuation is, for example, lack of efficacy, such patients may not be included in the application of the multiple imputation method. In DB4, a sensitivity analysis was performed without imputing missing values. There was very little difference between the primary analysis that utilized multiple imputation and the sensitivity analysis where just observed values were assessed.

In other programs, as demonstrated in this example, careful consideration should be made on the attributes of the efficacy estimands. Examples of other intercurrent events that could affect the efficacy estimand are concomitant therapy (see Section 14.2.2.4 for more detailed discussion). For example, in clinical trials for pain treatment, routine over-the-counter use of pain medication can affect a patient's measure.

14.3.1.2 Attributes of Estimands for Safety

In the DB3 and DB4 clinical trials, hyponatremia was defined as serum sodium $<= 120$ mmol/L with or without clinical symptoms or the serum sodium level between 126 and 129 mmol/L with clinical symptoms associated with hyponatremia (FDA Briefing Document 2016). There is some subjectivity in this definition; hence, hypothermia was also assessed using

TABLE 14.4

Effects Table for Noctivia

Effect	Study	Short Description	Noctivia 1.5 mcg	Placebo
Favorable Effect				
	DB3	ITT	N = 179	N = 186
		Adjusted mean change from baselines in nocturic episodes per night[a]	−1.6(0.1)	−1.4(0.1)
		Diff vs. placebo	−0.4(−0.6, −0.2)	
		50% reduction in nocturic voids	93/179(52%)	61/186 (33%)
		Absolute differences vs placebo	19%	
		P-value[b] (vs. placebo)	<0.001	
	DB4	ITT	N = 269	N = 260
		Adjusted mean change from baseline in nocturic episodes per night[a]	−1.5 (0.1)	−1.2 (0.1)
		Diff vs. placebo	−0.3 (−0.4, −0.1)	
		50% reduction in nocturic voids	120/260 (46%)	74/260 (29%)
		Absolute difference vs placebo	17%	
		P-value[b] (vs. placebo)	<0.0001	
Unfavorable effect				
	DB3 & DB 4	Safety	N = 448	N = 454
Hyponatremia			6/448 (1.3%)	1/454 (0.2%)
126–129 mmol/L		Incidence of nadir of serum sodium values	9/448 (2.0%)	0/454 (0%)
<= 125 Mmol/L		Incidence of the nadir of serum sodium values	5/448 (1.1%)	1/454 (0.2%)
Additional assessment (PRO endpoint)				
	DB4			
			N = 260	N = 260
		Adjusted mean change from the baseline in the PRO measure (scale ranged from 0 to 100)[a,c]	−14	−12
			−2.6	
		P-value	0.02	

[a] Change from the baseline was calculated using an ANCOVA model.

[b] P-values for treatment vs. placebo comparison using the Cochran–Mantel–Haenzel test.

[c] PRO measured impact of nighttime urination. Source: SER120 Nasal Spray: Advisory Committee Briefing Document (2016)

the patient's nadir of serum sodium assessment. It is to be noted that patients who were diagnosed with hyponatremia were to be discontinued in both studies. The population targeted by the safety question of interest is again adults with nocturia who are 50 years of age or older (see Section 14.3.1.1 for efficacy analysis population). In contrast, the safety analysis population was adults with nocturia who are 50 years of age or older, have received study treatment (Noctiva or placebo), and have a post-randomization assessment of safety. The safety (risk) variables of interest were whether or not a patient had hyponatremia while on treatment or whether or not a patient had a serum sodium assessment value below 130 mmol/L. The population-level summary endpoints were the proportion difference between the treatment arms in hyponatremia events and the proportion difference in the number of nadir serum sodium assessments that were < 130 mmol/L.

Hyponatremia definition is based on laboratory assessment; hence, if a protocol-specified scheduled assessment is missed, there is potential for missing an event. The missing assessment could be due to an intercurrent event or simply missing and not be related to any identified intercurrent event. Consider the intercurrent event early discontinuation of treatment. If early discontinuation of treatment results in the distribution of exposure times between treatment arms being different, it can have an impact on the safety estimand. Depending on the nature of the new therapy, both increasing and decreasing the exposure time for the treatment arm over the placebo (or active control) arm can affect the safety estimand. If the difference in exposure time is due to there being a greater number of treatment discontinuation in the treated arm, this can potentially result in undercounting the number of such events. In the Noctiva example, the discontinuation due to the AE was higher in the treated arm (FDA Briefing Document 2016). On the other hand, if patients are benefiting from treatment there is potential for patients in the new treatment arm to stay longer on treatment compared to the placebo (or active control)-treated patients, with increasing follow-up/exposure time increasing the chance of getting the risk event of interest. This is typically the case in oncology when the newer treatment is more efficacious than the active control. In such cases, a time-to-event analysis can reduce the impact of imbalances in exposure time in evaluating treatment differences and can be used as a primary or sensitivity analysis. For a more detailed discussion on the impact of differing exposure time on safety assessment see Section 14.2.2.5. One can also choose the population-level summary to be a hazard ratio instead of making it the sensitivity analysis. Another sensitivity analysis could be employed by creating a composite endpoint of hyponatremia or treatment discontinuation, especially if the reasons for discontinuation are expected to be related to treatment. One can also further limit the discontinuation events to be included in the composite endpoint to just discontinuation related to AEs that could be potentially due to the risk of interest (e.g., hypothermia). Finally, for the Noctivia example, for patients who missed lab assessment, it is also important to evaluate available laboratory assessment for trends and related safety records such as AEs. Additional discussion on safety intercurrent events can be found in Section 14.2.2.4.

14.3.1.3 Attributes of Estimand for BRA

In the Noctiva example, for efficacy (benefit), the primary analysis population for DB4 was non-responders in the placebo lead-in period of the two trials. This was called the modified intent-to-treat (mITT) population. The primary evaluation was changed to the ITT population (see Section 14.3.1.1 for the definition) at the request of the FDA. If the mITT population was used for the assessment of the efficacy, the difference between the safety and efficacy analysis populations would have been about 30%. Furthermore, since patients are not first placebo-treated before prescribing the treatment, the more appropriate analysis population for efficacy and BRA is the ITT analysis population. In contrast, using the full safety analysis population (see Section 14.3.1.2 for definition) helps in getting an accurate estimate of the difference in hyponatremia for both safety and BRA. In these two studies (DB3 and DB4), there was not much difference between the safety and ITT analysis populations. In general for population-level BRA, as the efficacy and safety target population are expected to be the same the resulting target BRA population will also be the same. However, the safety and efficacy analysis population may differ depending on the intercurrent events that are being taken into consideration when defining the analysis population for efficacy and safety. Hence, an efficacy estimate that is based on the analysis population used for estimating the efficacy estimand (ITT population in the Noctivia example) and the safety estimate that is based on the analysis population used for estimating the safety estimand (safety analysis population in the Noctivia example) can be used. In general, for population-level BRA, the efficacy and safety estimates and their associated analysis populations can be used, since both are selected to answer the appropriate efficacy and safety questions of interest taking into account intercurrent events that can potentially affect these estimands. However, there might be studies where there is a big difference between the efficacy analysis population and the safety analysis population. For example, in a given study, a number of patients may get randomized but not get treated or get treated by the wrong study drug. In these cases, it would be more appropriate to evaluate BR by using the safety analysis population for efficacy rather than the efficacy (ITT) as-randomized population. The rationale for this recommendation centers on the idea that the primary BR questions of interest are the questions that are of interest to the patients (see Table 3.1).

14.3.1.4 Other Considerations for Population-Level BRA

When employing population-level BRA such as the effects table, when possible, it is better to use the same type of variables for the risk and benefit estimand. For example, if the efficacy variable is a time-to-event variable and the risk is an AE of interest, the AE can also be summarized using the time to event analysis. Using the same type of variable for efficacy and safety makes the relative assessment of risk and benefit much easier and more transparent.

In evaluating the BR of the product, it is also important to look at sub-populations that might be at a higher risk for developing the safety risk of interest. In the Noctivia example, the indication is for subjects who are ≥ 50 years of age; furthermore, the incidence of hyponatremia with desmopressin is known to be numerically higher among those subjects who are ≥ 65 years of age compared to those 50-65 years of age (FDA Briefing Document 2016). Hence, in assessing the BR of the product , it is important to look at BR in this population. This is especially important for this indication since the ≥ 65 age group is a large segment of the population for this indication. The consideration of estimand for sub-population BRA can follow the same considerations as in the full-population BRA (see above). However, when looking at the sub-population for BRA, when the sample size is decreased it may result in increasing the discrepancy between the benefit (efficacy) analysis population and the risk (safety) analysis population.

Separate derivations of the benefit and risk estimands and assessing BR by quantitatively/ qualitatively looking at these risk and benefit estimands do not assess benefit in relation to the risk at the patient level. For example, are the patients that are benefiting more from the new treatment also more likely to have safety issues associated with the risk? In population-level BRA, these types of questions are answered via subgroup analysis while these questions can potentially be directly assessed using patient-level BRA. Such methods for assessing BR and the associated estimands are discussed in the next section.

Finally, BRA does not stop with drug approval. BRA is part of the lifecycle management of the drug, and regulatory documents (e.g., PBRER) need to include updated BR assessment. As safety surveillance is ongoing when the drug is on the market, there is more information on the risk profile of the product after the drug is on the market. In such assessments, the population would be patients treated with the drug, and the endpoint of interest would be whether or not a patient had a serious AE of interest. The summary would be the difference in this serious AE between the active treatment and other products that are available for the indication under consideration (published information). The issue with this estimand is that we do not have a good estimate of the denominator, and there may be a temporal difference in the information available on the competitor's drug. In such cases, the limitations of such comparisons should be clearly explained.

14.3.2 Estimand for Patient-Level BRA: Demonstrated in the Context of Patient's Perspective on BRA

Another way of assessing BR is by evaluating BR at the patient's level. For a given patient, the BR questions focus on three issues stated in Table 14.3. Patient-level BRA may be able to answer these three questions more directly and can help identify associations between benefit and risk endpoints of interest.

Chuang-Stein (1994) proposed deriving a risk-adjusted benefit score for each patient and then used this variable to evaluate treatment differences using a t-test. Pritchett and Tamura (2008) demonstrated how a global benefit-risk linear score can be used to design a trial. In their example, they categorize patient's outcomes into mutually exclusive categories that are based on both efficacy and safety. The incidence rate of each category is used to derive weighted linear statistics. This statistic is then used to evaluate treatment differences. A similar assessment based on a patient's safety and efficacy outcome was proposed by Evans et al. (2015). In the study by Pritchett and Tamura (2008), the two treatments' linear scores are compared using a t-test. In contrast, in the study by Evans et al. (2015) the treatment difference is evaluated using rank-based methods. Mt-Isa et al. (2013) also presented an exposition of both patient- and individual-level BRA measures. One such method they discuss is the quality-adjusted life year (QALY) indices. The application of QALY in the oncology therapeutic area is known as quality adjusted time without symptoms and toxicity (Q-TWiST). In Q-TWiST, for patients undergoing oncology therapy, three health states are assessed. These are, namely, the time when a patient is subjected to toxicity from treatment, the time a patient is free from toxicity and disease, and time of relapse until time of death.

Another method that can be used to assess BR at an individual level is the generalized pairwise comparison (GPC) statistical method that analyzes composite endpoints. A binary or a time to first event analysis of composite endpoints is unable to account for the importance of the variables that make up a composite endpoint. The GPC methods account for the importance of the components of the composite endpoint. In applying this method, one performs a comparison of every patient in the treatment arm with every patient in the control arm. Within each pair of comparison, one evaluates the different components of the composite outcome in the descending order of importance until the pair in the treatment arm is evaluated to have a better or worse outcome when compared with the pair in the control arm. If the patient in the treatment arm has a better outcome compared to the control patient, it is called a 'win'; if on the other hand, the control patient has a better outcome, it is called a 'loss'. If a 'win' or a 'loss' cannot be established, it is called a 'tie'. The difference between the two treatment arms in the pairwise comparison can be evaluated by using a statistic such as the Finkelstein–Schoenfeld test statistic which measures the win difference between the two treatment arms (Verbreeck et al. 2019). A scaled version of this statistic is the Buyse method or the total benefit statistics (Verbreeck et al. 2019). The win-ratio statistics was introduced by Pocock in 2012 (Verbreeck et al. 2019). In this case, the statistics of interest is the ratio between the total number of wins and losses in the treatment arm. A closely related statistics is the win odds (Dong et al. 2020a). The win-odds is a ratio of odds that adjusts for ties in the computation. Another related method is the adapted O'Brian method (Verbreeck et al. 2019). The GCP method can be applied to a vector of endpoints consisting of counts, time-to-event, ordinal, nominal, binary, and continuous endpoints. These vectors of endpoints can be selected to be benefit (efficacy) and risk (safety) endpoints of interest, allowing for a BRA at a patient level.

Assessing BR at the patient level can uncover a potential association between benefit (efficacy) and risk (safety) better informing developer of the drug, regulators, patients, and HTA agents on the safety profile of the product. Furthermore, it can help further summarize BR information in a unified way. Consider the theoretical example in Tables 14.5 and 14.6. If the benefit and risk are separately summarized using an effects table, the benefit (efficacy) advantage for the treatment is 20% more compared to the control-treated patients. In contrast, the risk increase is 19% for the treatment arm compared with the control arm. Table 14.6 shows that risk events only occur in patients that responded for efficacy. Further revealing that for responders in the treatment arm, the percentage for developing the risk event is 50% while this risk is 0% for control-treated patients, revealing the association between efficacy and safety for the treatment arm. Since patient-level BRA is better at detecting the interaction between treatment and BR outcomes, this patient-level BRA is better at answering the questions that are of interest to patients more directly.

Patient-level BRA can also be provided in CTD and BEPBR. However, as the BR assessment is evaluated at the patient level, the BRA is limited to studies that have both the benefit and the risk assessment. Patient-level BRA may also be useful for decision-making when the developer of the drug

TABLE 14.5

Effects Table

Effect being Measured	Treatment	Control
	N = 100	N = 100
Efficacy responder (binary endpoint)	40%	20%
Difference vs placebo	20%	
	N = 100	N = 100
Risk of interest (presence/absence)	20%	1%
Difference vs placebo	19%	

TABLE 14.6

Risk and Benefit Interaction Assessment

Treatment (N = 100)		
Efficacy	Presence of Risk	No Presence of Risk
Responder	20%	20%
Not a responder	0%	60%
Control (N = 100)		
Efficacy	**Presence of Risk**	**No Presence of Risk**
Responder	0%	20%
Not a responder	1%	79%

Note: Denominators for percentages are based on N of each treatment group.

evaluates the possibility of moving the product from one development phase to another. BR measures such as GPC can incorporate several endpoints and increase the sensitivity of the analysis while assessing the endpoints of interest in a unified manner.

14.3.2.1 Attributes of Estimand for Individual BRA

As patients' questions of interest are focused on outcomes of a potential treatment, if patient-level BRAs are being used to evaluate these questions, the target population of interest is patients with the indication of interest; for example, in the Noctivia example it would again be adults with nocturia who are 50 years of age or older. If the intercurrent events that affect the risk and benefit estimands are different, a separate analysis population may be defined for the risk and benefit endpoints of interest. Hence out of necessity, the patient-level BR analysis population will be defined by the intersection of these analysis populations. For example, for the DB4, Noctivia study, if the individual BRA is based on the change from the baseline in a daily reduction of nocturic events, the presence or absence of hyponatremia, and the PRO endpoint (see Table 3.2), then as the smallest of the three analysis sets is the PRO analysis set, the BRA will be based on this analysis population. If the difference between the different analysis populations is quite large and the intersecting set is small relative to, for example, the safety set, the resulting estimate may not be generalizable to the population of interest (target population). Such a situation can be avoided, to a certain extent, by carefully considering the intercurrent events that affect efficacy and safety estimands of interest and the impact on the BR estimand(s) at the study design phase and by ensuring the consistency of the various analysis populations through study monitoring so as to limit missing data.

The patient-level BR variable could be a single BR score derived from multiple benefits and risk endpoints (see for example Chuang-Stein (1994)). In this case, the population summary could be mean differences that could be assessed via a t-test. In contrast, for the GPC type of assessment, the patient-level variable would be a vector, and this vector consists of efficacy and safety endpoints of interest. In this case, the population summary of (unpaired) comparison of the prioritized BR outcomes can be summarized using win-ratio, win-odds, or the total benefit statistics (Verbreeck et al. 2019).

If the individual BRA variable is a single score derived from benefit and risk endpoints (Chuang-Stein 1994), and if one of the benefit or risk endpoints is missing, the BR score cannot be derived for a patient. Furthermore, as the number of the endpoints that are used to derive the BR endpoint increases, the number and type of intercurrent events that can affect the BR score increase, increasing the number of patients for whom a value cannot be directly derived. If such a situation arises, as a sensitivity or primary analysis one can use the multiple imputation method to handle some missing values. For example, if the reason for treatment discontinuation is lost-to-follow-up, multiple imputation can be considered in the primary or sensitivity analysis to handle such missing values. On the other hand, for a drug with a short half-life, if a patient discontinues treatment prior to the risk assessment and if the reason for discontinuation is the lack of efficacy, then the BRA variable can incorporate this as a low BR score. In contrast, if the GPC method is used to assess BR at an individual level, since the outcomes are ranked based on their importance, the impact of missing a benefit or risk endpoint may be lower compared to the approach where the BRA is based on a single BR score at a patient level. In applying the GPC method, if the risk or benefit outcome that is missing is ranked lower in importance, paired comparison can still be made over the higher-ranked variable. For example, for study DB4, if the GPC assessment is based on the two endpoints, the mean change from the baseline in a daily reduction of nocturic events and hyponatremia, it is reasonable to give the highest importance to the hyponatremia binary endpoint (presence/absence). The second prioritized endpoint would then be the change from the baseline in nocturic events. Hence, first, a patient in the treatment arm is compared with a patient in the control arm on the hyponatremia endpoint over the common follow-up period. Only if a 'win' or a 'loss' cannot be declared with this comparison, the two patients' mean change from the baseline value in the daily nocturic event is compared. Hence, such prioritized comparison by allowing to move from one endpoint to another can potentially reduce the impact of intercurrent events such as early termination of the study and missing data. However, GPC methods can still be impacted by missing and censored values. If a 'win' or a 'loss' cannot be declared over the hyponatremia endpoint and if one of the two patients have a missing change from the baseline nocturic endpoint, the comparison will result in a 'tie'. Furthermore, a large number of missing or censored values tend to result in a large number of 'ties'. Even if the number of missing or censored values is moderate, if the missing values are informative and are not at random, or for time-to-event variables the censoring is informative censoring, the estimand of interest can potentially be biased. In such cases, careful evaluation of the missingness and/or censoring patterns should be made, and sensitivity analysis that accounts for these should be considered.

As in the population-level BRA, for the individual-level BRA differences in exposure time can have a potential impact on the individual risk assessment and as a result on the BRA. Hence, the BRA method needs to account for exposure time. For example, if the risk assessment is based on the presence or absence of an AE of interest, the time to the event of interest would be a better method for comparing the two treatments for this risk. In the case of the GPC method, the time to the AE of interest can be used to compare the two treatments, or the number of times the AE occurs over the common exposure time for the two patients can be compared. However, if the censoring or the missingness is informative, this can still cause some bias in the parameter estimate (e.g., win-ratio). The impact of censoring on the win-ratio is discussed in the study by Dong et al. (2020b). Dong et al. (2020c) proposed a method for handling informative censoring when estimating the win-ratio.

14.4 Summary and Discussion

The structured BR assessment has become increasingly important. This structured evaluation should be enabled by not only the efficacy evaluation but also safety assessment. This is why the ASA safety working group has established two task forces: one on the aggregate safety assessment planning and another one on the BR assessment planning.

It is interesting to contrast the safety evaluation and the efficacy evaluation during clinical development. In general, the efficacy evaluation tends to be linear, targeted, and hypothesis-driven. On the other hand, the safety evaluation tends to be iterative, holistic, and dynamic. On the statistical analysis front, O'Neill pointed out that statistical methodology for safety monitoring has not been well developed to match that for efficacy (O'Neill, 2008). In recent years, global regulatory agencies including the US FDA and EMA have issued specific guidance to encourage fuller development of safety profiles and aggregate safety evaluation (FDA IND safety reporting guidance 2012 and 2015). These provide a great opportunity for biostatisticians and scientific communities to explore better qualitative, semi-quantitative, and quantitative methods for safety and BR evaluations.

In pursuit of thorough evaluations on safety and BR, it is always critical to clarify and get the questions right. The ICH E9 (R1) has laid out an estimand framework that is not only important for clinical trial design, but also very useful to clarify the questions, set the objective, and to plan and execute the safety and BR evaluation. These evaluations are commonly performed at a program level. In performing the program level safety, efficacy, and BR evaluation, it will be important to specify the treatment regimen, the population being analyzed, the endpoint definition, the intercurrent event and relevant clinical context, and the metrics for the treatment effect evaluation. The treatment regimen should be the same for efficacy and safety estimand. The analysis population should be largely the same – any major difference should be acknowledged. The determination for key safety endpoints should have the same level of rigor as that for the key efficacy endpoints. The metrics for the safety and efficacy estimand should also have alignment. In general, the treatment effect that can be expressed in absolute difference between the treatment groups can better help weigh the risk against the benefit. They are particularly useful for the number needed to treat and number needed to harm calculation. Regarding the handling intercurrent event and the choice of estimand strategies, it is always important to clearly understand the clinical context (e.g., concomitant medication) and the definition of intercurrent events (e.g., rescue medication). Sometimes, an intercurrent event for efficacy (e.g., discontinuation due to the AE) may not be the same for safety; therefore, the estimand strategy for efficacy may not always match that for safety; we need to clearly understand these differences. Sensitivity analyses will be useful in the BR decision-making. When the quantitative BR method (e.g., Multi-Criteria Decision Analysis or MCDA) is used, one should take those differences into consideration in determining the proper weights in the safety and efficacy components.

It should also be noted that the safety and BR evaluation is a continuous process throughout the product development life cycle. New safety and efficacy information does not always get collected and analyzed at the same time. While pre-clinical and early phase trials focus a lot on dose related safety and proof of concept efficacy in a limited population, the Phase III trials aim to confirm efficacy and evaluate BR in a larger scale and more representative population. Post-marketing focuses on safety monitoring based on accumulative safety with special attention on rare but critical safety information. More and more, effectiveness information is collected and may play a role in the continuous BR evaluation. All these dynamics call for a more principled and proactive approach. The estimand framework can become a useful tool in these efforts.

While it has been popular to evaluate safety and BR at the program and trial levels, there is an increasing trend to evaluate at the sub-group level and at the patient level. It is not only due to the importance for various decision-making by regulators and payers, but is also because of their importance for patients who will receive the treatment. In reality, patients themselves do not label what they have as efficacy and safety, patients get the treatment effect through their overall experiences. Therefore, having a patient-level BR evaluation can be a very meaningful way to get a deeper understanding of the drug safety and BR profile. Such methods as DOOR (Evans et al. 2015) have the potential to be used, not only in the post-hoc analysis, but also in the program/trial planning to build into the overall development strategy. We can use the estimand framework to enable patient-level BR evaluation.

Ultimately, the methodologies have to fit for the purpose. That's why we need to understand the context and get the questions right. With that, we can choose the estimand strategies for the right treatment regimen, in the targeted population, with properly defined endpoints as well as the appropriate treatment effect metrics that account for intercurrent events and relevant clinical situations. More research efforts are needed on these important topics.

References

Akacha M, Branson J, Bretz F, Dharan B, Gallo P, Gathmann I, Hemmings R, Jones J, Xi D, and Zuber E. Challenges in Assessing the Impact of the COVID-19 Pandemic on the Integrity and Interpretability of Clinical Trials, *Statistics in Biopharmaceutical Research*, 2020; 12(4), 419–426. DOI:10.1080/19466315.2020.17 88984

Akacha M, Bretz F, and Ruberg S. Estimands in clinical trials – broadening the perspective. *Statistics in Medicine*, 2016; 36(1).

Chuang-Stein C. A new Proposal for Benefit-Less-Risk Analysis in Clinical Trials. *Controlled Clinical Trials*, 1994; 15, 30–43.

Dong G, Hoaglin DC, Qiu J, Matsouaka RA, Chang YW, Wang J, and Vandemeulebroecke M. The Win Ratio: On Interpretation and Handling of Ties. *Statistics in Biopharmaceutical Research*, 2020a; 12(1), 99–106. DOI: 10.1080/19466315.2019.1575279

Dong G., Huang B, Chang Y, Seifu Y, Song J, and Hoaglin DC. The win ratio: Impact of censoring and follow-up time and use with non-proportional hazards. *Pharmaceutical Statistics*, 2020b; 19, 168–177. DOI:10.1002/pst.1977

Dong G, Mao L, Huang B, Gamalo-Siebers M, Wang J, Yu GL, and Hoaglin DC. The inverse-probability-of-censoring weighting (IPCW) adjusted win ratio statistic: an unbiased estimator in the presence of independent censoring. *Journal of Biopharmaceutical Statistics*, 2020c; 30(5), 882–899. doi: 10.1080/10543406.2020.1757692

European Medicines Agency (EMA). ICH E9 (R1) addendum on estimands and sensitivity analysis in clinical trials to the guideline on statistical principles for clinical trials. Step 5. 17 February 2020. Available at: https://www.ema.europa. eu/en/documents/scientific-guideline/ich-e9-r1-addendum-estimands-sensitivity-analysis-clinical-trials-guideline-statistical-principles_en.pdf. Accessed October 21, 2020.

Evans SR, Rubin D, Follmann D, Pennello G, Huskins WC, Powers JH, Schoenfeld D, Chuang-Stein C, Cosgrove SE, Fowler VG, Lautenbach E, and Chambers

HF. Desirability of Outcome Ranking (DOOR) and Response Adjusted for Duration of Antibiotic Risk (RADAR). *Clinical Infectious Diseases*, 2015; 61(5): 800–806. doi:10.1093/cid/civ495. Epub 2015 Jun 25. PMID: 26113652; PMCID: PMC4542892.

FDA Briefing Document. Bone, Reproductive and Urologic Drug Advisory Committee (BRUDAC). 2016. Available at https://www.fda.gov/media/100776/download. Accessed October 21, 2020.

Guidance for Industry, Good Clinical Practice: Human Gene Therapy for Rare Diseases. U.S. Department of Health and Human Services, Food and Drug Administration, Center for Drug Evaluation and Research (CDER), Center for Biologics Evaluation and Research (CBER), January 2020.

Guidance for Industry, Factors to Consider When Making Benefit-Risk Determination in Medical Device Premarket Approval and De Novo Classifications. U.S. Department of Health and Human Services, Food and Drug Administration, Center for Devices and Radiological Health (CDRH), Center for Biologics Evaluation and Research (CBER), August 2019.

Guidance for Industry, E6(R2) Good Clinical Practice: Integrated Addendum to ICH E6(R1). U.S. Department of Health and Human Services, Food and Drug Administration, Center for Drug Evaluation and Research (CDER), Center for Biologics Evaluation and Research (CBER), March 2018.

Guidance for Industry, M4E: The CTD-Efficacy. U.S. Department of Health and Human Services, Food and Drug Administration, Center for Drug Evaluation and Research (CDER), Center for Biologics Evaluation and Research (CBER), July 2017.

Guidance for Industry, E2C (R2) Periodic Benefit-Risk Evaluation Report (PBRER). U.S. Department of Health and Human Services, Food and Drug Administration, Center for Drug Evaluation and Research (CDER), Center for Biologics Evaluation and Research (CBER), July 2016.

Guidance for Industry, Considerations for the Design of Early-Phase Clinical Trials of Cellular and Gene Therapy Products. U.S. Department of Health and Human Services, Food and Drug Administration, Center for Drug Evaluation and Research (CDER), Center for Biologics Evaluation and Research (CBER), June 2015a.

Guidance for Industry, Safety Assessment for IND Safety Reporting, Food and Drug Administration, Center for Drug Evaluation and Research (CDER), Center for Biologics Evaluation and Research (CBER), December 2015b, https://www.fda. gov/regulatory-information/search-fda-guidance-documents/safety-assessment-ind-safety-reporting-guidance-industry Accessed December 2020.

Guidance for Industry, Bioavailability and Bioequivalence Studies Submitted in NDAs or INDs — General Considerations. S. Department of Health and Human Services, Food and Drug Administration, Center for Drug Evaluation and Research (CDER), Center for Biologics Evaluation and Research (CBER), March 2014Guidance for Industry, Development and Use of Risk Minimization Action Plans. U.S. Department of Health and Human Services, Food and Drug Administration, Center for Drug Evaluation and Research (CDER), Center for Biologics Evaluation and Research (CBER), March 2005a.

Guidance for Industry, Good Pharmacovigilance Practices and Pharmacoepidemiologic Assessment. U.S. Department of Health and Human Services, Food and Drug Administration, Center for Drug Evaluation and Research (CDER), Center for Biologics Evaluation and Research (CBER), March 2005b.

Guidance for Industry, Premarketing Risk Assessment. U.S. Department of Health and Human Services, Food and Drug Administration, Center for Drug Evaluation and Research (CDER), Center for Biologics Evaluation and Research (CBER), March 2005c.

Guidance for Industry, E9 Statistical Principles for Clinical Trials. U.S. Department of Health and Human Services, Food and Drug Administration, Center for Drug Evaluation and Research (CDER), Center for Biologics Evaluation and Research (CBER), September 1998.

Leverkus F, and Chuang-Stein C. Implementation of AMNOG: An industry perspective. *Biometrical Journal* 2016; 58(1), 76–88. DOI:10.1002/bimj.201300256

Mt-Isa S, Wang N, Hallgreen CE, Callreusse, T et al. Pharmacoepidemiological Research on Outcomes of Therapeutics by a European ConsorTium (PROTECT). *Review of methodologies for benefit and risk assessment of medication.* April 10, 2013. Available at http://protectbenefitrisk.eu/documents/ShahruletalReviewofmethodologiesforbenefitandriskassessmentofmedication May2013.pdf. Accessed October 21, 2020.

Meyer RD, Ratitch B, Wolbers M, Marchenko O, Quan H, Li D, Fletcher C, Li X, Wright D, Shentu Y, Englert S, Shen W, Dey J, Thomas L, Zhou M., Bohidar N, Zhao PL, and Hale M. Statistical Issues and Recommendations for Clinical Trials Conducted During the COVID-19 Pandemic, *Statistics in Biopharmaceutical Research*, 2020; 12(4), 399–411, doi: 10.1080/19466315.2020.1779122

Nilsson M, Crowe B, Anglin G, Ball G, Munsaka M, Shahin S, and Wang W. Clinical Trial Drug Safety Assessment for Studies and Submissions Impacted by COVID-19, *Statistics in Biopharmaceutical Research*, 2020; 12(4), 498–505, doi: 10.1080/19466315.2020.1804444

O'Neill RT. A perspective on characterizing benefits and risks. Derived from clinical trials: can we do more? *Drug Information Journal* 2008; 42, 235–245. doi:10.1177/009286150804200305

Potts J, Genov G, Segec A, Raine J, Straus S, and Arlett P. "Improving the safety of medicines in the European Union: from signals to action." *Clinical Pharmacology & Therapeutics*, 2020; 107(3), 521–529.

Prescribing Information (Labeling): Noctiva (desmopressin), Initial U.S. Approval, 2017. Available at https://www.accessdata.fda.gov/drugsatfda_docs/nda/2017/201656Orig1s000TOC.cfm. Accessed October 21, 2020.

Pritchett YL, and Tamura R. Global benefit-risk assessment in designing clinical trials and some statistical considerations of the method. *Pharmaceutical Statistics*, 2008; 7, 170–178.

Rockhold FW, Lindblad A, Siegel JP, and Molenberghs G. University of Pennsylvania 11th annual conference on statistical issues in clinical trials: Estimands, missing data and sensitivity analysis (morning panel session). *Clinical Trials*, 2019; 16(4), 350–362.

Unkel S, Amiri M, Benda N, Beyersmann J, Knoerzer D, Kupas K, Langer F, Leverkus F, Loos A, Ose C, Proctor T, Schmoor C, Schwenke C, Skipka G, Unnebrink K, Voss F, and Friede T. On estimands and the analysis of adverse events in the presence of varying follow-up times within the benefit assessment of therapies. *Pharmaceutical Statistics*, 2019; 18(2), 166–183. doi:10.1002/pst.1915. Epub 2018 Nov 20. PMID: 30458579; PMCID: PMC6587465.

Van Kerrebroeck P, Dmochowski R, FitzGerald MP, Hashim H, Norgaard JP, Robinson D, and Weiss JP. Nocturia research: Current status and future perspectives. *Neurourology and Urodynamics*, 2010; 29, 623–628.

Verbreeck J, Spit, Spitzer E, de Vries T, van Es GA, Aderson WN, Van Mieghem NM, Leon MB, and Molenberghs TJ Generalized pairwise comparison methods to analyze (non) prioritized composite endpoints. *Statistics in Medicine*, 2019; 38(30).

WHO, clinical considerations for evaluation of vaccines for prequalification, Points to consider for manufacturers of human vaccines, 2010, https://www.who.int/immunization_standards/vaccine_quality/clinical_considerations_oct10.pdf. Accessed November 29, 2020.

Yang F, Wittes J, and Pitt B. Beware of on-treatment safety analyses. *Clinical Trials*, 2019; 16(1), 63–70. doi:10.1177/1740774518812774. Epub 2018 Nov 16. PMID: 30445833.

Zhou Y, Ke C, Jiang Q, Shahin S, and Snapinn S. Choosing Appropriate Metrics to Evaluate Adverse Events in Safety Evaluation. *Therapeutic Innovation & Regulatory Science*, 2015; 49(3), 398–404. doi:10.1177/2168479014565470. PMID: 30222396.

15

Visual Analytics of Safety and Benefit-Risk from Clinical Trial Data

Melvin Munsaka
AbbVie, Inc., North Chicago, IL, USA

Kefei Zhou
Bristol Myers Squibb, New York, NY, USA

Krishan Singh
(retired) GlaxoSmithKline, Brentford, UK

Michael Colopy
UCB Biosciences, Morrisville, NC, USA

Chen Chen
UCB, Ponzano Vento, Italy

Meng Liu
AbbVie, Inc., North Chicago, IL, USA

CONTENTS

15.1 Introduction 352
15.2 Some General Considerations 353
    15.2.1 Principles of Graph Creation and Visualization.........................353
    15.2.2 Graph Construction Using the Grammar of Graphics.............353
    15.2.3 Question-Based Approach..354
    15.2.4 Regulatory Guidance on the Use of Graphs 356
    15.2.5 Defining Visual Analytics..356
15.3 Visual Analytics of Safety Data...356
    15.3.1 Complexity of Safety Data...356
    15.3.2 The Case for Visual Analytics in Safety Monitoring.................359
    15.3.3 Visual Analytics of Adverse Events Data....................................360
15.4 Visual Analytics of Benefit-Risk Data 363
    15.4.1 Overview..363
    15.4.2 Benefit-Risk Based on Patient-Level Data....................................365
    15.4.3 Questions of Interest in Benefit-Risk...365

DOI: 10.1201/9780429488801-19

15.4.4 The Case for Visual Analytics in Benefit Risk............................365
15.4.5 Some Examples of Visual Analytics Used in Benefit Risk 367
        15.4.5.1 Value Trees 367
        15.4.5.2 Forest Plots...367
        15.4.5.3 Norton Plots..368
15.5 Conclusion 368
References...370

15.1 Introduction

The preceding chapters covered a wide range of topics, including safety monitoring, data safety monitoring committees, quantitative methods for safety assessment, pragmatics trials, real-world evidence and methods, benefit-risk planning, and quantitative methods for benefit-risk assessment. Safety data from all sources present many challenges with regard to analysis and interpretation. The data have high variability in measurements and are multidimensional and interrelated in nature. The use of tabular outputs for safety data often results in large volumes of output leading to problems in generation, assessment, validation, assembly, comprehension, and communication of safety findings. Benefit-risk assessment of a drug is an integral component of drug assessment, review, and approval leveraging both qualitative and quantitative elements. It is well recognized that visual analytics present a useful alternative to tabular outputs for exploring safety data and present a great opportunity to enhance evaluation of drug safety. Graphs can play a big role in facilitating communication of safety results with regulators, investigators, data monitoring committees, and other stakeholders and help convey multiple pieces of information concisely and more effectively than tables. Similarly, the use of visualization throughout the benefit-risk process can help enhance transparency and clarity of tasks and decisions made. For both visualization of safety data and benefit-risk, it may be necessary and ideal to leverage multiple visual forms or types to communicate the intended message. The visualizations can be static and/or dynamic and interactive to allow for efficient exploration for communicating benefits and risks. This chapter will discuss the role of graphical methods in safety monitoring and benefit-risk. We will also discuss how to get maximum gain from using visual analytics in safety monitoring and reporting and benefit-risk planning and assessment, taking into account considerations that must be borne in mind for effective visualization. We will show via some examples how visual analytics can be used in safety evaluation of a drug and benefit risk and enhancements that can be applied to help in the assessment of safety monitoring and risk-benefit.

15.2 Some General Considerations

15.2.1 Principles of Graph Creation and Visualization

In order to obtain maximum gain from using visual analytics, some considerations must be borne in mind. These include accounting good principles of data visualization, frame the right question or questions, and use and design of visualization tools. Principles for graph construction to aid in the interpretation of data have been widely discussed in the literature, see for example, Duke *et al.* (2015). Some of these principles include considerations for graph content, communication, information, annotation, axes, and style. It is highly recommended that graph creation adheres to Tufte's (2001) principles including tell the truth (graphical integrity), do it effectively with clarity, precision (data ink ratio, data density, design principles color rules), and so on. Successful visualization of data can best be summarized in the context of information, a story, goal, and a visual form, as exemplified by, see, for example, McCandles (2021). Another consideration that must be borne in mind is with regard to what graphic is intended to be shown in terms of relationship, composition, distribution and comparisons, have a clear purpose, show the data clearly, and make the message obvious, see for example, Abela (2013), Schwabish and Ribecca (2020), Lundblad (2015, 2019a), and Krum (2020). Various other considerations pertaining to making better graphs and distribution detail level, and selection of the right graph are available for examples, in Duke *et al.* (2015), and Vandemeulebroecke *et al* (2019a, 2019b, 2019c).

15.2.2 Graph Construction Using the Grammar of Graphics

The concept of grammar of graphics in the construction of graphs has become very widely used due to its logical, structured, and highly efficient way in creating graphs and implementation in software tools, in particular ggplot. Within the context of the grammar of graphics, there are several components that make up the graphic starting with the data. As implemented in the ggplot package, after identifying the data that one would like to visualize, there is a need to specify the variables of interest. As an example, one may want to display one variable on the x-axis and another on the y-axis. The next step is to define the type of geometric object that one would like to utilize which can be anything from a bar plot to a scatter plot or any of the other existing plot types. These first three components are required. Without data, there is nothing to plot, and without axis definitions, there is nothing to plot either. The remaining components making up the grammar of graphics are optional and can be implemented to enhance the visualization. Facets refer to specifications of subplots, that is, plotting several variables within the data next to one another in separate plots. Statistical transformations mainly refer to the inclusion of summary statistics in the plot, such as the mean, median, or standard deviation. Coordinates describe the different coordinate systems available to you. The most used and default coordinate system is the Cartesian coordinate system. Depending on the structure of the data and what needs to be plotted, lesser used coordinate systems, such as the polar coordinate system which might provide a better way of visualizing the data. Finally, themes provide a variety of options to design all non-data elements of the plot, such as the legend, background, or annotations. There are many ways of visualizing the grammar of graphics. The additivity of these layers as well as the fact that they are building upon one another. For more details on the grammar of graphics, see, Wilkinson (2005). For additional details of grammar of graphics and implementation in the ggplot package, see for example, Wickham (2016), Moulik (2019), and Kassambara (2019).

15.2.3 Question-Based Approach

Safety data needs should be put into the context aligned with concepts of drug safety profiles. The goal of detecting safety issues early and accurately and the form of visualization should match with the data and the idea of what safety monitoring does. In the same spirit, to effectively use visual analytics in safety monitoring and reporting, it is recommended to begin with some questions with regard to safety data under consideration. One can also begin by asking the questions of interest and identifying the appropriate data that can address that question at hand. More specifically, safety monitoring and reporting should be driven by asking the question that can be addressed by the safety data at hand. Put differently, any analysis of safety can be associated with a specific safety concern or concerns. From this perspective, it makes sense to systematically tackle the analysis of safety data taking all these into account and employing various planning, analysis, and reporting suggestions that have been put forward such as those pertaining to the use of a tiered approach and multiplicity adjustments. The use of a question-based approach in analyzing safety data can be used in conjunction with the various proposals made to help improve the analysis of safety data in conjunction with medical discernment. The premise is that each safety concern can best be addressed by asking appropriate questions and then identifying the data that can be used to address the question followed by identifying a methodological approach (descriptive, graphical, analytical, or inferential) to address the question and ultimately reporting and medical decision-making associated with the findings. For the purposes of safety assessment and planning, it is recommended that the steps outlined in Table 15.1 are followed.

When one considers the many questions that one can ask in the safety monitoring setting to help in effective visualization of data and hence

TABLE 15.1

Question-Based Approach

Step	Task
1	What is(are) the safety question(a) that we want to address?
2	Determine what data source(s) will be used to address the question, or what sort of questions can be addressed with the available data source(s)
3	Safety question(s) will ultimately determine the analysis, tabular, and graphical outputs and delivery modality
4	Selection of the visual type or graph type or analysis may also be driven by the safety outcome, for example, in terms of adverse event (AE) tier categories, see Crowe (2009)
5	Safety question(s) will also dictate the tool to be used

identify potential concerns, the more evident it becomes to see that visualization types and settings can fit many subcategories. These subcategories range from graph types, graph complexity, graph usage, graph information type, and static and dynamic aspects for the graphs.

It is worthwhile to note that asking questions in connection to the assessment of safety data is not a new idea. Many authors have discussed questions that need to be considered and proposed some ways to address these questions when looking at safety data. For example, Durham and Turner (2008) discussed some patient-centric questions as part of the rationale for evaluating safety data in clinical trials. One question that they asked is: *how likely is that a patient will experience an adverse event reaction that is so serious that it may be life threatening*? Merz *et al.* (2014) laid out a set of questions when assessing liver safety data. For example, one question that they ask is: *Are there any Hy's law cases in the dataset*? Harrell (2005) also considered this approach in addressing safety issues. Among the questions he discussed are: *Who is having the selected AEs*?

Which AEs occur together? Which AEs tend to occur in the same patient? Within the context of subgroup analysis? Chow and Liu (2014) also posed the following questions among others: *Are the AE rates the same across a subgroup for patient taking the drug? Within subgroup levels, are AE rates the same across treatment groups? Is the time of occurrence of the AE the same across levels of subgroups*? A general discussion on this approach can also be seen in the study by Vandemeulebroecke *et al.* (2019a) and Wright (2019). For a detailed discussion on the question-based approach in the context of safety data, see Munsaka (2018). Ultimately, the set of questions and assessment will be a cross-functional and interdisciplinary effort and may be an iterative process. Although the above discussion focused on safety data, the concept of a question-based approach can easily be extended to and also applies to benefit-risk assessment.

15.2.4 Regulatory Guidance on the Use of Graphs

Although there is no specific regulatory guidance on the use of graphics in the analysis of safety data, the value and recommendations for incorporating some level graphical outputs are highlighted in some of the guidance documents. Some examples of these recommendations are highlighted in Table 15.2.

Additionally, there is overwhelming evidence pointing out and providing recommendations for using graphs in safety and benefit-risk assessment, see for example, Chakravarty *et al.* (2016), Wang *et al.* (2018), and Wen *et al.* (2016), the EMA Benefit-risk Methodology document (2010), and the PROTECT Work Package 5 documents (2013a), (2013b).

15.2.5 Defining Visual Analytics

Visual analytics can be viewed from two broad perspectives. On the one hand, it pertains to bringing static graphs to life through some enhancement, such as interactivity, drill down, animation, dynamic, and so on, that allows for interrogation of the data. On the other hand, visual analytics pertains to the implementation or incorporating statistical analysis and/or algorithms within graphical outputs that allow for enhanced quantitative assessment, allowing for more quantitative insights into the data. These considerations also help facilitate the development of tools for performing visual analytics. Figure 15.1 summarizes the various components that collectively constitute visual analytics in the present discussion.

15.3 Visual Analytics of Safety Data

15.3.1 Complexity of Safety Data

In comparison to efficacy data, safety data tend to be much more complex from an analysis perspective. Analyzing safety data with conventional statistical methods is difficult because many of the standard assumptions may not necessarily be satisfied. Furthermore, there are many pathological features frequently seen in safety data, including non-normal data, high variability, and heterogeneous sub-populations. For example, patients are deferentially prone to AEs depending on their prognosis, and two patients with the same prognosis can exhibit differences in their safety response and experience to treatment. Differences in the standard of care and clinical assessment can also contribute to high variability in a clinician's reporting and assessment of safety data. A further complication factor is that one often needs to look across several data domains and across difference studies to assess safety.

TABLE 15.2

Some Recommendations for the Use of Graphs in the Analysis of Safety Data

Reference	Description
Food and Drug Administration (FDA) Reviewer Guidance (2005)	**Page 27**: Applicants usually include displays and analyses designed to detect such outliers. The relevant data would come from shift tables, scatter plots, box plots, cumulative distribution displays, and tables providing incidence of patients across treatment groups who had a potentially clinically important deviation from normal on one or more laboratory parameters. **Page 46** Explorations of **Time-Dependency for Adverse Findings, Time of Onset**: A life table (Kaplan–Meier graph) describing risk as a function of duration of exposure (i.e., cumulative incidence). Plotting risk for discrete time intervals over the observation period (i.e., a hazard rate curve) reveals how risk changes over time.
FDA DILI Guidance (2009)	**Page 16**: Time course of serum enzyme and bilirubin elevations (consider tabular and/or graphical display of serial laboratory data)
ICH E3 (1996)	**Page 33**: A graph comparing the initial value and the on-treatment values of a laboratory measurement for each patient by locating the point defined by the initial value on the abscissa and a subsequent value on the ordinate. If no changes occur, the point representing each patient will be located on the 45 deg. line. A general shift to higher values will show a clustering of points above the 45 deg. line. As this display usually shows only a single time point for a single treatment, interpretation requires a time series of these plots for treatment and control groups. Alternatively, the display could show baseline and most extreme on-treatment value. These displays identify outliers readily (it is useful to include patient identifiers for the outliers).
CIOMS VI (2005)	**Page 97**: Uses of Statistics for Clinical Safety Data - The purpose of a statistical analysis is to present the data in a way that facilitates understanding of the effects of a drug and to make clear whether variation in results is likely to be due to chance or whether substantial effects might be associated with a drug. In such situations, the use of descriptive methods and well-designed graphics will be helpful in this process. **Page 98**: In addition to inferential statistical approaches, as mentioned above, descriptive statistical methods also play an important role in assessing data, particularly with the use of graphical and other displays, and some pointers on good practice are covered here. **Page 103**: Graphical displays, such as scatter plots of baseline versus later values for each trial participant, can help show both a shift from average and also draw attention to outlying values, both in terms of absolute levels but also large changes. **Page 113**: This and similar "survival" methods may be generally applied to adverse events, though their original use was in looking at death rates. The method is similar to that used for survival curves using a Kaplan–Meier estimate of survival. Kaplan–Meier curves start at 100% (everyone is alive) and move downward over time; AEs are best shown as cumulative hazard plots which move upward over time. **Page 115**: AEs that occur with sufficient frequency for formal analysis should be analyzed using "survival" type methods, and consideration should always be given to showing graphs of cumulative hazards. **Page 116**: It may be helpful to use graphical methods in meta-analyses, which can show similarities and differences for common as well as rare effects across different trials in a clear way. They can also be used to illustrate the uncertainty in effects so that apparently dissimilar results may be seen to be simply different by chance.

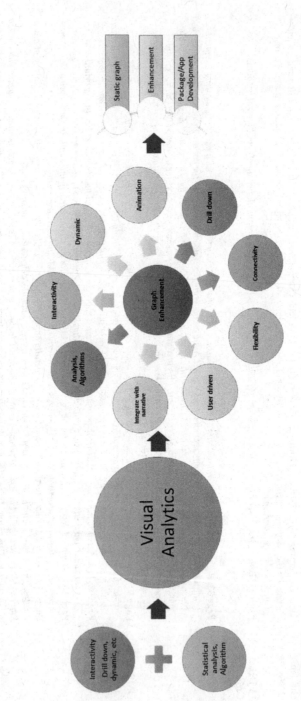

FIGURE 15.1
Defining visual analytics

The safety clinical trial data domains include AEs, clinical laboratory data, vital signs, and electrocardiograms. Specialty safety data based on indication and class of medication may also be collected to help assess certain drug adverse effects or AEs of special interest.

Another complicating factor is that certain adverse conditions may manifest themselves in different ways or may require several pieces of related safety information to conclusively ascertain harm and causal effects. Based on the standard analyses of safety data in clinical trial reports, it is evident that although a lot of safety data are collected, the overall treatment of safety data is not reflective of this, and it is probably and not necessarily the most appropriate data upon which to conclusively base safety decisions. In essence, the appropriate data to demonstrate the safety of a product will depend on the proposed indication, life-threatening potential, or quality of life enhancement, intended duration of use (one time versus short-term versus long-term versus intermittent versus recurrent use), and diversity of the patient population (age, race, gender, disease history, medication history, concomitant medication, concomitant disease, standard of care, genetic disposition, and many other factors). The resulting status quo is massive amounts of tabular outputs which may not render themselves useful in a thorough assessment of the safety profile of a drug. The tabular outputs lead to large volumes of outputs and as noted by Wittes (1996), *a plethora of tables that describe safety may bury some true signal in a cacophony of numbers.*

15.3.2 The Case for Visual Analytics in Safety Monitoring

The preceding discussion points out to a need to improve how safety data are analyzed. A well-known and widely and recognized alternative to tabulate outputs is the use of visual analytics to present safety data. Visualization of safety data can help convey multiple pieces of information concisely and more effectively than tables. It is worth noting that tabular outputs are also important, but as noted by Few (2004) *Numbers have an important story to tell. They rely on you to give them a voice.* With regard to graphs, Tukey (1977) pointed out that *The greatest value of a picture is when it forces us to notice what we never expected to see.*

Graphical exploration can substantially improve information gain from safety data. Visual analytics methods are useful for exploring safety data, and they present a great opportunity to enhance exploration and evaluation of drug safety. For example, Harrell (2005) pointed out that it is difficult to see patterns in tables and substituting graphs for tables can help increase the efficiency of review. McKain, Jackson, and Elko-Simms (2016) argued that traditional case reviews and use of tables and listings are not sufficient for safety surveillance principles. Vlachos (2016) argued that despite their potential, visual analytics methods are an underutilized resource in safety analysis. Graphs can be used to aid in inference and communicating safety results and to help display large amounts of safety data coherently and maximize the ability to detect unusual features or patterns. They can also play a big role in facilitating communication of safety results with regulators, investigators, data monitoring committees, and other stakeholders. Additional discussion on the case for using graphs in the analysis of safety data can be seen in, for example, Wang *et al.* (2020a, 2020b), Munsaka, Zhou, and Kracht (2017), and Amit, Heiberger, and Lane (2008) and the references therein.

15.3.3 Visual Analytics of Adverse Events Data

ICH-E6 defines an adverse event as *An adverse event (AE) is any untoward medical occurrence in a patient or clinical investigation subject administered a pharmaceutical product and which does not necessarily have a causal relationship with this treatment. An adverse event (AE) can therefore be any unfavorable and unintended sign (including an abnormal laboratory finding), symptom, or disease temporally associated with the use of a medicinal (investigational) product, whether or not related to the medicinal (investigational) product.* A host of considerations need to be taken into when analyzing events including timing/onset, seriousness, severity, relatedness to drug, frequency, outcome, and expectedness. The analysis of AEs tends to revolve around magnitude based on crude rates along with some assessment of aforementioned considerations with relying primarily on the tabular output. A complicating factor is the large number of possible events and incidence on placebo. A proper analysis of AEs should take into key considerations that should be well thought out in order to adequately and appropriately analyze and clinically assess AEs. Tremmel (1996) discussed some considerations including the type of event and clinical trial being conducted and duration of the trial. Taken together, these considerations should drive the appropriate metric, measure of risk, and consequently the method of analysis. The question-based approach can facilitate for a structured approach in analysis and assessment of AEs. Table 15.3 provides some samples of questions on AEs.

As an example, there may be interest in the safety question: *What are the frequencies, magnitudes, and differences between treatment and control in some pre-specified AEs?* One way to address this question is to look at the use of the risk plot described in the study by Amit, Heiberger, and Lane (2008). Both R and SAS codes for generating this plot with different analysis metrics are available at the CTSPedia website (2014). The programs generate static risk plots sorted by the size of the metric, largest risk difference (or relative risk or odd ratio) to smallest value of the metric. It may be of interest to explore options such as sorting the AEs by a different variable, for example, by one of the treatment arms or presenting the AEs by system organ class, and so on. An available tool and resource to do this is to make use of the R HH package developed by Heiberger (2020). An additional enhancement is the option within the package is to run an R Shiny app that allows for interaction with

TABLE 15.3

Some Safety Questions Pertaining to AEs

Category	Question
Incidence Rates	• What is the constellation of AEs that come with the drug? • Which AEs are elevated in treatment versus control and by how much? • What are the most common AEs in treatment?
Severity, Duration, and Relationship	• What is the severity of the AEs? • What is the duration of the AEs? • What is the relationship to the studied drug?
Temporal Relationship	• Is there a difference in the time to the first event across treatment groups? • What are the trends of time to the first event among different AEs? • Is the potential AE of interest increasing over time?
Dose relationship, Subgroups, and Risk Factors	• Is there evidence of a dose–response relationship? • Which AEs are elevated in patient subgroups? • What are the risk factors of the AE?
Concurrent and Intercurrent Events	• Is there a relationship with other AEs? • Is there a relationship with use of concomitant medications? • Are there withdraws and/or interruption due to AE of interest?

the risk plot using drop down menus and a check box. With the R Shiny app version, once can for example remove the tabular part, apply different sort orders, obtaining different sub-groupings, and so on. Hence using the app, one can use these various available options to address different questions, such as the most common AEs within each body system, or to obtain a different sort order for the risk plot. Figure 15.2 shows the interface of the R Shiny app of the HH package showing the various options available.

Some example code for generating the risk plot is available at the book resource at the URL: https://community.amstat.org/biop/workinggroups/safety/safety-home. An example of this R Shiny implementation of the HH package can be seen at the URL: https://visual-analytics.shinyapps.io/index/.

As an alternative to the forest plot, one can make use of an enhanced version of the static dot-line plot combination proposed by Salganik (2013). This graphic combines the dot plot and line plot with a sorting order based on the control rates. The line plot allows one to easily assess how the rates of the treatment compare with the control. The plot also incorporates Fisher's exact test for each preferred term comparing the treatment with the control. The preferred terms where treatment is greater than the control are highlighted in red and those which it is less are highlighted in green. An enhanced version

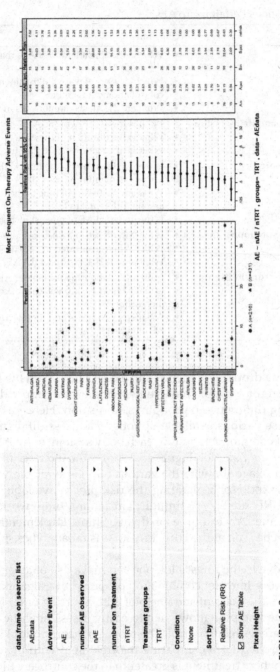

FIGURE 15.2
Forest plot

of the was developed and can be seen at the URL: https://visual-analytics. shinyapps.io/index/. In this enhancement, an R Shiny app has been built onto of the original R code. The R Shiny app extends the inference testing to include multiplicity adjustment methods implemented in the c212 package, see, Carragher and Robertson (2020). The c212 package includes methods for error controlling procedures for multiple hypothesis testing and Bayesian modeling approaches. Figure 15.2 shows the interface of the R Shiny app of the HH package showing the various options available (Figure 15.3).

Using the R Shiny app with the c212 implementation, one can explore AEs that are likely to be significantly different from the control while employing various methods controlling procedures for multiple hypothesis testing and Bayesian modeling approaches. Although the above discussions focused on AE data, the question-based approach can easily be extended to other data domains in clinical trial data and beyond. Below we discuss the question-based approach in the context of benefit risk

---

15.4 Visual Analytics of Benefit-Risk Data

15.4.1 Overview

The Benefit-risk balance of a drug/vaccine/device is considered positive/ favorable when the desired effects or *benefits* are greater than its undesired effects or *risks*. Chapter 5 introduced benefit-risk assessment, noting that it is an integral part of the iterative process of risk management along with developing and implementing tools to minimize risks and make adjustments as appropriate to risk minimization tools to further improve the benefit-risk balance. It was also noted that qualitative and quantitative characterization of the safety profile is essential for benefit-risk assessment throughout the product lifecycle. Qualitative methods referred to descriptive approaches that utilize summary statistics of key efficacy endpoints and key safety endpoints and include a structured approach on how this summary information weighs together into decision-making. Quantitative methods on the other hand integrate endpoints by quantifying their clinical relevance and trade-offs and require judgment in assigning weights to benefit and risk. Different frameworks including the FDA BRF, BRAT, and PrOACT-URL that can be used to describe and evaluate data in a structure were also highlighted. The concept of product benefit-risk value tree, a visual, hierarchical depiction of key benefits, and key risks was also noted as a foundation concept to the frameworks and quantitative methods. The multi-criteria decision analysis (MCDA) and its extensions the stochastic multicriteria acceptability analysis (SMAA) were highlighted. Another alternative to SMAA when

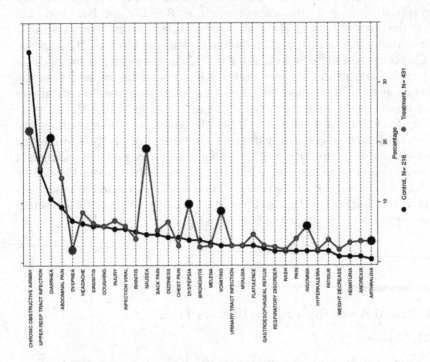

FIGURE 15.3
Dot plot-line plot

only accounting for uncertainty in the preference weight is the predicted choice probability analysis. Another alternative to the MCDA is the scale loss score method which helps one to avoid the recommendation of drugs with extreme risk or with no benefit and captures the usual preferences of the decision-makers. There are many additional methods that have been proposed in the literature, including modifications of aforesaid methods and/or Bayesian versions of these methods. For more details of these methods, see for example, Jiang and He (2015) and the URL: https://protectbenefitrisk. eu/methods.html.

15.4.2 Benefit-Risk Based on Patient-Level Data

Another important consideration in benefit-risk is that of patient-level analysis methods which focuses on evaluation of benefits and risks using patient-level information. This approach can provide valuable insights into the interaction of benefits and risks at the subgroup or patient level. Chung-Stein et al (1994) discussed one such method of composite benefit-less-risk score. Another method, the desirability of outcome ranking, was also suggested by Evans et al. (2015) and Evans and Follmann (2016). These methods use patient's experiences on efficacy and safety to derive an individualized benefit-risk measurement, which can be used to evaluate the overall benefit-risk comparisons. It is worth noting that although quantitative methods are well developed, they have not been widely used in regulatory settings. However, this may change as they become more widely acceptable along with readily available computational tools. Last but not least is the planning for benefit-risk assessment. Chapter 13 of this book discusses this aspect of benefit-risk focus on various considerations and recommendations when creating benefit risk plans.

15.4.3 Questions of Interest in Benefit-Risk

Just as in the case of safety data, the question-based approach can help facilitate for a structured approach in benefit-risk assessment and help drive the methods and graphs that can be leveraged in the assessment. Table 15.4 provides some sample questions on benefit-risk assessment

15.4.4 The Case for Visual Analytics in Benefit Risk

The use of data tables and certain displays makes this assessment difficult and challenging. In addition as noted by Levitan (2011), benefit-risk methodologies vary considerably both in formulation and complexity with most approaches requiring simultaneous consideration of information pertaining to multiple endpoints for multiple treatments. There is an opportunity to employ visual analytics to facilitate for a better understanding and

TABLE 15.4

Some Questions Pertaining to Benefit-Risk Assessment

Category	Question
The Indication	• Is it a rare disease?
	• How serious is the condition being treated?
The Benefits	• What are the benefits?
	• What is the evidence and uncertainty?
	• How clinically meaningful is the benefit and for whom?
	• How soon do the benefits occur and how long do they last?
The Risks	• What are the risks?
	• Are we using the right measure to characterize the risk?
	• How many risks are there?
	• What is evidence and uncertainty in the risks?
	• How frequent are the risks? How serious are the risks? Are the risks reversible?
	• How soon after treatment do the risks occur?
	• How long does the risk last?
	• Does the risk increase or decrease over time? Does risk persist after drug withdrawal?
	• What are the risks of the disease or condition if not treated?
Treatment Options	• What treatment options are available for the same indication?
	• Are there any other treatment options where the BR balance is more favorable
	• How well is the patient population's medical need being met by currently available therapies?
Risk Management	• How manageable is the risk?
	• Is (are) the risk(s) preventable or can it (they) be managed?
	• Can risk minimization measures be implemented and what are they?

assessment of information on multiple points, treatments, and preferences. The use of interactive and dynamic visualizations may also be useful when communicating benefits and risks visually and assist in the uncertainty assessment in benefit-risk models. The potential for using visual analytics methods has been widely recognized and continues to grow, see for example, Wen *et al.* (2016), Hallgreen *et al.* (2016), and Hughes *et al.* (2016). Due to the formulation and complexity of benefit-risk methodologies and the fact that the entire process requires simultaneous consideration of information pertaining to multiple endpoints for multiple treatments for both efficacy and safety endpoints, graphs utilized in benefit assessment are often specific to method and/or serve a specific purpose. Hence, it is not uncommon to employ more than one graph when assessing a drug product. In addition, there is not one single visual type that is consistently superior to others for the communication of benefit-risk information to various stakeholders. Thus when creating visuals for communication in benefit-risk assessments, it is important to consider the compatibility between a visual and its target audience. A multitude of visuals for benefit-risk assessment can be seen at the

URL: https://protectbenefitrisk.eu/visualisations.html. Below we highlight a few of these graphs and their use.

15.4.5 Some Examples of Visual Analytics Used in Benefit Risk

15.4.5.1 Value Trees

Tree diagrams in benefit-risk assessments are called value trees. They are a visual, hierarchical depiction of key ideas, values, or concepts used in decision, through an explicit visual map of the attributes or criteria of decisions that are of value to the decision-makers. Value trees have proven to be a very powerful graphic tool for eliciting and structuring the individual components that are essential for making a decision. They are a useful tool as they require decision-makers to clarify and make transparent the benefits and risks judged to be pivotal to the benefit-risk balance. The measures most relevant to a benefit-risk assessment will vary based on the indication, outcomes, data, and perspective or stakeholders and decision-makers. In essence, the value tree can be tuned to display or not display outcomes depending on the needs of the audience, the decision being made, and the data available. Value trees are thus essentially dynamic tools that are revisited and changed as new information is acquired over the lifecycle of a drug. Using value trees, some questions that can be addressed include: *What are the benefits?*, *What are the risks?*, *What measure are we using to characterize the benefits and risks?*. The creation of the value trees can also present a forum to ask additional questions such as, *Are we using the right measure or estimand to characterize the risk?* Additional details about value trees can be seen in: CIRS (2013) and at the URL: https://protectbenefitrisk.eu/bratvaluetree.html.

15.4.5.2 Forest Plots

Forest plots are helpful in communicating data on multiple benefits and risks simultaneously. These displays enable reviewers based on their own judgment to compare endpoints related to benefit-risk balance. A forest plot can instantly give the viewer a sense of the magnitude of difference and the variability or precision of a measure. The plots are constructed with dots as the central measures for each efficacy and safety criterion, with additional lower and upper limits. They are used to communicate summary measures such as the mean risk difference or risk ratio as well as the associated uncertainty, such as confidence intervals. Forest plots can also be used as means of benefit-risk communication to specialist audience such as physicians, regulators, and payers. Closely associated with the forest plot and in fact a precursor to forest plots are the benefit-risk summary tables. These are top-level representations of data in the framework that provide the key benefit-risk information organized for rapid interpretation. The rows of this table match the branches (outcomes and measures) in the value tree, and the columns of data contain summary measures of the most critical information that decision makers need as they provide decision makers with an understanding of the most important trade-offs behind the benefit-risk assessment. Some questions that can be addressed by forest plots include: *What is the evidence (or magnitude) and uncertainty in the benefits?* and *What is the evidence (or magnitude) and uncertainty in the risks?* Additional details about value trees can be seen in: CIRS (2013) and at the URLs: https://protectbenefitrisk.eu/bratvaluetree.html and https://protectbenefitrisk.eu/treebrassesment.html. Figure 15.4 shows a value tree and the corresponding and corresponding benefit-risk summary table and a forest plot.

15.4.5.3 Norton Plots

Although the most common benefit-risk methods involve comparisons of summary statistics for the entire sample, there are approaches where the comparison is done at the patient level. For example, one can classify each subject into a joint benefit and risk category, for example, assigning a score to each subject based on whether they experienced benefits only, risks only, both benefits and risks, or neither benefit nor risk. One such approach is to use Norton plots. Additional details of how to create Norton plots and the context can be seen in Norton (2011) and Jiang and He (2015). This plot can be used to answer the question: *What is the distribution of patients who are experiencing only benefits, only risks, both benefits and risks, and neither benefit nor risk.* An example of a Norton plot is provided in Figure 15.5.

15.5 Conclusion

The discussions in this chapter have provided some flavor of the type of visualizations that can be generated to help visualize safety and benefit-risk assessment. As noted earlier, a multitude of these visualizations exist with each addressing a question or questions of interest regarding safety or benefit-risk assessment. In benefit risk, some visualization are associated with a specific method and the same holds true for safety data as well. Also, just as in the case for clinical trial data, benefit-risk visualizations can be enhanced to allow for explorations of various considerations such as the effects of weights associated with benefits and risks, see for example, at URL: https://protectbenefitrisk.eu/interactivegraphics.html. Ultimately whether one is looking safety data or benefit-risk data, the visual type and tool used will depend on the question or questions under consideration. By considering various enhancements, such as interactivity and so on, one can select

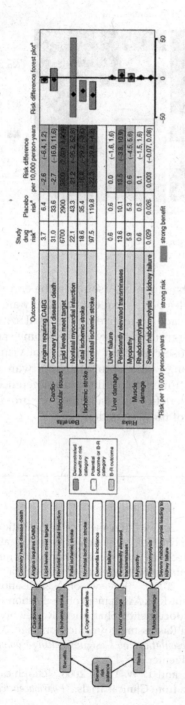

FIGURE 15.4

Value tree and corresponding benefit-risk summary table and forest plot.

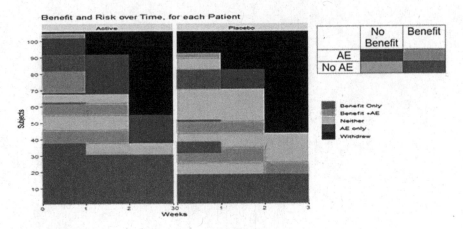

FIGURE 15.5
Norton plot

visualizations and tools that are most useful for the end-user. For additional visualizations and examples pertaining to both safety and benefit-risk along with available tools, the reader is referred to the URLs: https://community. amstat.org/biop/workinggroups/safety/safety-home and the references herein where additional use cases and examples of visualizations for both safety and benefit-risk along with resources abs software tools are provided. In addition, guidance documents and literature references for safety monitoring and planning and benefit-risk assessment are provided which are critical for execution of both of these considerations.

References

PROTECT Work Package. 2013a. "Review of Visualisation Methods for the Representation of Benefit-Risk Assessment of Medication: Stage 1 of 2." http://www.imi-protect.eu/documents/ShahruletalReviewofvisualisation methodsfortherepresentationofBRassessmentofmedicationStage1F.pdf.

PROTECT Work Package. 2013b. "Review of Visualisation Methods for the Representation of Benefit-Risk Assessment of Medication: Stage 2 of 2." https:// protectbenefitrisk.eu/documents/ShahruletalReviewofvisualisationmethods fortherepresentationofBRassessmentofmedicationStage2A.pdf.

Abela, A. 2013. *Advanced Presentations by Design: Creating Communication That Drives Action.* San Francisco: Pfeiffer.

Amit, O., R. M. Heiberger, and P. W. Lane. 2008. "Graphical Approaches to the Analysis of Safety Data from Clinical Trials." *Pharmaceutical Statistics* 7: 20–35.

Levitan, B. 2011. "A Concise Display of Multiple End Points for Benefit–Risk Assessment." *Clinical Pharmacology and Therapeutics* 89: 56–59.

Carragher, Raymond, and Chris Robertson. 2020. "C212: An R Package for the Detection of Safety Signals in Clinical Trials Using Body-Systems (System Organ Classes)." *Journal of Open Source Software* 5: 2706.

Chakravarty, A. G., R. Izem, S. Keeton, C. Y. Kim, M. S. Levenson, and M. Soukup. 2016. "The Role of Quantitative Safety Evaluation in Regulatory Decision Making of Drugs." *Journal of Biopharmaceutical Statistics* 26: 17–29.

Chow, S-C., and J. P. Liu. 2014. New York: Wiley.

CIRS. 2013. "The Brat Tool for Benefit-Risk Assessment: User's Guide to the Process, Version Beta 3.0." https://www.cirs-brat.org/.

Crowe, B. J., H. A. Xia, J. A. Berlin, D. J. Watson, H. Shi, S. L. Lin, J. Kuebler, et al. 2009. "Recommendations for Safety Planning, Data Collection, Evaluation and Reporting During Drug, Biologic and Vaccine Development: A Report of the Safety Planning, Evaluation, and Reporting Team." *Clinical Trials* 6: 430–40.

CTSPedia. 2014. "Most Frequent on-Therapy Ae Sorted by Risk Difference." https://www.ctspedia.org/do/view/CTSpedia/ClinAEGraph000.

Duke, S. P., F. Bancken, B. Crowe, M. Soukup, T. Botsis, and R. Forshee. 2015. "Seeing Is Believing: Good Graphic Design Principles for Medical Research." *Statistics in Medicine* 34: 3040–59.

Durham, T. A., and J. R. Turner. 2008. *"Introduction to Statistics in Pharmaceutical Clinical Trials."* London: Pharmaceutical Press.

E3, ICH Topic. 1996. "Structure and Content of Clinical Study Reports." https://www.ema.europa.eu/en/documents/scientific-guideline/ich-e-3-structure-content-clinical-study-reports-step-5_en.pdf.

EMA. 2010. "Benefit-Risk Methodology Project Work Package 2 Report: Applicability of Current Tools and Processes for Regulatory Benefit-Risk Assessment." https://www.ema.europa.eu/en/documents/report/benefit-risk-methodology-project-work-package-2-report-applicability-current-tools-processes_en.pdf.

Evans, S. R., and D. Follmann. 2016. "Using Outcomes to Analyze Patients Rather Than Patients to Analyze Outcomes: A Step Toward Pragmatism in Benefit: Risk Evaluation." *Statistics in Biopharmaceutical Research* 8: 386–393.

Evans, S., D. Rubin, D. Follmann, G. Pennello, W. C. Huskins, J. H. Powers, D. Schoenfeld, et al. 2015. "Desirability of Outcome Ranking (DOOR) and Response Adjusted for Duration of Antibiotic Risk (RADAR)." *Clinical Infectious Diseases* 61: 800–806.

Few, S. 2004. *Show Me the Numbers: Designing Tables and Graphs to Enlighten.* Oakland, CA: Analytics Press.

Guidance, FDA Reviewer. 2005. "Reviewer Guidance Conducting a Clinical Safety Review of a New Product Application and Preparing a Report on the Review." https://www.fda.gov/media/71665/download.

Hallgreen, C. E., S. Mt-Isa, A. Lieftucht, L. D. Phillips, D. Hughes, S. Talbot, A. Asiimwe, et al. 2016. "Literature Review of Visual Representation of the Results of Benefit–Risk Assessments of Medicinal Products." *Pharmacoepidemiology and Drug Safety* 25: 238–250.

Harrell, F. 2005. "Exploratory Analysis of Clinical Safety Data to Detect Safety Signals." https://www.fharrell.com/talk/gsksafety/.

Heiberger, R. M. 2020. *HH: Statistical Analysis and Data Display.* https://CRAN.R-project.org/package=HH.

Hughes, D., E. Waddingham, S. Mt-Isa, A. Goginsky, E. Chan, G. F. Downey, C. E. Hallgreen, et al. 2016. "Recommendations for Benefit–Risk Assessment Methodologies and Visual Representations." *Pharmacoepidemiology and Drug Safety* 25: 251–262.

Industry, FDA Guidance for. 2009. "Drug-Induced Liver Injury: Premarketing Clinical Evaluation." https://www.fda.gov/media/116737/download.

Jiang, Q., and W. He, eds. 2015. *In Benefit-Risk Assessment Methods in Medical Product Development: Bridging Qualitative and Quantitative Assessments.* Boca Raton: Chapman; Hall.

Kassambara, A. 2019. *GGPlot2 Essentials: Great Data Visualization in R.* Independently published.

Krum, R. 2020. "DataViz Reference Guide." https://coolinfographics.com/dataviz-guides.

Lundblad, P. 2015. "Third Pillar of Mapping Data to Visualizations: Usage." https://www.qlik.com/blog/third-pillar-of-mapping-data-to-visualizations-usage.

McCandles, D. 2021. "What Makes a Good Visualization." https://www.informationisbeautiful.net/visualizations/what-makes-a-good-data-visualization/.

McKain, L., T. Jackson, and C. Elko-Simms. 2016. "Optimizing Safety Surveillance During Clinical Trials Using."

Merz, M., K. R. Lee, G. A. Kullak-Ublick, A. Brueckner, and P. B. Watkins. 2014. "Methodology to Assess Clinical Liver Safety Data." *Drug Safety* 37: 33–45.

Moulik, T. 2019. *Applied Data Visualization with R and Ggplot2: Create Useful, Elaborate, and Visually Appealing Plots.* Birmingham: Packt.

Munsaka, M. 2018. "A Question-Based Approach to the Analysis of Safety Data." In *Biophar- maceutical Applied Statistics Symposium,* edited by S. Menon K. Peace D. G. Chen, 193–216. Singapore: Springer.

Munsaka, M. S., K. Zhou, and K. K. Kracht. 2017. *Enhancing Visual Analytics Approaches in Safety Monitoring.* JSM Proceeedings, Biopharm Section. https://ww2.amstat.org/MembersOnly/proceedings/2017/data/assets/pdf/593847.pdf.

Norton, J. D. 2011. "A Longitudinal Model and Graphic for Benefit-Risk Analysis, with Case Study." *Drug Information Journal* 45: 741–47.

Salganik, M. 2013. "Visualization of the Adverse Events Data." https://ww2.amstat.org/meetings/fdaworkshop/index.cfm?fuseaction=AbstractDetails&AbstractID=302825.

Schwabish, J., and S. Ribecca. 2020. "Visual Vocabulary: Designing with Data." https://ft-interactive.github.io/visual-vocabulary/.

Tremmel, L. 1996. "Describing Risk in Long Term Clinical Trials." *Biopharmaceutical Report* 4: 5–8.

Tufte, E. R. 2001. *The Visual Display of Quantitative Information.* Cheshire, CT: Graphics Press.

Tukey, J. W. 1977. *Exploratory Data Analysis.* Addison-Wesley.

Vandemeulebroecke, M., M. Baillie, D. Carr, L. Kanitra, A. Margolskee, A. Wright, and B. Magnusson. 2019a. "Graphics Principles." https://graphicsprinciples.github.io/.

Vandemeulebroecke, M., M. Baillie, D. Carr, L. Kanitra, A. Margolskee, A. Wright, and B. Magnusson. 2019b. "How Can We Make Better Graphs? An Initiative to

Increase the Graphical Expertise and Productivity of Quantitative Scientists." *Pharmaceutical Statistics* 18: 106–14.

Vandemeulebroecke, M., M. Baillie, A. Margolskee, and B. Magnusson. 2019c. "Effective Visual Communication for the Quantitative Scientist." *CPT: Pharmacometrics and Systems Pharmacology* 8: 705–19.

VI, CIOMS. 2005. "Management of Safety Information from Clinical Trials: Report of Cioms Working Group Vi." https://cioms.ch/publications/product/management-of-safety-information-from-clinical-trials-report-of-cioms-working-group-vi/.

Vlachos, P. 2016. "Graphical (and Analytical) Tools for the Systematic Analysis of Safety Data in Clinical Trials."

Wang, W., M. Nilsson, R. Revis, and B. Crowe. 2020a. "Interactive Visualization for Clinical Safety Data Review." *Biopharmaceutical Report* 27: 2–7.

Wang, W., R. Revis, M. Nilsson, and B. Crowe. 2020b. "Clinical Trial Drug Safety Assessment with Interactive Visual Analytics." *Statistics in Biopharmaceutical Research*, 13, 1–12.

Wang, W., E. Whalen, M. Munsaka, J. Li, M. Fries, K. Kracht, M. Sanchez-Kam, K. Singh, and K. Zhou. 2018. "On Quantitative Methods for Clinical Safety Monitoring in Drug Development." *Statistics in Biopharmaceutical Research* **10**: 85–97.

Wen, S., W. He, S. R. Evans, G. Quartey, J. D. Norton, C. Chuang-Stein, Q. Jiang, H. Ma, and M. Bhattacharya. 2016. "Practical Analysis and Visualization Tools for Benefit-Risk Assessment in Drug Development: A Review of Recent Fda Advisory Committee Meeting Packages." *Statistics in Biopharmaceutical Research* 8: 409–16.

Wickham, H. 2016. *Ggplot2: Elegant Graphics for Data Analysis*. New York: Springer-Verlag. https://ggplot2.tidyverse.org.

Wilkinson, L. 2005. *The Grammar of Graphics (Statistics and Computing)*. Berlin, Heidelberg: Springer-Verlag.

Wittes, J. 1996. "A Statistical Perspective on Adverse Event Reporting in Clinical Trials." *Bio- pharm. Report* **4**: 5–10.

Wright, A. 2019. "Show Me the Question: Putting the Questions Front and Center." RSS Conference.

Index

Page numbers in *Italics* refer to figures; page numbers in **bold** refer to tables

A

Active Risk Identification and Analysis (ARIA), 235
ADAPTABLE, 258
adverse drug reaction (ADR), 85
    quantitative analysis of, 90–92
adverse events (AEs), 106, 121, 134, 169, 188, 319, 360–363
adverse events of special interest (AESI), 83
Aggregate Assessment, 17
aggregate data, signals, 47
    adverse events, 48–50
    laboratory data, 50–51
    post-marketing, 51–52
aggregate safety assessment
    emergence of, 3–13
    monitoring data, 15–41
Aggregate Safety Assessment Planning (ASAP), 287
    background, 16–18
    communication of safety information, 29
    data analysis approaches, 22–24
    hurdles toward implementing, 33–34
    key gap analysis, 24
    as modular template-drive approach, 30
    Ongoing Aggregate Safety Evaluation (OASE), 25
    pooling strategy, 20, 22
    responding to COVID-19, 30–32
    safety topics of interests, 20–22
    trends influencing direction of, 40–41
    value proposition and governance, 19
Akaike's information criterion (AIC), 152

American Statistical Association (ASA), ix, 256
ARIA sufficiency, 248–250

B

background rates, 37–39
Bayesian approach, safety monitoring, 105–107
Bayesian confidence neural network (BCPNN), 108
Bayesian hierarchical models (BHMs), 149
Bayesian safety methodology
    blinded data, 141–148
        BDRIBS method, 145–148
        event rate methods, 142–143
        framework, 142
        methods for proportion of subjects with event, 144–145
        Schnell-Ball method, 146–148
    borrowing historical information, 156–167
    hierarchical modeling approaches, 148–156
        Bayesian hierarchical models, 150–156
        literature review, 149–150
    statistical methods in safety analysis, 134
    statistics, 135–139
        hierarchical models, 140–141
        illustrative analysis, 138–140
benefit-risk assessment, 286–287
    origins of, 291–293
Benefit-Risk Assessment Planning (BRAP), xiv, 291
    industry perspective, 304–319
    during medical product lifecycle, *293*

published recommendations applicable to, 295–304

questions, **298–299**

recent developments applicable to, 292–295

suitability of real-world data, 299–301

bias, 52–54

binding, design consideration, 218–219

blinded, dynamic methods, 27–28

blinded safety assessment, 105

blinded safety monitoring, 121–122
  advantages, **123**
  Bayesian method, **126**
  challenges in, 122–123
  consideration in implementing, 129
  defining, 122
  disadvantages, **123**
  frequentist method, **126**
  illustrative example of, 130–134
  inference approaches, 124–125
  likelihood method, **126**
  linking with unblinded safety monitoring, 129–130
  quantitative methods for, **126–128**
  related regulatory consideration, 123–124
  software for, 130
  statistical methods for, 125–126, **127–128**

blinded, static methods, 27

Blood Surveillance Continuous Active Surveillance Network (BloodSCAN), 236

Bradford Hill causation criteria, 60–66
  analogy, 64
  biologic gradient, 62
  coherence, 63
  consistency, 61
  experimental evidence, 64
  plausibility, 63
  specificity, 61
  strength of association, 60
  temporality, 61

C

cardiovascular safety outcome trials (CVOTs), xiii

causal association, 54–59

CBER Biologics Effectiveness & Safety (BEST), 236

Center for Biologics Evaluation and Research (CBER), 290–291

Center for Devices and Radiological Health (CDRH), 240, 291

Center for Drug Evaluation and Research (CDER), 290–291

China NMPA regulations, 9–11

chronic hepatitis C viral (HCV), case study, 256–257

CIOMS IV, 81

CIOMS VI, 123

classification and regression tree (CART), 262–263

CMaxSPRT, 177

Cochran-Mantel-Haenszel, 109

cohort identification and descriptive analysis (CIDA), 239–240

common graphical display
  adverse event review, 66, *67–70*
  laboratory data review, 70–71
  other resources for, 71–73

Community Building & Outreach Center (CBOC), 237

comparator selection, design consideration, 219

confidence intervals (CIs), 125

confounding, 52–54, 258–266
  methods, **265–266**

constant continuity correction, **198**

continuous glucose monitors (CGMs), 273

controlled design, signal identification, 244

Council for International Organizations of Medical Sciences (CIOMS), 3, 188, 288–290

COVID-19
  general effect on data collection, 30–31
  unanticipated infections of subjects in the clinical trials, 31
  using ASAP to respond to, 32

CTSPedia, 71

cumulative incidence function, *68*

cumulative meta-analysis (CMA), 204

D

data monitoring committee (DMC), 78, 141
  composition of, 81–83
  evaluating need for, 80–81
  operations, 83
  role of, 78–80
  scope, 83–92
Data warehouse, 36
DeMets, D. L., 207
Designated Medical Events, 17–18
design tools, design consideration, 221–222
Development Safety Update Report (DSUR), 6
deviance information criterion (DIC), 152
direct acting antivirals (DAAs), 256
directed acyclic graphs (DAGs), 262
discrete choice experiment (DCE), 112
dot plot, *67*
dot plot-line plot, *364*
Duke Margolis Center for Health Policy, 304
DuMouchel, W., 106, 108, 152
Dutch STRIDER, 240
dynamic evaluation, 109–110

E

electronic health record (EHR), 94, 108
eMAP prior, 158–159
empirical Correction, **198**
endpoints, design consideration, 220
estimand framework, 22–23, 318–319
  patient-level BRA, 340–347
  population-level BRA, 334–339
estimand framework for safety assessment, 320–332
  intercurrent events, 323, **324,** 325–327
  missing data, 327
  population, 321–322
  summary measure, 328, **329–330,** 331–332
  treatment, 321
  variable (outcome), 322–323

European Medicines Agency (EMA), 79, 255, 290
  benefit-risk perspective, 295–296
  effects table, **291**
  PrOACT-URL framework, **296**
  regulations, 7–8
exposure-adjusted adverse event rate (EAER), 125
exposure-adjusted incidence rate (EAIR), 125
external evidence method (EEM), 301

F

false discovery rate (FDR), 25, 107, 149
FDA adverse event reporting system (FAERS), 241–242, 320
Federal Food, Drug, and Cosmetic Act, 102
Food and Drug Administration Amendments Act (FDAAA), 233–236
Food and Drug Administration (FDA), 17, 291
  analysis considerations, 254–256
  benefit-risk guidance, 110–113
  CVOTs, 222–231
  DMC definition, 79
  IND safety reporting guidance, 123–124
  meta-analysis, 191
  regulations, 5–7
  statistical methodologies, 102–103
forest plots, 367–368
Foundational Analysis Rules, 22–24
French imputability method, 60
frequentist approach, safety monitoring, 106–107

G

generalized boosted models (GBMs), 263
generalized linear mixed effects model, **197**
generalized pairwise comparison (GPC), 340
Good Vigilance Practice (GVP), 8

Grading of Recommendations
  Assessment, Development, and
  Evaluation (GRADE), 256
graphical assessment, 94–96
Greenberg Report, 78
Guidelines for Good
  Pharmacoepidemiology
  Practices (GPP), 256

H

health-related quality of life (HRQoL),
  301–304
Hepatic Safety Explorer, 72
hepatocellular carcinoma (HCC), 257
histogram, *69*
Hochberg, Y, 183
hyper-parameters, 136

I

ICH E2C, 123
ICH E9, 82
ICH E9(R1), 318–319
important medical events, 47
independent data monitoring committee
  (IDMC), 19, 78–80
INdividual Data ANalysis of
  Antihypertensive intervention
  trials (INDANA), 203
industry perspective, BRAP
  overview, 304–305
  sBRA utilization, 306, **307**, 308
  survey results
    coordination, 305–306
    data/information sources, 306
Institutional Review Boards
  (IRBs), 85
integrated safety evaluation, 110
International Council for Harmonization
  (ICH), 3
International Society for
  Pharmacoeconomics and
  Outcomes Research (ISPOR),
  111
investigational medical product (IMP),
  320

J

Japan, PDMA regulations of, 8–9

L

likelihood-based methods, safety
  monitoring, 168–171
  examples, 179, **180**
    LongLRT, 182
    rotavirus vaccine monitoring,
      181–182
    vaccines, 181
  general framework, 171–174
    misleading evidence, 173–174
    randomized studies, 171
    risk/evidence levels, 172–173
    universal bound, 174
  software, 182–184
  statistical methods, 175–179
    CMaxSPRT, 177
    LongLRT, 178–179
    MaxSPRT, 176–177
    SeqLRT, 178
    SGLRT, 179
    SPRT, 176
likelihood ratio test (LRT), 175–176
logistic Regression, **197**
LongLRT, 178–179
lower level terms (LLT), 36

M

Mantel–Haenszel method, 193, **197**
Markov Chain Monte Carlo (MCMC),
  138
maximum tolerated dose (MTD), 320
MaxSPRT, 176–177
Medical Device Innovation Consortium
  (MDIC), 301
meta-analysis, drug safety assessment,
  188–189
  analysis tools/software, 206–208
  cumulative meta-analysis, 204
  multivariate meta-analysis, 201–204
  planning, 189–191
  rare events, 195–200

reporting, 206
sequential meta-analysis, 204–205
summarized results, 191–195
visualizing, 206
meta-analytic predictive (MAP), 106
monitoring, state of, 169–170
multi-criteria decision analysis (MCDA),
111–113, 363
multi-item Gamma Poisson shrinker
(MGPS), 108
multi-source data, synthesis of, 266–267
case studies, 271–273
external control arms, 269–270
linking data sources, 268
meta-analysis of safety, 270–271
multivariate Bayesian logistic regression
(MBLR), 106, 149
multivariate meta-analysis, 201–204
myocardial infarctions (MIs), 195

N

Naranjo method, 60
National Medical Products
Administration (NMPA),
regulations, 9–11
Nissen, S. E., 198
non-arteritic anterior ischemic optic
neuropathy (NAION),
241–242
non-vitamin K oral anticoagulants
(NOACs), 263–264
Norton plots, 368

O

Ongoing Aggregate Safety Evaluation
(OASE), 25
optimum information size (OIS), 205

P

patient experience data, suitability of,
301–304
patient-focused drug development
(PFDD), 301–304
patient-level data, analyses from, 110

Patient Preferences in Benefit-Risk
Assessments during the Drug
Life Cycle (PREFER), 302
Periodic Adverse Drug Experience
Report, 4
periodic benefit-risk evaluation report
(PBRER), 332
Periodic Safety Update Report, 4
Peto method, 193, **197**
Pharmaceuticals and Medical Devices
Agency (PMDA), regulations,
8–9
Pharmacovigilance Risk Assessment
Committee (PRAC), 257
post-marketing PV evaluation, 108–109
post-marketing requirements (PMRs),
248–250
Post-market Rapid Immunization Safety
Monitoring (PRISM), 236
pragmatic clinical trials (PCTs), 93–94
pragmatic trials, 215–217
cardiovascular outcome trials,
215–235
CVOTs, 222–231
pragmatic trial, 224–231
regulatory guidance, 222–223
design considerations, 217–222
PRECIS wheel, 225
preclinical findings, signals, 46
pre-marketing evaluation, 108–109
PROACT-URL framework, 111
Problem framing, Objectives, Action,
Consequence and Tradeoff
(PROACT), 111
product risk
analogy, 64–66
association strength, 60
biologic gradient, 62–63
coherence, 63–64
consistency, 61
experimental evidence, 64
plausibility, 63
specificity, 61
temporality, 61–62
Program Safety Analysis Plan (PSAP),
19
proportional reporting ratio (PRR), 108

propensity score, 258–260
    considerations in using, 260–264
    methods, **260**
protopathic bias, 54

Q

quantitative benefit-risk assessment
    (QBRA), 112
quantitative safety evaluation, xiii

R

randomization, design consideration,
    218
randomized clinical trial/real-world
    evidence (RCT/RWE), xiii
randomized clinical trials (RCTs),
    215–217, 319
randomized controlled trials (RCTs),
    189, 254–256
rare events, dealing with, 195–200
real-world data (RWD), xiii, 254–256
    using for benefit-risk evaluation,
        299–301
    multi-source data, 266, 270
real-world evidence (RWE), 254–256, 296
    case study involving, 256–257
    confounding, 258–266
    multi-source data synthesis, 266–274
real-world health data, 12–14
relative risk (RR), 125
risk difference (RD), 125
risk management, 110–114
Robust MAP (rMAP) prior, 158
rotavirus efficacy and safety (REST)
    study, 181–182

S

safety assessment committee (SAC), 7,
    78, 124
Safety Assessment Entity (SAE), 89–90
safety evaluation, 220–221
Safety Explorer Suite, 72
safety monitoring, *104*
    blinded, 121–134

Likelihood-based methods for,
    167–184
multiple drugs, 92–93
perspective of statistical methods in,
    103–110
pragmatic trials, 93–94
stakeholders, *103*
statistical methodologies for,
    101–114
safety monitoring reporting (SMR), 102
Safety Scientific Working Group (SWG),
    256
safety signals
    governance structure for evaluation,
        20
    quantitative frameworks for
        screening, **27**, 28
Safety topics of interest, 18, 20, **21**, 22
Safety Working Group (SWG), viii
self-controlled design, signal
    identification, 244
Sentinel operating center (SOC), 236
Sentinel system (FDA), 234
    ARIA sufficiency, 248–250
    common data model, 237–239
    drug utilization using, 239–240
    history/role of, 234–236
    organizational structure of, 236–237
    signal identification system in,
        240–244, **243**
    signal refinement, 245–248
SeqLRT, 178
sequential meta-analysis (SMA),
    204–205
serious adverse event (SAE), 3–4, 25
serious adverse events (SAEs), 16, 79,
    188
SGLRT, 179
signals
    available tools for, 66–73
    detecting, 46–52
    determining product risk, 59–66
    evaluating, 52–59
Simpson's paradox, 53
single-case reports, signals, 46–47
single-exposure group, signal
    identification, 244

spaghetti plot, *70*
SPRT, 176
stochastic multi-criteria acceptability
      analysis (SMAA), 111–113,
      363–364
Strengthening the Reporting of
      Observational studies in
      Epidemiology (STROBE), 256
structured benefit-risk assessment
      (sBRA), 286
   current approaches to, 289–292
swimmer plot, *69*

T

Three-level Bayesian Hierarchical
      Mixture Model, 151–156
treatment arm correction, **200**
treatment emergent adverse events
      (TEAEs), 321
TreeScan, 242
Tufts Center for the Study of Drug
      Development, 254
21st Century Cures Act, 254, 291
type 2 diabetes mellitus (T2DM),
      222–223

U

unblinded safety assessment, 105
up-versioning, 36
US Food and Drug Administration (FDA)

benefit-risk framework, **291**
benefit-risk perspective, 297–298
Sentinel system, 234–246
U.S. Kefauver Harris Amendment, 102

V

vaccine adverse event reporting system
      (VAERS), 320
value trees, 367
visual analytics, 352
   adverse event data, 360–363
   benefit-risk data, 363–370
      case for, 365–366
      examples, 366–367
      patient-level data, 365
      questions of interest, 365, **366**
   defining, 356, *358*
   grammar of graphics, 353–354
   graph creation principles, 353
   question based approach, 354–356
   recommendations, **357**
   safety data, 356–363
   safety monitoring, 359–360
volcano plot, *67*

W

Women's Health Initiative (WHI), 225
workstreams, xiii–xv
World Health Organization (WHO),
      255

Printed in the United States
by Baker & Taylor Publisher Services